PHILOSOPHICAL CONSEQUENCES
OF QUANTUM THEORY

Studies in Science and the Humanities
from the
Reilly Center for Science, Technology, and Values

Volume II

PHILOSOPHICAL CONSEQUENCES
OF QUANTUM THEORY

Reflections on Bell's Theorem

JAMES T. CUSHING AND
ERNAN McMULLIN, EDITORS

UNIVERSITY OF NOTRE DAME PRESS
NOTRE DAME, INDIANA

N. David Mermin, "Quantum Mysteries for Anyone," is re-printed with permission from *Journal of Philosophy* 78 (1981): 397–408.

Bas van Fraassen, "The Charybdis of Realism: Epistemological Implications of Bell's Inequality," is reprinted with permission of Kluwer Academic Publishers from *Synthese* 52 (1982): 25–38. Copyright © 1982 by D. Reidel Publishing Company.

Figure 5 in James T. Cushing, "A Background Essay," is Figure 1 from A. Peres and W. H. Zurek, "Is Quantum Theory Universally Valid?" *American Journal of Physics* 50 (1982): 809–810, and reprinted with permission. Copyright is held by the American Association of Physics Teachers.

Cover: figure from John S. Bell's Collège de France lecture, "Bertlmann's Socks and the Nature of Reality," delivered to an audience of philosophers and physicists in 1980. The reader not already aware of the profound implications of Bertlmann's socks for the nature of reality should consult Bell's elegant lecture, originally published in the *Journal de Physique,* 1981, and re-printed in his anthology, *Speakable and Unspeakable in Quantum Mechanics,* 1987.

Library of Congress Cataloging-in-Publication Data

Philosophical consequences of quantum theory : reflections on
Bell's theory / James T. Cushing and Ernan McMullin, editors.
   p.   cm.—(Studies in science and the humanities from the
Reilly Center for Science, Technology, and Values ; 2)
   Bibliography: p.
   Includes index.
   ISBN 0-268-01578-3   —   ISBN 0-268-01579-1 (pbk.)
   1. Quantum theory.   2. Physics—Philosophy.
3. Philosophy.   I. Cushing, James T., 1937–
II. McMullin, Ernan, 1924–   . III. Title: Bell's theory.
IV. Series: Studies in science and the humanities from the
Reilly Center for Science, Technology, and Values ; v. 2.
QC174.12.P43   1989
530.1′2—dc19   89-40014

For JOHN BELL
In honor of his sixtieth birthday

CONTENTS

PREFACE

On October 1–3, 1987, a conference entitled, "Philosophical Lessons from Quantum Theory," was held at the University of Notre Dame. The conference brought together a sizeable number of those philosophers and physicists who have been concerned with the challenges raised by quantum mechanics for traditional philosophical issues of causality, explanation, and objectivity. These challenges were, to some degree at least, implicit in the very first formulations of the quantum ideas in the early days of the century. But it was only with John Bell's formulation of his now celebrated theorem in 1964 that the full measure of the challenge came to be appreciated. Though the conference had not originally been intended to focus on Bell's theorem only, it is an indication of the importance now accorded to it that virtually all the talks delivered at the conference took it as their starting-point.

This volume consists of papers deriving from the conference. Each of the original speakers was asked to shape the final version of his or her paper in the light of the conference discussions. Two are represented here in part by essays already in print, either as an appendix or with an appendix. The conference ended with a panel discussion. Those panelists who were not also presenting papers at the conference were invited to extend their remarks into full-length essays. Since Jon Jarrett's formulation (1984) of the consequences of Bell's theorem had been frequently cited during the course of the conference, it was agreed that he also should be asked to contribute a paper to the conference volume.

These essays provide a detailed analysis of, as well as reflection on, the consequences of the quantum formalism for our understanding of the world. These consequences have not yet been fully appreciated outside the small circle of those who have directly concerned themselves with them. This is why every effort was made in the original conference, as well as in this volume, to make the presentations as accessible as possible to an audience unfamiliar with the forbidding technicalities of quantum theory. Philosophers will find that the developments chronicled here constitute as significant a challenge to our ideas of natural order as anything that has occurred since the "new science" of Galileo overturned the age-old certainties of the Aristotelian worldview.

The order in which the essays appear is deliberately chosen. After the stage has been set for discussion (Cushing, Shimony), the implications of the puzzling correlations discovered between certain classes of quantum events occurring at a distance from one another are drawn out in a variety of ways, and a number of important distinctions are debated (Mermin, Jarrett, Wessels, van Fraassen, Butterfield, Redhead). The matter of "explaining" these correlations is then discussed, some of the essays outlining possible ways to proceed, others arguing that no explanation may be needed (Stapp, Fine, Hughes, Teller, Howard). Two retrospective essays return to earlier history, one speculating as to how Bohr might have reacted to these recent developments, and the other suggesting analogies between the present situation and the debate prompted by Newton's account of planetary motions (Folse, McMullin).

The book ends with a bibliography of all the works that bear on the philosophy of quantum mechanics. Although it does not purport to be exhaustive, being restricted to works actually cited in the essays published here, it affords a useful guide to a fast-growing literature.

The conference was sponsored by the Reilly Center for Science, Technology, and Values, as well as by the Program in History and Philosophy of Science and the Department of Philosophy at the University of Notre Dame. Support for the conference was provided by the National Endowment for the Humanities, whose aid is gratefully acknowledged by the conference organizers.

As this volume was in preparation, Abner Shimony reminded the editors that its publication date would coincide pretty nearly with John Bell's sixtieth birthday. As an example of serendipity—and, after all, is not the famous theorem a testimony to serendipity on a wider scale?—this presented an unexpected and welcome opportunity. And so this volume is dedicated to John Bell, with the enthusiastic concurrence of all those whose essays appear here.

CONTRIBUTORS

Jeremy Butterfield, Faculty of Philosophy, University of Cambridge

James T. Cushing, Department of Physics, University of Notre Dame

Arthur Fine, Department of Philosophy, Northwestern University

Henry J. Folse, Department of Philosophy, Loyola University, New Orleans

Don Howard, Department of Philosophy, University of Kentucky

R. I. G. Hughes, Department of Philosophy, Yale University

Jon P. Jarrett, Sage School of Philosophy, Cornell University

Ernan McMullin, Program in History and Philosophy of Science, University of Notre Dame

N. David Mermin, Laboratory of Atomic and Solid State Physics, Cornell University

Michael L. G. Redhead, Department of History and Philosophy of Science, University of Cambridge

Abner Shimony, Departments of Philosophy and Physics, Boston University

Henry P. Stapp, Lawrence Berkeley Laboratory, University of California, Berkeley

Paul Teller, Department of Philosophy, University of Illinois at Chicago Circle

Bas C. van Fraassen, Department of Philosophy, Princeton University

Linda Wessels, Department of History and Philosophy of Science, Indiana University

A BACKGROUND ESSAY

James T. Cushing

The purpose of the papers in this volume is to discuss the hard questions and hard choices that recent quantum physics has presented for philosophy in general, not just for the philosophy of science. The authors examine what has been established, what options are still available, and what revisions, radical or otherwise, may be necessary in our philosophical views. Since the volume is intended for philosophers in general, and not just for experts in the foundational problems in quantum theory, the papers are not thick with technical details. A central development to which all of these papers are in some way related is Bell's theorem. There is an enormous literature on the technical aspects of Bell's theorem and on the foundational problems of quantum mechanics (see, for example, Ballentine 1987). My task here is to provide some introductory material that will help the reader new to this subject to understand the subsequent papers. Let me be explicit in stating that the tale which follows is not always chronologically faithful to the historical record nor is it in all details a literal transcription from the original papers cited. By way of orientation, I begin with a general and somewhat loose overview of the subject and then proceed to define terms and concepts more precisely.

A little history

While it is true that interest in the interpretative problems of quantum mechanics received major impetus from the seminal paper of John Bell (1964) on the Einstein-Podolsky-Rosen (EPR) paradox, there was life in the field before Bell—and before EPR too (see, for example, Wheeler and Zurek

Partial support for this work was provided by the National Science Foundation under Grants No. SES-8606472 and No. SES-8705469. I thank especially Don Howard, Ernan McMullin, and Abner Shimony for comments on an earlier draft of this essay.

1

1983). As early as 1913, *before* Bohr's paper of that year on the semiclassical model of the hydrogen atom had appeared in print, Rutherford pointed out a problem for causality in Bohr's model. Bohr had postulated that the frequency ν of light emitted by an electron in its transition from an initial energy level E_m to a final level E_n (figure 1) is given by

$$E_m - E_n = h\nu.$$

To Rutherford, it appeared as though the electron would have to know to what energy level it was going before it could decide what frequency it should emit (Hoyer 1981, 112). By 1917, Einstein wanted to know how, in Bohr's model, the photon decided in what direction it should move off (figure 2). Schrödinger attempted a largely classical interpretation of his own equation, but Max Born (1926) proposed a consistent statistical interpretation of quantum mechanics. Determinism, in the sense of our being able to predict *the* unique outcome of a measurement on an event-by-event basis, was gone from the formalism, although Einstein and Schrödinger struggled (Przibram 1967) against what became codified as the ''Copenhagen'' interpretation of quantum mechanics. True, the majority of physicists (if they chose to think about the issue at all) *believed* that atomic events could not, even in principle, be predicted on an event-by-event basis. Still, one could (and some notables did) question the completeness of quantum mechanics, asking whether there might not exist a successor theory which *could,* in principle, make such event-by-event predictions. In fact, von Neumann ([1932] 1955, 313–328) offered a ''proof'' that such ''hidden-variables'' theories could not exist. Much later, Bell (1966) did address the question of the relevance of that ''proof.''

In 1935, Einstein, Podolsky and Rosen (EPR) published a paper in which they questioned the completeness of quantum mechanics. That is, they asked whether one could be certain, on physical grounds, that more could be specified (or known) about a system than could be predicted with certainty by the formalism of quantum mechanics. By means of a specific thought experiment, they argued that the incompleteness of quantum mechanics was entailed

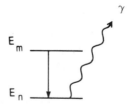

Figure 1.

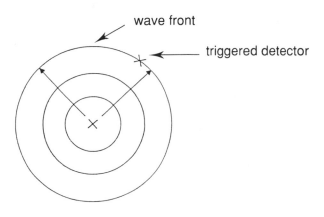

Figure 2.

by the formalism of quantum mechanics itself, along with entirely plausible assumptions excluding action at a distance (''locality'') and about the reality (or definiteness) of a physical quantity independent of our choice to observe it. Their argument has been lucidly discussed by Shimony (1978). We do not consider the original EPR thought experiment here. For pedagogical purposes, there is a simpler one due to Bohm (1951). The EPR paper did not offer any alternative theory to quantum mechanics, nor did it mention hidden variables. Nevertheless, the additional parameters that would be necessary to give a complete specification of the state of a system have subsequently come to be referred to as ''hidden variables'' and any theory encompassing such parameters as a ''hidden-variables theory.''

Figure 3 is a schematic representation of Bohm's thought experiment. At the center is a source (such as an atom) which decays and emits two electrons (or photons)[1] in opposite directions as indicated in the figure. For simplicity of discussion, we assume that the spin of the atom (''source'') is zero, both before it emits the two ''electrons'' and after as well.[2] Conservation of angular momentum then requires that the spin of electron 1 must be oppositely directed to the spin of electron 2. That is, if the first electron is observed to have spin ''up'' in some direction, then the second electron's spin, if observed along that same direction, would be found to have spin

---

[1]Most of those experiments actually performed to date have used photons. Other quantum systems (such as electrons, protons and kaons) have also been employed (or at least proposed). I use the term 'electrons' here only because the reader may have an easier time picturing what is going on.

[2]For the present introductory discussion it does no real harm for the reader to picture the spin of an atomic system as one would the spin of a ball or of a planet about an axis through its center.

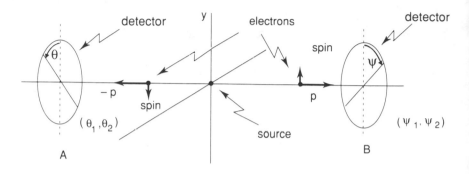

two "choices" (i = θ_1 ,θ_2)
two possible outcomes (x = r_{A_k} = ± 1)

two "choices" (j = ψ_1 , ψ_2)
two possible outcomes (y = r_{B_k} = ±1)

Figure 3.

"down." We choose units such that the observed spin of the electron in any given direction is either +1 or −1.[3] At stations A and B there are instruments (labeled "detector" in the diagram) which can be set to measure the spin of an electron along an axis transverse to the line of flight of the electrons. In each detector this transverse axis can be chosen or set in either of two orientations: θ_1 or θ_2 at station A, and ψ_1 or ψ_2 at station B.[4] It is an empirical fact that, whenever the spin of an electron is measured along a given axis, one always finds that spin to point *either* "up" (+1) *or* "down" (−1) along that axis, but *never* to have some fractional, intermediate value for its projection along an axis. That is, each individual measurement or observation yields either +1 or −1, never any other value.[5]

---

[3]The magnitude of the spin of an electron is $\frac{1}{2}\hbar$. Here $\hbar = h/2\pi$, where h is Planck's constant. We *could* choose a system of units in which $\frac{1}{2}\hbar = 1$. Or, equivalently, we can express all spin projections in terms of multiples of ($\frac{1}{2}\hbar$).

[4]The use of a Greek ψ, here and later in this essay, to denote an angle of orientation should cause no confusion with the use of psi to denote a wavefunction. Not only would the distinction be clear from the context, but in the body of this essay we neither write down any explicit wavefunctions nor use any symbol for them.

[5]This behavior of the measured value of the spin projection of an electron is not peculiar to *this* particular experimental arrangement. It is a feature of nature whenever an electron's spin projection is measured by *any* means.

In summary, then, the experimenter (or observer) at station A has two choices (θ_1 or θ_2) for instrument settings, and for each instrument setting a result (or datum) $+1$ or -1 is possible for the outcome of a measurement. Similarly, at station B the choices are ψ_1 or ψ_2 for each of which an outcome $+1$ or -1 is possible. This experiment, or sequence of observations, can be repeated as many times as we wish. Each repetition is a run, which we label with an index k, where $k = 1, 2, \ldots N$, with N being the total number of runs. Let us denote by r_{A_k} the result at station A for the kth run. The quantity r_{B_k} has a similar meaning for station B. It is important to appreciate that, for any given run, r_{A_k} can have *only* the value $+1$ or -1 (no other), and similarly for r_{B_k}. For the probability (or distribution) of the outcome r_{A_k} (at A when the setting is θ) and r_{B_k} (at B when the setting is ψ), we use a notation suggested by David Mermin:

$$p^{AB}(r_{A_k}, r_{B_k} | \theta, \psi). \tag{1}$$

More generally, $p^{AB}(x, y | i, j)$ represents the joint distribution of obtaining a result (or outcome) x at station A when the setting (or parameter) i has been chosen by the experimenter at A, while the result y obtains at station B when setting j has been chosen there. The spirit of this thought experiment is that the experimenter at station A can make a free choice of the setting (or parameter), i, *independently* of the free choice, j, his colleague makes at B.

2. Bell's theorem

Prior to Bell's 1964 paper, the question of whether or not there could exist a deterministic hidden-variables theory with no instantaneous action at a distance seemed incapable of resolution. Of course, no one had succeeded in writing down an empirically adequate example of one. But, that did not prove that one could not exist. After all, if a student fails to solve a difficult homework problem, the reason could be that he or she lacks the wit to do it or, indeed, it could be a problem with no solution. In the absence of a successful deterministic, local, hidden-variables theory, discussion of the *possibility* of such a theory could appear to be little more than idle argument appropriate only for a free Saturday afternoon or for cocktail parties. Bell's paper changed that in a dramatic fashion. The strength of a theorem is inversely proportional to the strength of the assumptions it makes. That is, if you assume a lot and prove a little, no one is particularly impressed. But if you (apparently) assume practically nothing and obtain a remarkable result, that is impressive.

In effect, Bell (1964) argued that determinate (i.e., predetermined prior to the measurement) projections for the spins of the electrons and locality are incompatible with the (spin) correlations predicted by quantum mechanics.

Once his argument had ruled out determinate values, then it also ruled out the possibility of a local, deterministic theory. Subsequently, Bell (1971) gave an argument that did not involve any type of determinism (or even determiniteness). Nevertheless, for simplicity of presentation here, I take the license of speaking of his proof in terms of the restrictive framework of a deterministic, local theory which could account for the known outcome of a simple (thought) experiment. Here, as previously, 'deterministic' means just that it is in principle possible to specify enough about the system so that the outcome of the experiment could be predicted on an event-by-event basis. Again, 'local' here serves to characterize the absence of instantaneous action at a distance. Such locality would seem to be required by special relativity (figure 4). At this stage of the discussion, locality requires a first signal principle according to which two events separated by a distance l cannot affect each other before a time $t = l/c$ has elapsed, where c is the speed of light. Certainly, such determinism and locality would be granted as unproblematic by anyone inclined toward a classical ontology. The idealized experiment considered by Bell was simple and its "known" outcome not in serious doubt.

If we let λ denote those parameters (the "hidden" variables) necessary to give a complete specification of the state of a system, then, again in Mermin's notation, $p_\lambda^{AB}(x,y|i,j)$ is the corresponding joint probability. Here λ

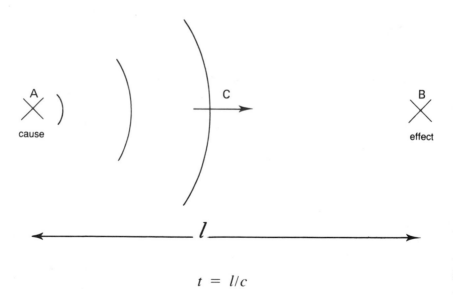

$$t = l/c$$

Figure 4.

stands collectively for *all* λ_1, λ_2, λ_3, . . . , necessary for a complete state specification (of the source and of the electrons as they emerge from the source). (Not all of these need be *hidden* variables.[6]) In terms of a density function $\varrho(\lambda)$ which assigns appropriate weighting to the λ, the joint distribution (Eq. [1]) of the experimental results is given as

$$p^{AB}(x,y|i,j) = \int p_\lambda^{AB}(x,y|i,j)\varrho(\lambda)d\lambda. \qquad (2)$$

On the basis of locality, Bell argued that these $p_\lambda{}^{AB}$ could be factored as

$$p_\lambda^{AB}(x,y|i,j) = p_\lambda^A(x|i)p_\lambda^B(y|j). \qquad (3)$$

From this factorization (or "factorizability," a term due to Fine [1981, 536 ff]) assumption it follows, essentially by algebraic manipulations alone (Clauser and Horne 1974), that the observed distributions for the experiment of figure 3 must be bounded as:

$$\begin{aligned} -1 \leq\ & p^{AB}(++|\theta_1,\psi_1) + p^{AB}(++|\theta_1,\psi_2) + p^{AB}(++|\theta_2,\psi_2) \\ & - p^{AB}(++|\theta_2,\psi_1) - p^A(+|\theta_1) - p^B(+|\psi_2) \leq 0. \end{aligned} \qquad (4)$$

Here $p^A(+|\theta_1)$ is the probability of observing an electron with spin up along the θ_1 direction at station A irrespective of what the spin measurement outcome is at station B.[7] The term $p^B(+|\psi_2)$ has a similar meaning for station B.

In fact, the actual experiment (in a real laboratory with real equipment) is much more difficult to do than my rather glib characterization in figure 3 might lead one to expect. A detailed discussion of the experimental situation can be found in the comprehensive review article by Clauser and Shimony (1978) and in Redhead (1987b). There also exists a general, less technical review by Shimony (1988). Such experiments have been carried out, some of the latest and most convincing being those by Aspect, Grangier, Dalibard, and Roger (1981, 1982) in Paris. The empirical results are representable, well within the limits of experimental error, by the simple distributions:[8]

---

[6]In all of the papers in the volume (*except* Redhead's), this λ can include the quantum-mechanical state-vector specification (often denoted by Ψ).

[7]The *marginal* $p^A(+|\theta_1)$ is recovered from the $p^{AB}(++|\theta_1,\psi)$ and $p^{AB}(+-|\theta_1,\psi)$ as

$$p^A(+|\theta_1) = p^{AB}(++|\theta_1,\psi) + p^{AB}(+-|\theta_1,\psi).$$

That is, we sum over *both* possible outcomes at station B. The careful reader may wonder why the left side of this equation is not written as $p^A(+|\theta_1,\psi)$. We shall discuss this point below (see Eq. [8]). For now let us simply state that this is an *empirical* fact about the marginals of this experiment. This can also be seen to follow from Eqs. (5) below.

[8]It is also a fact that the formalism of quantum mechanics yields the results of Eqs. (5). However, that has nothing to do with a discussion of the significance of the Bell inequality. For our purposes now, we can take Eqs. (5) as phenomenological fits to the data—discovered perhaps by a modern-day Ptolemy! For the present discussion, they need not be considered as having been derived from some theory.

$$p^{AB}(++ \mid \theta, \psi) = p^{AB}(-- \mid \theta, \psi) = \frac{1}{2} \sin^2 \left(\frac{\theta - \psi}{2} \right), \qquad (5a)$$

$$p^{AB}(+- \mid \theta, \psi) = p^{AB}(-+ \mid \theta, \psi) = \frac{1}{2} \cos^2 \left(\frac{\theta - \psi}{2} \right), \qquad (5b)$$

$$p^{A}(+ \mid \theta) = p^{B}(+ \mid \psi) = \frac{1}{2}. \qquad (5c)$$

Although we shall return to Eqs. (4) and (5) in more detail below, one can already see the crux of the conflict by choosing, for example,

$$\theta_1 = 60°, \ \theta_2 = 180°, \ \psi_1 = 0, \ \psi_2 = 120° \qquad (6)$$

in Eq. (4) and then using the *empirical* distributions of Eqs. (5) to obtain[9]

$$-\frac{1}{8} \geq 0. \qquad (7)$$

The logical skeleton of the argument is that the assumptions of locality and determinism, plus the actual experimentally observed distributions of the real world, have produced the contradiction of Eq. (7). Although one can, in principle, attempt to undermine the empirical leg of the triad upon which this argument rests (cf. Clauser and Shimony 1978), each successive experiment forecloses more such possible loopholes and makes such a line of attack ever less plausible. So, the arrow of *modus tollens* appears more reasonably directed at the assumptions of locality and/or determinism. We have purposely not gone into the details of the argument by which Bell passed from Eq. (3) to the (Bell) inequality of Eq. (4) because we want to focus on the logical structure of the argument. In the appendix to this essay, the reader can find a simple proof of a contradiction like Eq. (7). So, Bell's remarkable result, or theorem, is that no deterministic, local hidden-variables theory can account for the empirical result of the experiment. It is worth emphasizing that these types of correlations are a pervasive feature of the quantum world. They are not peculiar to the Bohm-EPR class of experiments alone. However, the Bohm-EPR configuration is in a sense the "simplest" one yet known which exhibits these "mysterious" quantum correlations.

---

[9]That is, if, for the sum of the six distributions in the middle term of the inequality of Eq. (4), the expressions of Eqs. (5) are used with the values specified in Eq. (6), then these six terms sum to $-9/8$. Thus, Eq. (4) becomes

$$-1 \leq -\tfrac{9}{8} \leq 0,$$

which implies Eq. (7).

Let me stress two points here. First, Bell never wrote down a single local, deterministic theory. Rather, he proved, without ever having to consider any dynamical details, that *no* such theory can in principle exist. The entire class was killed at a stroke—a classic "no-go" theorem. Second, Bell's theorem really depends in no way upon quantum mechanics. It refutes a whole category of (essentially) classical theories without ever mentioning quantum mechanics. And it turns out that the experimental results not only refute the class of local, deterministic theories but also agree with the predictions of quantum mechanics. (That is, a straightforward application of the rules of quantum mechanics does lead to the results of Eqs. [5].) Abner Shimony (1984b, 35) has appropriately given the name "experimental metaphysics" to this type of definitive empirical resolution of what appears to be a metaphysical question.

3. Some distinctions

In my presentation thus far, I have been rather cavalier in oversimplifying the issues and in conflating terms that must be carefully distinguished. So I now turn to the purpose of the subsequent papers in this volume and to some of the work that the authors have done in recent years. Today, when one looks back at Bell's original paper and at some of the early responses to it, one is struck by at least two facts. First, the paper contains a modicum of mathematical formalism. Depending upon one's level of mathematical sophistication, the proof may not be immediately transparent and one can wonder whether something has gone awry in those pages and symbols. After all, the result is *so* remarkable: it forces us to face indeterminism and/or nonlocality *in principle*. Could the proof be flawed? As often happens with great discoveries, proofs are subsequently fashioned which make the important result seem almost self-evident. Bell's theorem was no exception. Eventually, there were picture proofs and nonmathematical discussions (d'Espagnat 1979; Mermin 1981a, 1985) of Bell's result and of the quantum-mechanical riddles it makes us face. While such discussions are nontechnical, they can remain rather long and involved. The reader's eyes may glaze over before the end. However, if one is willing to pay the price of a little algebra—really, only about six lines of arithmetic—one can immediately go from Bell-type premises and a requirement of empirical adequacy to a contradiction like $1 > 2$ (Stapp 1971, 1979; Redhead 1987a). The mathematics is *so* simple and brief you are certain no error has been made. You think you understand it all! (The details of such a proof are given in an appendix to this paper.)

So then, first, the formalities or manipulations in the proofs were greatly simplified. But then, the second, and in many ways more difficult, phase began—unpacking the assumptions and the meanings of the terms used in these proofs and coming to some understanding of just what the implications are. This is a job that philosophers are particularly well equipped to do. The terms 'reality', 'determinism', and 'causality' cannot be used interchangeably and one must be especially careful to distinguish between locality and separability. Perhaps a few sketchy definitions will help for a start:[10]

> *reality* — existence of an objective, observer-independent
> world (often closely related to determinate values)
> *determinism* — sufficient information at t_o allows prediction of a
> specific result at a later time t
> *causality* — a specific preceding event (or ''cause'') for every
> effect — a concept familiar from prequantum,
> classical theories
> *locality* — no influence transmitted faster than light
> *separability* — spatially separated systems always have
> independently definable properties and existence
> (and these properties exhaust the description of any
> system made up of these subsystems).

Arthur Fine (1984a, 1984b) and Don Howard (1985, 1987) have provided a useful perspective for several of these issues by their careful and enlightening historical reconstructions of Einstein's views on locality and separability, bringing out essential differences here between Einstein and Bohr. Henry Folse (1985), Don Howard (1986), and Dugald Murdoch (1987) have done similar work in reconstructing Bohr's philosophy of science. Furthermore, as we indicated previously in Eq. (3), a crucial mathematical step in the usual proof of Bell's theorem is the factorization of a certain expression for joint probabilities. A long debate has arisen as to the physical warrant for this step. This factorizability (or ''Bell'' locality or statistical independence) is not implied by the first signal principle of relativity (''Einstein'' locality). Michael Redhead (1983) and Linda Wessels (1985) have analyzed in detail the assump-

---

[10]In the present context, the term 'determinism' is usually predicated of a *theory*, as in a *deterministic* theory. In quantum field theory, 'causality' is used in a sense rather different from (but related to) the classical cause-effect one. (See Cushing, 1986, for a fuller discussion of the meaning of the term 'causality' in modern theoretical physics.) The reader should be warned that the terms 'locality' and 'separability' are the most problematic as far as universally-agreed-upon definitions are concerned. The ones I give here alert the reader to a distinction between these terms. However, each author below must be checked carefully for his or her own precise use of these terms. It is also true historically that the evolution of an explicit distinction between those two terms was a long time in coming. (See Howard [1985] and Folse [this volume] for careful discussions of this issue.) Furthermore, we distinguish among different types of locality and nonlocality. Finally, d'Espagnat (1984) treats the issues of reality and of separability carefully and at great length.

tions made in various proofs of Bell's theorem. (Actually, there are by now many similar results that are generically referred to as "Bell theorems.") In an important analysis, Jon Jarrett (1984) pointed out that the Bell locality (or factorizability) used in the proofs of these theorems is the logical conjunction of two other conditions, violation of either of which leads to some form of nonlocality. Abner Shimony (1984a) has termed these violations controllable and uncontrollable nonlocality. The first would violate the first signal principle of special relativity, but the second is innocuous in this regard. Since quantum mechanics exhibits only uncontrollable nonlocality, special relativity and quantum mechanics seem capable of a peaceful coexistence (Shimony 1978; Redhead 1983, 1987b).

Let me summarize these important distinctions that Jarrett (1984) has made. By "*Jarrett locality*" (or what Shimony [1984a] terms "*parameter independence*"), we mean that the single distribution for outcome x at station A (given the state specification λ) is independent of the choice j made at station B:

(I) $p_\lambda^A(x|i,j) = p_\lambda^A(x|i)$.   (8)

A similar condition follows for $p_\lambda^B(y|i,j)$. If this condition is respected (as it is, for instance, by quantum mechanics [Shimony 1984a]), then the experimental determination of such a distribution cannot be used to send a signal from station B to station A (e.g., to transmit information faster than the speed of light). However, if condition (8) were violated, then such a signal could be sent[11] and Shimony has referred to such a circumstance as "*controllable nonlocality*." Controllable nonlocality (which obtains if Eq. [8] is violated) would produce a conflict with the first signal principle of special relativity. By "*Jarrett completeness*" (or what Shimony terms "*outcome independence*"), we mean that the single distribution for outcome x at station A (given the state specification λ) is independent of the measurement outcome y at station B:

(II) $p_\lambda^A(x|i,j,y) = p_\lambda^A(x|i,j)$.   (9)

If this condition fails (as it does for quantum mechanics) while (I) holds, then an explanation in terms of (past) common causes is not possible (cf. Shimony 1984a and Jarrett, this volume). However, the violation of Eq. (9) does *not* allow the measurement of such a distribution to be used for superluminal

---

[11]It is important to note that the *observable* (or experimentally relevant) distributions are the integrated $p^A(x|i,j)$ or $p^A(x|i)$ (cf. Eq. [2]), *not* the $p_\lambda^A(x|i,j)$ or the $p_\lambda^A(x|i)$ of Eq. (8). The point is that not all of the λ need be under the control of the experimenter (even in principle). Even if the $p_\lambda^A(x|i,j)$ did depend upon j, it does not follow that the integrated quantities $\int p_\lambda^A(x|i,j)\varrho(\lambda)d\lambda$ will still *necessarily* depend upon j. It is only these integrated quantities which are directly related to experiment. Jarrett's argument that a violation of Eq. (8) would allow signaling assumes an idealized situation (cf. his paper in this volume).

signaling. For this reason, Shimony has termed a violation of condition (9) *"uncontrollable nonlocality."* Such uncontrollable nonlocality produces no conflict with the first signal principle of special relativity. (Shimony [1984a] gives a general proof that the uncontrollable nonlocality of quantum mechanics does not produce a violation of the first signal principle of special relativity. This is what he calls "peaceful coexistence" between quantum mechanics and special relativity. See also Shimony [1986] for further discussion of these implications).

The importance of Jarrett's (1984) clarifying analysis in terms of the conditions (I) and (II) of Eqs. (8) and (9) respectively is that (I) and (II) *together* imply the factorizability of Eq. (3) and hence allow derivation of a Bell inequality. In Mermin's notation, Jarrett's argument can be summarized as follows. From the very meanings of joint and conditional probabilities, it follows that

$$p_\lambda^{AB}(x,y|i,j) = p_\lambda^A(x|i,j,y)p_\lambda^B(y|i,j). \tag{10}$$

If we now assume that both (I) and (II) hold, then Eq. (10) reduces to the factorizability condition of Eq. (3). Thus, if we can block either move (8) or move (9), we can block the standard derivation of a Bell-type inequality.

Another insightful observation about the meaning of the Bell inequality was made by Fine (1982b). He argued that Bell inequalities of the type in Eq. (4) above are the necessary and sufficient conditions for the existence of a deterministic hidden-variables model which will produce the joint distributions for the Bohm (EPR) experiment of figure 3. But the existence of such a complete set of state variables λ is equivalent to a common-cause explanation (in the common past of the parts of the system to be observed) for these distributions or experimental outcomes. Knowing that there is such an empirically applicable test for the possibility of a common-cause explanation will prove important for the discussions which follow in subsequent papers in this volume.

4. Philosophical implications

We can now ask just what the implications of all of this are for our view of the physical world. Thus far we have pointed out certain restrictions on allowable world views (or representations of reality) that are demanded by quantitative relations (the Bell inequalities) containing only empirically measurable distributions of experimental results. In a sense, the tone has been negative since we have stressed what type of theories or explanations are *not* possible. Must we, for example, abandon belief in an observer-independent reality? Or, as David Mermin has put it, "Is the moon there when nobody looks?'' We have shown what cannot work rather than exploring some theory or explanatory framework that is successful in reproducing the results of experiment. Of course, we *do* have an empirically adequate theoretical framework within which to organize the observational data—namely, quantum mechanics. However, this enormously empirically successful theory has difficult interpretative problems associated with it. Henry Stapp (1979, 14) makes a point similar to Mermin's when he characterizes our immediate reaction to a literal acceptance of some of the more extreme interpretations of quantum mechanics:

> One objection to this view is that it seems excessively anthropocentric, at least if consciousness is reserved for human beings and higher creatures. Before the appearance of such creatures the world would be synthesizing endless superposed possibilities, with nothing actual or real, waiting for the first conscious creature to occur among the possibilities. Then a gigantic collapse would occur. Similarly, the Martian landscape would be nothing but superimposed possibilities until Mariner landed and some observer in Houston viewed his TV screen. Then suddenly the rocks and boulders would all snap into their observed places. This view seems to assign a role to such observers that is out of proportion to their place in the world they create.

That is, our most successful theory of processes at the microlevel, namely quantum mechanics, poses serious problems for scientific realism (which requires roughly and at a minimum that our scientific theories are to be taken as giving us literally true descriptions of the world).[12]

Bas van Fraassen (1982a; this volume) has argued that the experimental violation of the Bell inequality tells against scientific realism. That is, if scientific realism does not work at the microlevel, then it cannot be generally valid. In a provocative article, Asher Peres (1985) has posed yet another quantum paradox ''as a challenge to those physicists who claim that they are realists'' (p. 201). His conclusion at the end of that article (p. 205) is that ''Any attempt to inject realism in physical theory is bound to lead to inconsistencies.''[13]

At the other end of the spectrum from van Fraassen (1980) or Peres on views of scientific realism, we find Ernan McMullin (1984) who points to the great success *structural* theories have enjoyed in several sciences (such as chemistry, astrophysics, geology, and genetics) in taking a starkly realistic

---

[12]John Bell (1987) has recently published his own collected essays on the philosophy of quantum mechanics, and on interpretations of it other than the standard Copenhagen one.

[13]At the 1986 Quantum Measurement Theory Conference (Greenberger 1986) in New York City, I mentioned to Peres that his position appeared to be an instrumentalist one. He replied with no apparent discomfort that others had told him that before. For a physicist's statement on an instrumentalist interpretation of quantum mechanics, see Peres (1988).

view of the entities contained in those theories. It is in regard to the interpretation of the ontologies underlying *mechanical* theories (whether classical or quantum) that problems most often arise (McMullin 1989). McMullin recommends treating these theories as a special class and considers as inappropriate the demand for a realistic interpretation of *force* or *field* that would be unproblematic for *molecule* or *gene* or *galaxy*. He is willing to put aside for the present certain difficult questions of a realistic interpretation for mechanics (that is, classical mechanics, quantum mechanics, quantum field theory—all of what would seem to many to be the foundations of physics): "Because of its many special features, mechanics is quite unsuitable as a paradigm of science generally, though philosophers are wont to overlook this" (1984, 10). Rather than being *the* paradigm of natural science, much of physics becomes, at least in the context of this issue, an anomaly. It appears that McMullin restricts consideration to cases that satisfy, in some broad sense, Newton's Rule III of Reasoning in Philosophy in Book III of the *Principia* (Newton [1726] 1934):[14] "The qualities of bodies, which admit neither intensification nor remission of degrees, and which are found to belong to all bodies within the reach of our experiments, are to be esteemed the universal qualities of all bodies whatsoever." This rule is often taken as saying that we may extrapolate general features of the macroworld to the microworld. To some, McMullin's circumscription of mechanics may be too costly a move to make on behalf of scientific realism. Somewhere between van Fraassen and McMullin, we find Heisenberg (1958, 185) with his suggestion that we must admit a new class of physical entity into our theories: *potentia* (Shimony 1978, 1986; Stapp 1979, 1985a). Bell (1984) has made a similar suggestion in speaking of the "*be*ables" of quantum field theory.

One of the most interesting philosophical questions, perhaps, concerns the relations among empirical adequacy, explanation, and understanding for quantum phenomena. Are explanation and understanding really possible when a detailed causal explanation is in principle impossible? In Bas van Fraassen's (1985) terms, are the EPR correlations a mystery? Paul Teller has suggested that relational properties of physical objects may not simply supervene wholly upon nonrelational properties of localizable individuals, but that a type of "*relational holism*" is essential in which the objects have inherent relations among themselves. He claims that this is "a holism we can understand" (Teller 1986, 73). But is it? A central issue is whether or not we can truly *understand* such descriptions of our world. These problems are forced upon us, of course, when we take the present formulation of quantum mechanics as exactly correct, needing no modification.

---

[14]I thank André Goddu for this insight, although he is not responsible for my interpretation of it.

In the same vein, we can even ask whether all the *desiderata,* which we may want in a theory that accords with the phenomena of the real world, can be mutually compatible. Peres and Zurek (1982) set up a triad (figure 5) involving the three "wishes" of determinism, verifiability, and universality and argue that no theory can, *even in principle,* satisfy simultaneously all three demands. (By "verifiability" here they mean the freedom of choice of an observer or experimenter to fix a given setting on, or orientation of, a measuring device to test the predictions of a theory.) We can have at best just any two of the three. In the end, quantum mechanics may be the best theory it is possible to have.

We might question the value of discussing the implications that essentially *nonrelativistic* quantum mechanics (say, the usual Schrödinger equation) has for such issues, since the more complete (and problematic) theory in use today is relativistic quantum field theory. I have argued elsewhere in some detail (Cushing 1988) that the root of quantum paradoxes is the superposition principle and *that* remains in any quantum theory. Michael Redhead and Paul

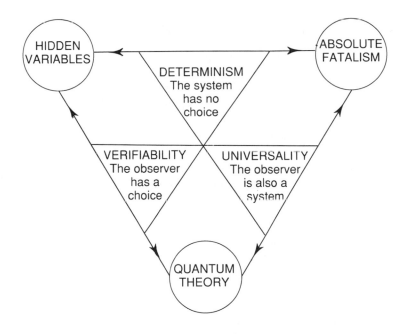

Figure 5.

Teller (cf. Brown and Harré 1988) do believe that quantum field theory introduces new philosophical problems which must now be faced. These would, of course, be in addition to, and not solutions of, the interpretative difficulties already presented here.

This introductory essay may at least establish a *prima facie* case for the relevance of quantum mechanics to general philosophical issues related to epistemology and ontology. There are serious problems here, not simply questions of mathematical formalism. Hence, the rationale for this volume. Several years ago John Bell (1975, 98) made an observation about the understanding of our world which quantum theory gives us:

> The continuing dispute about quantum measurement theory is not between people who disagree on the results of simple mathematical manipulations. Nor is it between people with different ideas about the actual practicality of measuring arbitrarily complicated observables. It is between people who view with different degrees of concern or complacency the following fact: so long as the wave packet reduction[15] is an essential component, and so long as we do not know exactly when and how it takes over from the Schrödinger equation, we do not have an exact and unambiguous formulation of our most fundamental physical theory.

In our search for a new understanding, we face the challenge characterized by Costa De Beauregard (1983, 515–516) in this way:

> Hard paradoxes . . . are resolved only by producing a new and adequate paradigm, in Kuhn's words. In physics, this implies the production of a new mathematical recipe (e.g., Copernicus's heliocentrism, or Newton's inverse-square law) and tailoring an explanatory discourse exactly fitting the mathematics (e.g., Einstein's interpretation of the Lorentz-Poincaré formulas; or still better, Minkowski's).
>
> This sort of "explanation" is usually felt (and often for a long time) as itself paradoxical. Newton's action at a distance, Einstein's "reciprocal" interpretation of the Lorentz contraction, have very often been deemed "hardly explanations at all."

---

[15]The reader unfamiliar with the expression "wave packet [or state] reduction" can take (an example of) this to be the following. A system in a state ψ, which is a linear superposition of eigenstates (φ_j) of an operator (observable) A *prior to* a measurement of that observable, is in a specific eigenstate (say φ_f) immediately *after* the measurement. The (discontinuous) "transition"

$$\psi = c_1\varphi_1 + c_2\varphi_2 + c_3\varphi_3 + \ldots \rightarrow \psi' = \varphi_f$$

is (an example of) state reduction. It is commonly the case that $\psi' \neq \psi$. This reduction is *not* governed by the Schrödinger equation. See d'Espagnat (1979) for a full discussion of reduction.

What occurs in the "paradox and paradigm" peripateia (or, in Kuhn's words, in a "scientific revolution") is a victory of formalism over modelism. In the EPR case we do have, since many years, the formalism. We are at home with it for performing calculations, but not yet for viewing our world, and our relation to it.

The papers in this volume are attempts to fashion an explanatory discourse with a view to producing an understandable view of our world. The ultimate goal is to construct a framework that is empirically adequate, that explains the outcomes of our observations, and that finally produces in us a sense of understanding how the world can be the way it is. These are three linked but distinct goals. It remains an open question whether all of these are simultaneously attainable.

5. *Conference papers*

This essay has been an informal and somewhat loose summary of the situation prior to the Notre Dame Conference in October of 1987. Let me conclude with come comments on the remaining papers in this volume and on the interrelations among them.

Abner Shimony discusses the various components of a consistent world-view and assesses the general implications of quantum theory for our world-view *provided* we take the present formulation of quantum mechanics to be a complete one. His paper sets the stage for the detailed analyses of the subsequent papers. Shimony argues against a (mere) instrumentalist interpretation of quantum mechanics and in favor of a realist interpretation. As one possible means of producing a coherent realist picture (to "close the circle" between metaphysics and epistemology), he suggests that the formalism of quantum mechanics may have to be modified (in a way which he specifies). This could serve as an example of adjusting a physical theory to meet certain metaphysical requirements.

David Mermin (1981a) has given a clear, nontechnical representation of the essential features of the Bohm-EPR "experiment." (This is reprinted as an appendix to his conference paper.) He shows convincingly that the emitted "particles" cannot simply be considered as individually carrying the information that determines the outcome (i.e., which color light flashes) once the particles interact with a particular detector that has been set. He argues that the particles cannot possess definite values for the measured quantities (prior to the measurement). In his conference paper, Mermin then goes on to give the novice an acute sense of the mystery of these correlations by suggesting the (apparently obvious or at least innocuous) validity of his "Strong Baseball

Principle.'' The reader feels the rug has been pulled out from under him or her when the ''data'' refute this principle.

Jon Jarrett in his paper also uses a ''Mermin device'' for purposes of illustration. He first defines the general framework of *local realism* and shows that it could not survive the ''empirical'' results which would be produced by a (real) laboratory version of a Mermin device. He then discusses his important (and by now well-known) result that locality (parameter independence) and completeness (outcome independence) together imply factorizability (which in turn implies a Bell inequality). Jarrett draws out more of the philosophical implications of the Bell-EPR analyses and experiments.

Linda Wessels spells out in careful detail the assumptions (both explicit and implicit) that are made in derivations of a Bell inequality. Given the failure of that inequality, she examines which of those assumptions most likely need to be given up. In the end Wessels singles out three of these assumptions, one or more of which is probably the culprit.

Bas van Fraassen argues that the outcomes of Bell-EPR type experiments cannot, even in principle, be given a common-cause explanation. He discusses in detail the notion of a causal mechanism and the reasons some philosophers have given for requiring such in any explanation that allows us to understand correlations between events (or ''outcomes''). In an appendix to his paper, van Fraassen analyzes various types of explanation that can be employed to account for statistical correlations. Since the concept of a common cause figures prominently in this and in several of the later papers in this volume, let me here indicate the basic idea involved. Suppose that, for some sample, the outcomes x and y are *not* statistically independent,

$$p(x,y) \neq p(x)\, p(y). \tag{11}$$

Suppose further, though, that when we condition (or restrict) the sample upon another factor z then the correlation does factorize as

$$p(x,y|z) = p(x|z)\, p(y|z). \tag{12}$$

In such a case, z is said to be the *common cause* of x and y.

In Jeremy Butterfield's paper, as in Michael Redhead's which follows, the reader encounters the concept of *screening-off*. The basic idea is this. If two events x and y are correlated, as in Eq. (11) above, but there exists a third event z such that conditionalization upon it produces stochastic independence, as in Eq. (12), then z ''screens off'' x and y. Butterfield focuses on the principle of the common cause according to which, if two events are correlated and one does not cause the other, then a common cause for both events can be found in their common past. His contention is that a principle of common cause and special relativity are *both* required to obtain Jarrett locality and/or Jarrett completeness. He presents locality and completeness each as screening-off conditions. Butterfield shows that, even under the weaker assumption of an extended domain for the common past of the state of a system, the Bell inequality follows. That is, even this extended (weakened) version of the principle of the common cause is refuted by the EPR correlations.

Michael Redhead introduces the criterion of *robustness* as a necessary condition for events to be connected even by stochastic causality (by which we mean that Eq. [12] can be made to obtain).[16] The motivation for such a requirement is that the events (or outcomes) x and y should be stable against some class of perturbations affecting the conditions surrounding x or y. Explicit calculation shows that the Bohm-EPR correlation (for the experiment of figure 3 above) is *not* robust. This result and special relativity together allow Redhead to rule out direct causal links and/or common causes as accounting for these correlations. He suggests using Shimony's expression, "passion at a distance," for this property of the quantum-mechanical state responsible for the correlation.

A central issue on which several of our authors divide is whether or not EPR-type correlations stand in need of an explanation, as opposed to being simply irreducible brute facts about nature. Henry Stapp begins by considering, and by and large rejecting, several possible ontologies currently on offer as underlying quantum phenomena. But he does see part of the explanation of quantum correlations as lying in the nonlocal connections of those phenomena. Stapp's thesis is that the quantum phenomena themselves require nonlocal influences in nature. Much of his argument is concerned with justifying assumptions (free from counterfactual definiteness) that can warrant this conclusion. Here counterfactual definiteness refers to a claim that the actually occurring outcome would still result under certain different conditions, had they instead obtained. Stapp's requirement is weakened from what *would* still occur in wing A when a change is made in wing B to what *could* still occur under those conditions.

Arthur Fine begins with a summary of his work on the conditions under which a common-cause explanation is (in principle) possible for a given set of data. Like van Fraassen, he forecloses the possibility of a causal explanation for the EPR correlations. He also comments upon the effectiveness of "would" and "could" distinctions of the type made in Mermin's Baseball Principle and in Stapp's analysis (respectively). But, then Fine turns the tables on the reader and makes the challenging suggestion that, in a truly indeter-

---

[16]Stochastic (local) hidden-variables theories have the advantage of avoiding any assumption of determinism (and this is desirable since deterministic local hidden-variables theories have been refuted by experiment). But, a Bell inequality follows for these stochastic theories as well (because only the factorizability of the joint probability is required for that; cf. Eq. [3]). Redhead's considerations hold for deterministic theories too.

ministic world, such correlations stand in no more need of an explanation than does a random string of outcomes of measurements made at a single (fixed) location. This is a dramatic and radical move that takes the correlations as primitive givens which may themselves be employed as basic elements in explaining other phenomena.

R. I. G. Hughes is one of those who feel that the EPR correlations *do* require an explanation. He proposes a "*structural*" explanation based on the terms used in the (mathematical) models which the general theory of the phenomena (i.e., quantum mechanics) employs. Hughes analyzes the (mathematical) components common to a general quantum-mechanical framework and then describes the EPR experiments in terms of four "theses" he has identified. He also discusses possible objections to the structural explanation sketched there.

Paul Teller's basic thesis is, like Arthur Fine's, a radical one, but in a very different direction from Fine's. Teller offers an explanation for (or perhaps better, a lesson to be drawn from) the EPR correlations in terms of an *ontological* relational holism. He takes the nonseparability of the quantum-mechanical formalism quite seriously and interprets it as a warrant for a modified ontology in which not all relational properties need simply "supervene" (or be impressed upon) those of independently specifiable individual entities. Teller argues for the need for such a move on the basis of an analysis of the role relativity plays in proofs of a Bell inequality. For him, the source of our difficulty in understanding the implications of the failure of the Bell inequalities is our implicit (perhaps unconsciously made) assumption of *particularism* (roughly, separability in the EPR context). Teller trades in particularism for relational holism to rid us of the interpretative conflicts we otherwise face.

Don Howard provides careful statements of the separability principle and of the locality principle, and then offers a proof of the Bell inequality based on locality and separability conditions. He examines the role these concepts played in Einstein's view of physical reality. His paper indicates that the tension between quantum theory and relativity becomes more severe when the general, rather than just the special, theory of relativity is considered. In fact, Howard questions the consistency of quantum field theory and of quantum mechanics itself as a *fundamental* theory.

Henry Folse's essay is a retrospective study of how Bohr's ideas (and to some extent, Einstein's as well) relate to the discussion surrounding the EPR correlations and Bell's theorem. He sketches Bohr's position on realism and argues that Bohr should not be seen as a positivist. Folse shows us that Bohr appreciated early on the conflict between a space-time description of physical phenomena and quantum reality. We also learn the philosophical basis of the differences between Bohr and Einstein, with separability being a key issue.

This type of philosophically sensitive analysis of the writings of Bohr and of Einstein on questions associated with quantum theory and physical reality has recently engaged the attention of several philosophers of science (Fine 1986; Folse 1985, 1987; Howard 1985; Krips 1987; Murdoch 1987).

The volume concludes with Ernan McMullin providing some perspective, from the history of science, on how scientific practice and regulative principles (to use current terminology) have in the past interacted to produce an explanation within an accepted (but possibly evolving) worldview. He shows that there are precedents in science for the present situation in which empirical adequacy obtains without any accompanying causal explanation. McMullin focuses on the history of planetary motion and argues that prior metaphysical commitments (or constraints) were responsible, at least in part, for the sense of bafflement which existed until certain of this baggage was jettisoned. We see that there are precedents for the type of crisis resolution suggested by, say, the dramatic moves made by Teller or by Howard in their essays.

APPENDIX

In order to illustrate the algebraic simplicity of some later proofs of Bell-type inequalities and the gradual realization and unpacking of the significance of the assumptions made, let us consider Stapp's 1971 proof. Although algebraically even simpler proofs are now available (e.g., Peres 1978; Redhead 1987a), Stapp's has the virtue of being one of the earliest of these "obvious" proofs and it has had an interesting subsequent history of its own. In accord with the variables employed in Eq. (1) of this introduction, let us agree upon the notation:

$r_{A_k}(\theta,\psi)$—the result (or outcome) at station A for the kth run ($k = 1, 2, 3, \ldots, N$) when the parameter ("choice") θ has been selected at station A and the parameter ("choice") ψ has been selected at station B.

$r_{B_k}(\theta,\psi)$—the result (or outcome) at station B for the kth run ($k = 1, 2, 3, \ldots, N$) when the parameter ("choice") θ has been selected at station A and the parameter ("choice") ψ has been selected at station B.

The choice (θ) at A and the choice (ψ) at B are made *independently* of each other. Each choice could be made so late (i.e., just before the "electrons" in figure 3 arrived at their respective detectors at stations A and B) that a light signal (or causal influence) could not travel from A to B (or vice versa) before each detector registered its determination of the spin of the electron passing through it. That is, we have already stated that the first signal principle of special relativity (cf. figure 4 again) is sometimes formulated as the requirement that an event at A cannot produce an effect at B (a distance l away) before a time of at least l/c has elapsed. Such a principle would seem to warrant the "locality" requirements that r_{A_k} can depend only upon the choice θ made at A (but not upon the choice ψ made at B) and, similarly, that r_{B_k} can depend only upon ψ (but not upon θ):[17]

$$r_{A_k}(\theta,\psi) = r_{A_k}(\theta), \tag{A.1}$$

$$r_{B_k}(\theta,\psi) = r_{B_k}(\psi). \tag{A.2}$$

The experimentally determined correlation $\langle r_A r_B \rangle$ is defined as

$$< r_A r_B \ (\theta, \ \psi) > \ \equiv \ \frac{1}{N} \sum_{k=1}^{N} r_{A_k} r_{B_k}. \tag{A.3}$$

In terms of the joint distributions (or probabilities) $p^{AB}(r_A, r_B | \theta, \psi)$, this can also be expressed as

$$\langle r_A r_B(\theta,\psi) \rangle = p^{AB}(++|\theta,\psi)(+1)(+1) + p^{AB}(+-|\theta,\psi)(+1)(-1)$$
$$+ p^{AB}(-+|\theta,\psi)(-1)(+1) + p^{AB}(--|\theta,\psi)(-1)(-1). \tag{A.4}$$

With the *empirical* distributions of Eqs. (5), we can express Eq. (A.4) (by means of an elementary trigonometric identity) as

$$\langle r_A r_B(\theta,\psi) \rangle = -\cos(\theta - \psi). \tag{A.5}$$

Stapp's argument is now essentially the following. We have from Eqs. (A.1), (A.2), (A.3) and (A.5) for the choice $\theta = \psi$

$$\frac{1}{N} \sum_{k=1}^{N} r_{A_k}(\theta) r_{B_k}(\theta) = -1. \tag{A.6}$$

Since each r_{A_k} and each r_{B_k} can be only *either* $+1$ *or* -1, Eq. (A.6) can be true if and only if

$$r_{B_k}(\theta) = -r_{A_k}(\theta), \text{ for all } k \text{ (for any given } \theta). \tag{A.7}$$

That is, we can (for a fixed θ) exchange a station index B for a station index A at the price of a minus sign. In simple succession, we then obtain at once the sequence of equalities and inequalities:

---

[17]The requirements of Eqs. (A.1) and (A.2) are in the spirit of (but *not* identical with) the Jarrett locality (Shimony's parameter independence) of Eq. (8).

$$\left| < r_A r_B \,(\theta, \, \psi_1) > \, - \, < r_A r_B \,(\theta, \, \psi_2) > \right|$$

$$\equiv \left| \frac{1}{N} \sum_{k=1}^{N} \left[r_{A_k}(\theta) r_{B_k}(\psi_1) - r_{A_k}(\theta) r_{B_k}(\psi_2) \right] \right|$$

$$= \left| \frac{1}{N} \sum_{k=1}^{N} r_{A_k}(\theta) r_{B_k}(\psi_1) \left[1 - r_{B_k}(\psi_1) r_{B_k}(\psi_2) \right] \right|$$

$$\leq \frac{1}{N} \sum_{k=1}^{N} \left| \left[1 - r_{B_k}(\psi_1) r_{B_k}(\psi_2) \right] \right| = \frac{1}{N} \sum_{k=1}^{N} \left[1 - r_{B_k}(\psi_1) r_{B_k}(\psi_2) \right]$$

$$= \frac{1}{N} \sum_{k=1}^{N} \left[1 + r_{A_k}(\psi_1) r_{B_k}(\psi_2) \right] = 1 + < r_A r_B \,(\psi_1, \psi_2) > .$$

Therefore, we have the general inequality among the correlations

$$\left| \langle r_A r_B(\theta, \psi_1) \rangle - \langle r_A r_B(\theta, \psi_2) \rangle \right| \leq 1 + \langle r_A r_B(\psi_1, \psi_2) \rangle. \qquad (A.8)$$

For the experimental correlations of Eq. (A.5), this inequality becomes

$$\left| \cos(\theta - \psi_1) - \cos(\theta - \psi_2) \right| \leq 1 - \cos(\psi_1 - \psi_2). \qquad (A.9)$$

For the choices

$$\theta = 0, \; \psi_1 = 45°, \; \psi = 135°, \qquad (A.10)$$

this reduces to

$$\sqrt{2} \leq 1. \qquad (A.11)$$

It may seem reasonable to claim that this proof demonstrates that experiment (i.e., the empirical correlations of Eq. [A.5]) requires nonlocality since only locality (but not determinism) is assumed (cf. Eqs. [A.1] and [A.2]). However, Stapp himself (1971) pointed out that statements such as those of Eqs. (A.1) and (A.2) are counterfactual definite ones, since they are equivalent to

$$r_{A_k}(\theta, \psi_1) = r_{A_k}(\theta, \psi_2), \qquad (A.12)$$

which is a claim that r_{A_k}, which has some specific numerical value when $\psi = \psi_1$ is chosen at B, *would* have the same value (or outcome) if the choice $\psi = \psi_2$ had instead been made on the kth run. However, there is no way empirically to verify a claim of this sort since the kth run can never be repeated, although, of course, the $(k + 1)^{\text{st}}$ run can be performed. *If* the underlying theory were a deterministic one, then it could serve as a warrant for the statements of Eqs. (A.1) and (A.2).

While there are other difficulties with these (algebraically) "simple" proofs of Bell-type inequalities, we do not pursue them here. (See Redhead, 1987a, for an engaging discussion of the subtle difficulties involved in these "self-evident" proofs; also Redhead 1987b, 90–96; Clauser and Shimony 1978, pp. 1898–1900.) Our purpose in this example has been to indicate the type of unpacking of assumptions that has followed upon the simplification of the mathematical technicalities of the Bell arguments. Without using either determinism or counterfactual definiteness or any other ideas alien to orthodox quantum theory, Stapp (1985a, 1988a, 1988b, 1988c, and the contribution to this volume) has recently presented arguments that one can deduce from the simple arithmetic contradiction given above (or an equivalent one) the incompatibility of quantum theory with the locality claim that causal influences cannot act outside the forward light cone.

SEARCH FOR A WORLDVIEW WHICH CAN ACCOMMODATE OUR KNOWLEDGE OF MICROPHYSICS

ABNER SHIMONY

1. Contribution of the natural sciences to a worldview

In a sense every human being with rudimentary intelligence has a worldview, that is, a set of attitudes on a wide range of fundamental matters. A philosophical worldview must be more than this, however. It must be articulate, systematic, and coherent. A philosophical worldview may be naturally subdivided into at least the following parts: (1) a metaphysics, which identifies the types of entities constituting the universe (possibly organizing them into a hierarchy, with some fundamental and others derivative), and which in addition asserts fundamental principles, such as those of cause and chance, that govern these entities; (2) an epistemology, concerned with the assessment of human claims to knowledge or justified belief; (3) a theory of language; and (4) a theory of ethical and aesthetic values. For reasons of professional training and preoccupation, and also of relevance to this collection of essays, I shall confine my comments almost entirely to metaphysics and epistemology.

A necessary condition for the coherence of a philosophical worldview is the meshing of its metaphysics and its epistemology. The metaphysics should be capable of understanding the existence and the status of subjects like ourselves who are capable of deploying the normative procedures of epistemology. And the epistemology should suffice to account for the capability of human beings to achieve something like a good approximation to knowledge of metaphysical principles, in spite of the spatial and temporal limitations of our experience and the flaws and distortions of our sensory and cognitive apparatus. This meshing can be briefly called "the closing of the circle."

In my opinion the twentieth century is one of the golden ages of metaphysics, probably surpassed by the fourth century B.C. in conceptual innovations, but probably surpassing all previous ages in the control and precision of the best metaphysical thinking. At least four important factors can be cited to account for the flowering of twentieth-century metaphysics. (1) There has been an unparalleled sharpening of logical, mathematical, and semantical analysis, which has provided powerful tools for philosophical criticism. The positivists used some of these tools to challenge the legitimacy of the enterprise of metaphysics, but this challenge has been a stimulus to careful constructive work.[1] (We learned from Carnap that in every neighborhood of a real problem there is a *Scheinproblem,* arising from abuse of language; but we also learned—from whom? maybe Plato—that in the neighborhood of every *Scheinproblem* there is a real problem awaiting intelligent extraction.) (2) The methodology of metaphysics has been greatly extended by the use of the hypothetico-deductive method—often tacitly, but in some cases (like Whitehead's *Process and Reality*) self-consciously and explicitly. There are, of course, classical instances—in the work of Newton, for example, and even earlier in the Greek atomists—of renouncing the assumption that a fundamental principle concerning the world has to be known with greater certainty than derivative principles. Aristotle already made the famous distinction between "things which are more obscure by nature but clearer to us" and "those which are clearer and more knowable by nature,"[2] but the full exploitation of his profound perception required the replacement of his own method of intuitive induction by the indirect procedures of the hypothetico-deductive method. (3) Parts of the natural sciences have been deepened to the point where evidence can be brought to bear in a controlled way upon problems traditionally classified as metaphysical. The results of relevant investigations in the natural sciences make it possible to use the hypothetico-deductive method specifically and fruitfully. (4) Phenomenological investigations have provided refined reports of ordinary experience, which are relevant to metaphysics. Although I am skeptical that phenomenology provides direct access to metaphysical principles, I believe that phenomenological reports are valuable supplements to typical scientific evidence in metaphysical investigations using the hypothetico-deductive method.[3]

Of the foregoing, I am most preoccupied with (3) and can cite several spectacular examples. General relativity has shown that the space-time field

---

[1] A fine example is Hao Wang, *Beyond Analytic Philosophy* (Cambridge, Mass.: MIT Press, 1986). See especially his remark about metaphysics on p. 131.

[2] Aristotle, *Physics* 184a 20–1, translated in *The Basic Works of Aristotle,* ed. R. McKeon (New York: Random House, 1941).

[3] See, for example, A. N. Whitehead, *Adventures of Ideas* (New York: Macmillan, 1933), chap. 15.

interacts dynamically with matter, thereby behaving more like a substance than did the fixed space-times of Newtonian kinematics or of special relativity. Molecular biology has provided, in some detail, an explanation in terms of chemical cybernetics of the crucial organic properties of teleological behavior, spontaneous morphogenesis, and reproduction, thereby eliminating the need for postulating nonphysical vitalistic elements in primitive biology. We can discern that elementary particle physics and cosmology are in the process of making contributions to metaphysics, but it is probably premature to say exactly what these are. The metaphysical implications of quantum mechanics will be discussed in some detail below.

It is legitimate, in view of all these rich results, to speak of the enterprise of *experimental metaphysics*. A warning is needed, however, against possible misunderstanding of this term. One should not anticipate straightforward and decisive resolution of metaphysical disputes by the outcomes of experiments. We know, in fact, that laboratory tests alone do not settle without careful conceptual analysis even those problems which are commonly classified as scientific, and *a fortiori* such analysis is indispensable in coming to grips with metaphysical problems.

2. *Metaphysical implications of quantum mechanics*

Whether quantum mechanics has definite metaphysical implications depends upon two conditions: (a) the legitimacy of interpreting the quantum state of a physical system "realistically," as something independent of the knowledge of an observer, or indeed of all observers; and (b) the completeness of the quantum-mechanical description of a physical system. If these two conditions are satisfied, then we are obliged to accept a literal understanding of the peculiarities of the quantum formalism. The fact that in any pure quantum state there are physical quantities that are not assigned sharp values will then mean that there is *objective indefiniteness* of these quantities. The fact that the outcome of a measurement of such an indefinite quantity is not determined by the quantum state of the object, even in conjunction with the quantum state of the measuring apparatus, means that there is *objective chance*. The fact that the quantum state does determine (with due consideration for statistical fluctuations) the frequencies in an ensemble of measurements means that there is *objective nonepistemic probability* (sometimes called "propensity"). The fact that there exist quantum states of two-body systems which cannot be factorized into products of one-body quantum states means that there is *objective entanglement* of the two bodies, and hence a kind of *holism*. And the fact that spatially well-separated bodies can be entangled means that there is a peculiar kind of *quantum nonlocality* in nature, which is manifested in correlations of the outcomes of measurements performed upon the separated bodies.

The challenge to condition (b) is the program of "hidden variables," which are putative properties of a physical system not specified by the system's quantum state. The simplest type of hidden-variables theory is one in which a complete state of a system assigns to every proposition (or "eventuality") of the system a definite truth value (so that the quantum-mechanical indefiniteness of a quantity is interpreted epistemically as due to ignorance or to practical limitations upon measurement). It was proved by Gleason and others[4] that in almost all interesting cases such a simple hidden-variables theory is mathematically excluded by the algebraic structure of the quantum formalism. Bell (1966) noted, however, that Gleason's theorem does not preclude a more complex type of hidden-variables theory, called "contextual," according to which a complete state assigns a definite truth value to a proposition only relative to a specified context. The success of a contextual hidden-variables program would permit all of the peculiarities of the quantum-mechanical formalism to be interpreted epistemically, thereby avoiding the metaphysical innovations implied by a realistic interpretation.

The encouragement which Bell provided to advocates of a more conservative metaphysics by exhibiting the family of contextual hidden-variables theories was undermined by a great theorem which he proved under restrictive conditions in 1964 but much more generally in 1971: Any hidden-variables theory which obeys a certain "locality condition" will imply that the correlations of certain quantities of two spatially separated systems must satisfy a certain inequality, and that under specified conditions the predictions of quantum mechanics violate that inequality. Bell's theorem precludes the interpretation of quantum mechanics by a contextual hidden-variables theory satisfying the locality condition. The discrepancy exhibited in Bell's theorem has the further consequence that in principle an experimental test can be made between quantum mechanics and the entire family of physical theories satisfying the locality condition. At least thirteen tests have been performed, of which all but two confirm the predictions of quantum mechanics and violate Bell's inequality. And there are good reasons for suspecting systematic errors in the two anomalous cases.[5] The most spectacular of these experiments is that of Aspect and his collaborators, in which the choice of analyzers used to examine the polarization of each of a pair of photons in a quantum-mechan-

---

[4]Gleason (1957); Bell (1966); Kochen and Specker (1967); Belinfante (1973).

[5]A detailed survey of experiments up to 1978, and also of variant proofs of Bell's theorem up to that date, is given by Clauser and Shimony (1978). A tabulation of experiments up to 1987 is given by Redhead (1987b); the experiment with time-varying analyzers is reported by Aspect, Dalibard, and Roger (1982).

ically entangled state is made too rapidly for information to be conveyed subluminally from the apparatus used for one of the photons to the apparatus used for the other; consequently, the violation of Bell's locality condition by the correlations exhibited in this experiment cannot be explained (except by rather contrived and desperate assumptions) as instances of information transfer in accordance with relativity theory.

The foregoing summary of Bell's theorem and the experiments which it inspired is extremely condensed, but details may justifiably be omitted since I have given detailed expositions elsewhere (1984a; 1984b; 1986) and many of the papers in this volume are devoted to detailed analysis. The following remarks will suffice for my purposes here.

1. Jarrett (1984) made the valuable discovery that Bell's locality condition is equivalent to the conjunction of two conditions, which he calls "locality" and "completeness" (for which I have suggested the names "parameter independence" and "outcome independence"). The former holds if the probability of an outcome of an observation on one particle is independent (given the complete state of the particle pair at the moment of emission) of the parameter chosen for the analyzer of the other particle; the latter holds if the probability of an outcome of an observation on one particle is independent (again given the complete state of the pair) of the outcome of the observation of the other particle. The violation of Bell's inequality by the experimental correlations implies by *modus tollens* that Bell's locality condition is violated in the experimental situation, and therefore one or the other of Jarrett's locality and completeness conditions must be false. But which one?

Since the experimental results not only violate Bell's inequality but confirm with accuracy the quantitative predictions of quantum mechanics, it is reasonable to answer this question by consulting quantum mechanics. It is easily seen that the quantum predictions of an entangled state can violate completeness (outcome independence), but analysis also shows[6] that the quantum predictions do not violate locality (parameter independence). Jarrett demonstrated that the violation of locality would in principle permit a message to be sent between the two pieces of apparatus with superluminal velocity. By contrast, even though a message can be sent by exploiting a violation of completeness, it would be subluminal (Jarrett 1984; Shimony 1986). Consequently, even though there is evidently some tension between relativity theory and the violation of completeness by quantum mechanics (and, by the argument just given, by the experimental results), there is nevertheless a kind of "peaceful coexistence" between quantum mechanics and relativity theory.

2. Because of peaceful coexistence, the tension between quantum me-

---

[6]Eberhard (1978); Page (1982); Ghirardi, Rimini and Weber (1980).

chanics and relativity theory is not a crisis, resolvable only by the retrench-
ment of one or both. Nevertheless, a deeper understanding of the relation
between these two fundamental parts of physical theory is desirable. One
possibility is that success in quantizing space-time structure (perhaps more
radically than envisaged in current programs of quantizing general relativity)
will throw some unexpected light upon peaceful coexistence. Specifically,
modification of the topology of space-time in the small may yield a new
interpretation of nonlocality. Another possibility is that illumination will
come more from conceptual analysis than from new physics. The failure of
completeness (outcome independence) indicates that the classical concept of a
localized event needs to be broadened and that the classical concept of
causality must be modified. Investigations in both of these directions are
reported in the papers by Henry Stapp, Don Howard, and Paul Teller in this
volume.

   3. Someone might propose to bypass Bell's theorem and the experiments which it inspired by accepting a hidden-variables theory that violates
Bell's locality condition. To be sure, in an experimental situation like Aspect's such a strategy would have been unpalatable to Einstein, but from the
standpoint of a conservative metaphysician it might be attractive: one could
exorcize the quantum-mechanical innovations of objective indefiniteness, objective chance, objective probability, and entanglement by paying the price of
accepting nonlocality. In my opinion, the crucial weakness of this proposal is
the unlikelihood that a nonlocal hidden-variables theory can be devised which
achieves the peaceful coexistence with relativity theory that quantum mechanics enjoys.[7] Unless revolutionary experiments overthrow the relativistic prohibition of superluminal signals, such peaceful coexistence remains a de-
sideratum for any physical theory.

   4. In the foregoing reasoning about the significance of Aspect's experiment several loopholes have been pointed out, the importance of which
should be assessed.

   (a) In Aspect's experiment, the choice of analyzers for examining the
polarization of each photon of a pair is made by switches which are periodically switched off and on. Even though relativity precludes the direct
transfer of information from one switch at a given moment in the laboratory

---

[7]The hidden-variables model of Bohm (1952) is so designed that it replicates all the predictions of standard quantum mechanics, though details of a relativistic version are not worked out.
It would seem, therefore, that if standard quantum mechanics can achieve peaceful coexistence
with relativity theory, so can this model of Bohm. However, Bohm himself was not content with
a completely phenomenological postulation of a ''quantum potential,'' and in later work he spoke
of objectively real fields which are the locus of the quantum potential (e.g., 1957, 112–114). The
question of peaceful coexistence of his models with relativity theory is discussed in the much
more recent work by Bohm and Hiley (1984) and Bell (1987).

frame of reference to the other switch at the same moment, it does not preclude sufficiently clever hidden variables located at one of the switches from using inductive logic to infer the contemporaneous state of the other switch. This criticism can be partially answered by pointing out that two switches are driven independently and have independent phase slippage from time to time, and furthermore that there is no obvious mechanism for the hypothetical exercise in inductive logic. Nevertheless, a revised experiment in which the switches are operated stochastically rather than periodically is desirable.

(b) Most of the pairs which pass through the analyzers on the two sides of the experiment are not detected in experiments performed to date. There is a hidden-variables model which reproduces the quantum-mechanical predictions of observed correlations because the hidden variables not only determine passage or non-passage of a particle through an analyzer but also detection or non-detection; and furthermore, all the operations postulated by this model are local.[8] The model requires, however, a limit upon the efficiency of the detectors, and hence this loophole can be blocked by performing a correlation experiment in which both the analyzers and the detectors are sufficiently efficient (Lo and Shimony 1981; Mermin 1986).

(c) Although the actions of the two switches for one pair of particles in Aspect's experiment are events with spacelike separation, and furthermore with spacelike separation from the event of particle emission, there nevertheless exists a region of overlap of the three backward light cones extending

---

[8]Clauser and Horne (1974). This model yields quantum-mechanical counting rates (provided the detectors are sufficiently inefficient) for any choice by the experimenter of analyzer orientations. The experimenter's choice does not affect the rules of operation of the model, but only specifies initial and boundary conditions, as in any physical experiment. Fine (1982d) has presented two models of a genre which he calls "prism models," in which there is response or nonresponse by the two parts of the apparatus, subsequent to the specification of the orientations of the analyzers only for a subspace of the space of states. His two models are much simpler than that of Clauser and Horne, but each has the limitation of being tailored to a specific choice of two alternative orientations of one analyzer and two alternative orientations of the other analyzer. For instance, in the "maximal" model on p. 287, the response function $A_i(\lambda)$ when the first analyzer is put in the ith orientation ($i = 1$ or 2 only), depends upon θ and θ', which in turn depend on the two alternative orientations chosen for the second analyzer. Although Fine says that there exist many more prism models, he has not explicitly exhibited one in which the response function $A_i(\lambda)$ for all the infinitely many values of i are defined in a way that is independent of the choice of two orientations of the second analyzer to be used in an actual experiment; and likewise for the response functions $B_j(\lambda)$ for all the infinitely many values of j of the second analyzer. If a different prism model has to be provided for each choice of the two pairs of orientations in an actual experiment, this would be tantamount to endowing the experimenter not only with the power to choose initial and boundary conditions but also with the ability to influence the laws of physics. Consequently, Fine has not justified his claim: "There is, therefore, no question of the consistency of prism models, at least in principle, with the quantum theory" (1986, 52).

from the switching and emission events; and it is logically possible that events in this overlap region can control the emission and analysis of each pair in a manner fully consistent with relativity theory, but in such a way that the quantum-mechanical correlations are achieved. There is, however, no known or even proposed mechanism for achieving this result. Consequently, general considerations of scientific methodology suggest that this loophole not be taken seriously. After all, any experimental result which seems to disconfirm a favored hypothesis could be attributed to a conspiracy with an unknown mechanism, but at the price of generic skepticism about experimental investigations of nature.

In spite of the reservations which have been indicated, the case is strong for rejecting hidden-variable theories and accepting the completeness of the quantum-mechanical description of physical systems. Before taking the next step and accepting the metaphysical implications of a realistic interpretation of quantum mechanics, a discussion is required of the realistic interpretation, labeled above in Section 2 as condition (a). There is an important philosophical tradition, extending from Berkeley to Mach to some of the positivists, according to which any scientific theory is nothing more than an instrument for organizing and predicting human experience. The usual arguments for an instrumentalist interpretation of scientific theories are independent of the content of the theories and apply just as well, or just as poorly, to Newtonian particle mechanics as to quantum mechanics. There are semantical arguments, to the effect that terms putatively referring to unobserved entities are meaningless; epistemological arguments, advocating maximum caution in performing inductions; and methodological arguments, espousing Ockham's principle.[9] But how good is the case for instrumentalism? The *a priori* arguments of Berkeley and his successors against physical realism rest upon unjustifiably restrictive and arbitrary semantical, epistemological, and methodological theses, as a vast body of critical literature has attempted to demonstrate.[10] If the *a priori* arguments fail, then instrumentalism must be judged *a posteriori* in competition with various versions of physical realism, preference being given to the point of view with the greatest explanatory power. It is particularly difficult to incorporate instrumentalism into a coherent philoso-

---

[9]I discovered, with great delight, Machs' [sic] General Store in Pawlet, Vermont; it reveals the Yankee storekeeper within the instrumentalist: ''What we have is on the shelves; what you don't see, don't exist; and we don't give credit.''

[10]Three papers from this literature which I have found particularly illuminating are C. S. Peirce's review of Fraser's edition of the works of George Berkeley, *Collected Papers*, vol. 8, ed. A. W. Burks (Cambridge, Mass.: Harvard University Press, 1958), 9–38; J. F. Thomson's article on Berkeley in *A Critical History of Western Philosophy*, ed. D. J. O'Connor (New York: Free Press, 1964), 236–252; and H. Stein (1972), 367–438, especially pp. 370–373 and pp. 405–409.

phy which "closes the circle" between metaphysics and epistemology, if the metaphysics takes cognizance of the evidence that human beings evolved only very recently in the history of the universe.[11]

3. *Reduction of the wavepacket and stochastic variants of quantum mechanics*

Although the content of a scientific theory is irrelevant to the usual arguments for interpreting it instrumentally, there is a peculiarity in quantum theory that supplies an entirely novel argument for instrumentalism. The peculiarity is the problem of measurement (also known as "the problem of the reduction of the wavepacket" and as "the problem of the actualization of potentialities"), and it arises from the linearity of quantum dynamics. Here is a highly idealized formulation of the problem, which suffices for present purposes. Suppose that u_1 and u_2 are normalized vectors representing states of a microscopic object in which a physical quantity A has distinct values a_1 and a_2; that v_0 is a normalized vector representing the initial state of a measuring apparatus; that v_1 and v_2 are normalized vectors representing states of the measuring apparatus in which a macroscopic quantity B (such as the position of a pointer needle) has distinct values b_1 and b_2 respectively; and finally that the interaction between the microscopic object and the measuring apparatus is such that

$$u_i \otimes v_0 \rightarrow u_i \otimes v_i \ (i = 1 \ or \ 2),$$

where the arrow stands for temporal evolution during an interval of specified duration t. Under these circumstances, if it is known that initially the microscopic object was either in the state represented by u_1 or the state represented by u_2, but it was not known which, then the missing information can be obtained simply by examining the quantity B of the apparatus at time t after the interaction commenced and ascertaining whether the value of this macroscopic quantity is b_1 or b_2. Furthermore, the observation of B will permit the inference of the value of A both at the beginning and at the termination of the measurement. Thus far there is no conceptual difficulty.

Suppose, however, that initially the microscopic object is prepared in the state represented by the superposition $c_1 u_1 + c_2 u_2$, where the sum of the absolute squares of c_1 and c_2 is 1, and neither c_1 nor c_2 is zero. States of this

[11]This argument is often presented in works on evolutionary epistemology. One version may be found in Shimony (1971). Mach himself was an ardent advocate of the theory of evolution; there was evident tension between this part of his thought and his dictum that "the world consists only of our sensations," *The Analysis of Sensations* (New York: Dover, 1959), 12.

kind are physically possible, according to the formalism of quantum mechanics, and indeed it is often experimentally feasible to prepare such states. The linear dynamics of quantum mechanics implies then that

$$(c_1 u_1 + c_2 u_2) \otimes v_0 \rightarrow c_1 u_1 \otimes v_1 + c_2 u_2 \otimes v_2.$$

In the state represented by the sum on the right hand side of this process, the macroscopic quantity B does not have a definite value. This fact in itself is peculiar, because our ordinary experience indicates that macroscopic physical quantities always have definite values. Furthermore, there is a conceptual difficulty in understanding the quantum formalism. The standard interpretation of the superposition $c_1 u_1 + c_2 u_2$ is that the quantity A has an indefinite value, but in the event that A is actualized, there is a probability $|c_1|^2$ that the result will be a_1 and a probability $|c_2|^2$ that it will be a_2. Now if the quantum dynamics precludes a definite measurement result, what sense does it make to speak of the probabilities of various outcomes? A literal and realistic interpretation of the quantum dynamics undermines the literal and realistic interpretation of the quantum state! This problem evidently is dissolved, however, if the quantum formalism is interpreted instrumentally. For then the quantum state is interpreted as a shorthand for a procedure which permits the reliable anticipation of a statistical distribution in observations of a certain kind. The superposition $c_1 u_1 + c_2 u_2$ can be used effectively for this purpose, and one should avoid the *Scheinproblem* of demanding the values of quantities A or B apart from their role in human experience.

My opinion is that a realistic interpretation of the quantum formalism, and thereby the metaphysical implications of quantum mechanics summarized earlier, can be salvaged by a change of physics: *viz.*, the abandonment of linear dynamics. This proposal—admittedly conjectural—is the antithesis of my attempt to draw philosophical consequences from scientific results, for it indicates rather a reliance on philosophical considerations to supply the heuristics for a scientific investigation. Working physicists are happy with standard quantum dynamics (the time-dependent Schrödinger equation and its relativistic surrogates), and there are no puzzling phenomena at present which point to a modification of this dynamics. The motivation for this suggestion is that of saving physical realism. The title of my paper—"Search for a worldview which can accommodate our knowledge of microphysics"—in a way is not quite appropriate, since I am proposing to modify the microphysics in order to achieve a coherent worldview.

One way to modify linearity is to maintain a differential equation governing dynamic evolution of a state, but to replace the linear differential equation of Schrödinger by a nonlinear one. The hope would be that the nonlinear equation would agree very closely with the Schrödinger equation for systems with few degrees of freedom, where the latter has been very well confirmed, but would deviate sharply from it for systems with many degrees of freedom. Specifically, a superposition of states in which a macroscopic quantity has distinct values should spontaneously and rapidly evolve into a state in which that quantity has a sharp value. The requirement "rapidly" is imposed to agree with the phenomenology of laboratory measurements, which are often completed in time intervals of the order of microseconds or less. No proposals have been made concretely which satisfy all of these *desiderata,* and it may be mathematically impossible to satisfy them simultaneously.

More promising is the family of stochastic modifications of the Schrödinger equation, in which the initial state does not determine the final state unequivocally, and hence a role of chance in the outcome is explicitly postulated—and such modifications have been proposed.[12]

The theory of Ghirardi, Rimini, and Weber (1986) seems to me the most promising to date, and indeed they have sketched a very impressive unification of microdynamics and macrodynamics. They postulate that during all but a discrete set of instants the Schrödinger equation governs the evolution of the quantum state of a physical system; however, at instants which themselves are selected stochastically, there is a spontaneous reduction of the spread of the quantum state in configuration space. The rate of such stochastic reductions of spread is slow for systems with few degrees of freedom and very fast (automatically from their formalism, with no additional postulation) for systems with many degrees of freedom. One anticipates, therefore, that even if a measurement should result in a state which is a superposition of different positions of a pointer needle, which is a macroscopic system with a number of the order of 10^{23} of degrees of freedom, there would be a spontaneous transition within a microsecond or so into a well-localized state of the needle.

A number of objections may be raised against the theory of Ghirardi, Rimini, and Weber. It implies that energy and momentum are not conserved, though the amount of nonconservation is small. The theory is nonrelativistic. It is very much tied to the position representation, even though experimental considerations may indicate more desirable choices of the representation in which reduction takes place. It may not suffice to explain the rapidity of results in most measurement situations, for as Albert and Vaidman pointed out,[13] the excitation of a few molecules, with few degrees of freedom, by an incident particle is the crucial precipitating step of a typical measurement. My strongest objection, however, is that the theory permits the formation for a

---

[12]Pearle (1976; 1979; 1982); Gisin (1981; 1984); Gisin and Piron (1981); and Ghirardi, Rimini, and Weber (1986).

[13]D. Albert and L. Vaidman, "On a theory of the collapse of the wave-function," unpublished.

short time of monstrous states of macroscopic objects—states in which the pointer needle has an indefinite position, or Schrödinger's cat is neither dead nor alive—and then it rapidly aborts the monstrosity. It would be much better to have a stochastic theory which provides contraception against the formation of such a monstrosity, by destroying unwanted superpositions in the earliest stages of the interaction of the microscopic object with the macroscopic apparatus. I have no suggestions at present for the details of such a theory, but it is my preoccupation and the subject of my current research.

A stochastic modification of quantum mechanics will not be acceptable unless it has clear experimental consequences which are confirmed. At present I have no way of saying what those consequences might be. My hope is that some of them will concern macromolecules, which occupy a strategic position between microscopic and macroscopic bodies. The quantum mechanics of macromolecules has provided essential ingredients in the triumphs of molecular biology of the last three decades: the stability of DNA, the folding of proteins to form characteristic globular shapes, and the stereospecific recognition of a substrate by an enzyme. I am not convinced, however, that current quantum dynamics is entirely satisfactory for biological purposes. It is biologically important that an enzyme behave like a switch, which is definitely off or definitely on with regard to mediating a chemical reaction. It would be troublesome to have macromolecular analogues of Schrödinger's cat—enzymes which are in a superposition of switching on and switching off. Pragmatically, molecular biology pays no attention to the possibility of such superpositions, which seems to indicate that they seldom if ever occur. It is not clear to me, however, that the nonoccurrence of such superpositions is well understood within the framework of standard linear quantum dynamics, whereas a stochastic modification of this dynamics might provide a natural explanation (see Pattee 1967).

4.   The mind-body problem

The conjectured stochastic modification of quantum dynamics would contribute greatly to "the closing of the circle," which is one of the chief requirements for a coherent worldview, for it would explain in principle how there are definite outcomes of experiments—e.g., definite registrations on photographic plates and definite firings of Geiger counters—and these outcomes provide the evidence upon which the immense superstructure of physical theory is based. But unless a physicalistic explanation of mental events is correct—which seems incredible to me, for reasons that are naive but strong—the definite physical outcomes of experiments constitute only a necessary but not a sufficient condition for human experience and inference. The greatest obstacle to "closing the circle" is the ancient one which haunted Descartes and Locke—the mind-body problem. Does contemporary microphysics have any implications concerning this problem?

One very interesting implication is negative. Whitehead proposed to solve the mind-body problem by a modern version of Leibniz's monadology, according to which the fundamental constituents of natural things are endowed with protomental characteristics. (Whitehead's actual occasions differ from Leibniz's monads by being short-lived and possessing "windows.") In ordinary matter the protomental characteristics of neighboring occasions are "incoherent," and therefore the descriptions of matter by physical theory are statistically very accurate; but a sketch of an explanation is also provided for the possibility of emergence of high level mentality, when the protomental characteristics of a society of occasions achieve "coherence."[14] It is, however, very difficult to reconcile Whitehead's theory with the great body of evidence favoring the intrinsic identity of electrons, for his postulated protomental features at the subphysical level would endow each electron with unequivocal individuality.

On the other hand, the metaphysical innovations of quantum mechanics— objective indefiniteness, objective chance, objective probability, and entanglement—have obvious analogies to some phenomenological features of mentality. It is premature to judge the significance of these analogies. Nevertheless, I conjecture that a world in which these metaphysical principles hold is somehow more hospitable to a dualism of mind and body than a world governed by the metaphysical principles associated with classical physics. I certainly do not want to suggest that quantum mechanics by itself provides a resolution to what Whitehead calls "the bifurcation of nature," but it may provide a framework within which a resolution can be sought.[15] In making this conjecture I run the risk that my position may be conflated with that of Maharishi Mahesh Yogi, who expounds the thesis that macroscopic quantum coherence is exhibited by a meditating cohort (the "Maharishi effect"). I suspect, however, that the Maharishi is not as cognizant as I of the need to determine with precision the mathematical structure of the space of mental states or to design careful and informative experiments.

---

[14]A. N. Whitehead, *Adventures of Ideas,* 266–267.

[15]See R. Faber, *Clockwork Garden* (Amherst, Mass.: University of Massachusetts Press, 1986), chaps. 10–11. This book is reviewed by Shimony, *Foundations of Physics* 17 (1987): 1041–1043.

CAN YOU HELP YOUR TEAM TONIGHT
BY WATCHING ON TV?
MORE EXPERIMENTAL METAPHYSICS
FROM EINSTEIN, PODOLSKY, AND ROSEN

N. David Mermin

A few years ago I described (1981a)[1] a simple device that reveals in a very elementary way the extremely perplexing character the data from the Bohm-Einstein-Podolsky-Rosen experiment assumes in the light of the analysis of J. S. Bell. There is a second, closely related, form of that *Gedanken* demonstration,[2] which I would like to examine for several reasons.

1. It is simpler: there are only two (not three) settings for each switch.
2. The *Gedanken* data resemble more closely the data collected in actual realizations of the device.
3. None of the possible switch settings produce the perfect correlations found in the first version of the *Gedanken* demonstration, where the lights *always* flash the same color when the switches have the same setting. Since absolutely perfect correlations are never found in the imperfect experiments we contend with in the real world, an argument that eliminates this feature of the ideal *Gedanken* data can be

Some of the views expressed here were acquired in the course of occasional technical studies of Einstein-Podolsky-Rosen correlations supported by the National Science Foundation in a small corner of grant No. DMR 86–13368.

[1]Reprinted as an appendix below. An only slightly more technical but significantly more graceful version appeared a few years later (Mermin, 1985).

[2]What follows is my attempt to simplify some reformulations of EPR and Bell by Henry Stapp (for example, 1985a), but the interpretation I give differs from his, and any foolishness in what follows is entirely my own.

applied to real data from real experiments. (If you believe, however, along with virtually all physicists, that the quantum theory gives the correct ideal limiting description of all phenomena to which it can be applied, then this is not so important a consideration.)

4. Because the ideal perfect correlations are absent from this version of the *Gedanken* demonstration, one is no longer impelled to assert the existence of impossible instruction-sets. To establish that the new data nevertheless remain peculiar, it is necessary to take a different line of attack, which has again intriguing philosophical implications, but of a rather different character.[3]

1. The modified demonstration

In the modified *Gedanken* demonstration, there are only two switch settings (1 and 2) at each detector. Otherwise the setup is unchanged: there are two detectors (*A* and *B*) and a source (*C*), and the result of each run is the flashing of a red or green light. If one had actually built such a device according to the quantum-mechanical prescription, it could be transformed to run in this modified mode simply by readjusting the angle through which certain internal parts of each detector turned as the switch settings were changed.[4]

In its new mode of operation, the device produces the following data:

(*i*) When the experiment is run with both switches set to 2 (22 runs), the lights flash the same colors only 15% of the time; in 85% of the 22 runs different colors flash.

(*ii*) When the experiment is run with any of the other three possible switch settings (11, 12, or 21 runs) then the lights flash the same colors 85% of the time; in only 15% of these runs do different colors flash.

As in the earlier version of the *Gedanken* experiment, RR and GG are equally likely when the lights do flash the same colors, and RG and GR are equally likely when different colors flash. Also as earlier, the pattern of colors observed at any single detector is entirely random. There is no way to infer

---

[3]There are more orthodox ways of extracting the peculiar character of these data. The route I take here requires fewer formal probabilistic excursions, and leads to a rather different philosophical point, though I suspect a careful analysis of the use of probability distributions in the conventional arguments might uncover something quite similar.

[4]Physicists might note that if setting 1 at detector *A* corresponds to measuring the vertical spin component, then setting 2 at *A* measures the component at 90° to the vertical; setting 1 at *B*, 45° to the vertical, and setting 2 at *B*, −45° to the vertical, all four directions lying in the same plane. (In the earlier version the three switch settings at either detector corresponded to 0°, 120°, and −120°.) The fraction 85% is just $\cos^2(22.5°) = \frac{1}{4}(2 + \sqrt{2})$.

from the data at one detector how the switch was set at the other. Regardless of what is going on at detector B, the data for a great many runs at detector A is simply a random string of R's and G's, that might look like this:

Typical Data at Detector A

A: R G R R G G R G R R R G G G R G R R R G R G R G . . .

The choice of switch settings only affects the *relation* between the colors flashed at *both* detectors. If, for example, the above data had been obtained at detector A when its switch was set to 2, and in all those runs the switch at B had also been set to 2, then, as noted above, the color flashed at B would have agreed with that flashed at A in only 15% of the runs, and the lights flashed at both detectors together might thus have looked like this:[5]

Data from a Series of 22 Runs

A2: R G R R G G R G R R R G G G R G R R R G R G R G . . .
B2: G R G G G R G R G G R G R R R G R G G R R R R G R . . .

Although the list of colors flashed at either detector remains quite random, the color flashed at B is highly (negatively) correlated with the color flashed at A. In the overwhelming majority (85%) of the runs the detectors flash different colors. Only in a few (15%) of the runs do the detectors flash the same colors.

On the other hand, for any of the other switch settings (take 21 as an example) the comparative data would have looked something like this:

Data from a Series of 21 Runs

A2: R G R R G G R G R R R G G G R G R R R G R G R G . . .
B1: R G R G G G R R R R G G R G G R R R G G G R G . . .

Again we have two lists of colors, each entirely random, but they now agree with each other in 85% of the runs, disagreeing in only 15%.

There are various ways to run the modified *Gedanken* demonstration, but let me focus on the following procedure, which it seems to me makes a rather striking contribution to Abner Shimony's field of experimental metaphysics. Suppose we do a long series of runs in each of which both switches are set to 1:

---

[5]The numbers after A and B denote the fixed setting of their switches throughout the sequence of runs. In contrast to some earlier versions of the *Gedanken* demonstration, we now try out various fixed switch settings, rather than randomly resetting the switches after each run.

Data from a Series of 11 Runs

A1: R G G G R G R G R R G R R G R G R R R G R G G . . .
B1: R G G R R G G R R R G R R G R G R G R R G G G G . . .

About 85% of these 11 runs will produce the same colors, and 15%, different. Now because there are no connections of any kind between the detectors at *A* and *B*, it seems clear that whatever happens at *A* cannot in any way depend on how the switch was set at *B*, and vice versa. Let us elevate this commonsense remark into a principle, which I shall call the Baseball Principle. Before examining the implications of the *Gedanken* demonstration for the Baseball Principle, let us discuss it in the context from which its name derives, where it assumes (at least for me) an especially vivid character.

2. The Baseball Principle

I am a New York Mets fan, and when they play a crucial game I feel I should watch on television. Why? Not just to find out what is going on. Somewhere deep inside me, I feel that my watching the game makes a difference—that the Mets are more likely to win if I am following things than if I am not. How can I say such a thing? Do I think, for example, that by offering up little prayers at crucial moments I can induce a very gentle divine intervention that will produce the minute change in trajectory of bat or ball that makes the difference between a hit or an out? Of course not! My feeling is completely irrational. If you insisted that I calm down and think about it, I would have to admit that the outcome of the game does not depend in the least on whether I watch it or not. What I do or do not do in Ithaca, New York, can have no effect on what the Mets do or do not do in Flushing, New York. This is the Baseball Principle.

Now a pedant comes along and says, "What do you really mean by that Baseball Principle?" And then, being a pedant, he tells me what I really mean. What I really mean is this: If we examined a great many Mets games and divided them up into those I watched at least part of on TV and those I did not watch at all, and if my decision to watch or not was entirely independent of anything I knew about the game—made, for example, by tossing a coin—then we would find that the Mets were no more or less successful in those games I watched than in those games I did not.

Now I reply, "That's very nice, but I mean something much simpler. I mean that in each individual game, it doesn't make any difference whether I watch it or not. Tonight, for example, whatever the Mets do, will be exactly the same, whether or not I end up watching the game."

"C'mon," says the pedant, "that's silly. Either you watch the game or you don't. You can't say that what happens in the game in the case that didn't happen is the same as what happens in the case that did, because there's no way to check. What didn't happen *didn't happen*."

I say to the pedant: "Who's being silly here? Are you trying to tell me that it *does* make a difference in tonight's game whether I watch it or not?"

"No," says the pedant, "I'm saying that your statement that it doesn't make any difference whether or not you watch an individual game can only be viewed as a very convenient construct to summarize the more complex statistical statement about correlations between watching and winning over many games. All of its statistical implications are correct, but it has no meaning when applied to an individual game, because there is no way to verify it in the case of the individual game, which you cannot both watch and not watch."

But is it *wrong* to apply the Baseball Principle to an individual game?

3.   *The Strong Baseball Principle*

Let us call the claim that the Baseball Principle applies to each individual game the Strong Baseball Principle. The Strong Baseball Principle insists that the outcome of any particular game does not depend on what I do with my television set—that whatever it is that happens *tonight* in Shea Stadium will happen in exactly the same way, whether or not I am watching on TV.

As a rational person, who is not superstitious, and does not believe in telepathy or the efficacy of prayer on the sporting scene, I am convinced of the Strong Baseball Principle. True, there is no way to verify it, since I cannot both watch and not watch tonight's game, and am therefore unable to compare how the game goes in both cases to make sure nothing changes. Nevertheless, deep in my heart, I do believe that because there is no mechanism connecting what I do with the TV at home to what happens in Shea Stadium, the outcome of tonight's game genuinely does not depend on whether I watch it or not: the Strong Baseball Principle. Try as you may to persuade me that the Strong Baseball Principle is meaningless, in my heart, I know it is right.

Remarkably, when run in the second mode, the *Gedanken* demonstration provides us with a case in which if it really does make no difference whether or not I watch the game, then it is not only meaningless, but demonstrably *wrong* to assert this principle in the individual case. If the Baseball Principle is right for the device, then the Strong Baseball Principle must be wrong, not merely because it naively compares possibilities only one of which can be realized, but because it is directly contradicted by certain observed facts. Such an experimental refutation of the Strong Baseball Principle would have been impossible before the discovery of the quantum theory; you cannot get into trouble using the Strong Baseball Principle in classical physics, and it can, in fact, be a powerful conceptual tool.[6] I believe that those who take the view that an experimental refutation is of no interest since reasoning from the Strong Baseball Principle was impermissible all along miss something of central importance for an understanding of the character of quantum phenomena.

4. The device and the Baseball Principle

We return from ball games to the device. There are no connections between the detectors or between the source and either detector. The Baseball Principle therefore applies, and asserts that what goes on at detector A does not depend on how the switch is set at detector B, and vice versa. This is readily verified in the statistical sense insisted on by the pedant. Keep the switch at A set to 1. Do a great many runs with the switch at B set to 1. Then, keeping the switch at A at 1, do a second series of runs with the switch at B set to 2. Compare the data at A in the two cases. It will have exactly the same character—namely a featureless sequence of R's and G's like the series of heads and tails you get by repeatedly flipping a coin. There is nothing in the outcome at A to distinguish between the runs in which B was set to 1 or to 2.

But what about the Strong Baseball Principle? Given the lack of any connection between the detectors, can we not also assert that what goes on at one detector in any *individual* run of the experiment does not depend on how the switch is set at the other detector? Granted, there is no way to test this stronger assertion, but surely, for the same reason, there is also no way to refute it. But here, remarkably in my opinion, we have a case in which the Strong Baseball Principle is directly contradicted by the data. Consider what happened when the device was run with both switches set to 1:

Actual Data from a Series of 11 Runs

A1: R G G G R G R G R R G R R G R G R R R R G R G G . . .
B1: R G G R R G G R R R G R R G R G R G R R G G G G . . .

If there are really no connections between A and B, and no spooky actions at a distance, then what happens at detector A cannot depend on how the switch is set at detector B (and vice versa). The Strong Baseball Principle takes this to

---

[6]In a deterministic world in which the future can be calculated from present conditions, the Strong Baseball Principle can be given an unambiguous meaning.

mean that in the first run of this sequence (in which both lights flashed R) the light at detector *A* would have flashed R even if the switch on detector *B* had been set to 2 instead of 1, and similarly, for every other run in the series, if *B* had been set to 2 nothing would have changed at *A*. In no individual run can the outcome at *A* depend on how the switch was set at *B*. (Compare this with "In no individual baseball game can the outcome at Shea Stadium depend on how the switch was set on my TV.")

Well, if that is so, we can say something about what would have happened if the run had been 12 (*A*1 and *B*2) rather than 11 (*A*1 and *B*1)— namely the outcomes at *A* would have been exactly the same as before:[7]

The 11 Runs and What the Strong Baseball Principle Can Say About What Would Have Happened Had They Been 12 Runs

B2: ? ...
A1: R G G G R G R G R G R R G R R G R G R R R R G R G G ...

A1: R G G G R G R G R G R R G R R G R G R R R R G R G G ...
B1: R G G R R G G R R R G R R G R G R G R G R R G G G G ...

Note that in this application of the Strong Baseball Principle we make no commitment at all to what colors flashed at *B* in the case that did not take place (with the switch at *B* set to 2) since, after all, that did not happen. We merely assert that whatever might have taken place at *B* in that unrealized experiment, nothing would have turned out any differently at *A*.

We can also say the same thing about what would have happened at *B*, if we had set the switch differently at *A*. This gives us one more pair of rows:

The 11 Runs and What the Strong Baseball Principle Can Say About What Would Have Happened Had They Been 12 Runs Or What Would Have Happened Had They Been 21 Runs

B2: ? ...
A1: R G G G R G R G R R G R R G R G R R R R G R G G ...

A1: R G G G R G R G R R G R R G R G R R R R G R G G ...
B1: R G G R R G G R R R G R R G R G R G R R G G G G ...

B1: R G G R R G G R R R G R R G R G R G R R G G G G ...
A2: ? ...

---

[7]This does not imply determinism—indeed, I am not convinced that what happens in a baseball game *is* deterministic; it simply says, in the baseball case, that whatever it is that does happen is not going to depend on what a television set 300 miles away is doing.

Consider now what we have laid out here. The middle two (third and fourth) rows show what actually happened: both switches were set to 1, and the first run gave RR, the second, GG, the third GG, the fourth GR, etc. The top two rows (first and second) express the Strong Baseball Principle in the form that asserts that the outcome of *each individual* run at A does not depend on how the switch is set at B. The bottom two (fifth and sixth) express it as an assertion that the outcome of each run at B does not depend on the switch setting at A.

Now what about the question marks? They appear in the top (first) and bottom (sixth) rows because those rows represent what would have happened at B and A had the switches there been other than what they actually were. Evidently *some* sequence of R's and G's would have been produced in either case,[8] but we have no way of telling which. Experience with the device, however, tells us some of the features these sequences would have had, if the runs had been 12 or 21 runs rather than the 11 run that actually took place. An acceptable sequence of R's and G's for the first ($B2$) row, must agree with the sequence of R's and G's in the second ($A1$) row in about 85% of the positions, since that is the way 12 runs always work. Similarly a sequence of R's and G's replacing the question marks in the sixth row must agree in about 85% of the positions with the sequence in the fifth row, since that is what always happens in 21 runs. These considerations cut down on the number of ways of replacing question marks with R's and G's, but many different possibilities are still allowed.

A final application of the Strong Baseball Principle can be made to restrict these possibilities further. Suppose both switches had been set to 2 rather than 1. We can regard this 22 series of runs either as a modification of a 21 series (modified by changing the switch setting at B without changing anything at A) or as a modified 12 series (in which the switch was changed at A without anything having been done at B). We do not know, of course, what would have happened at B in the hypothetical 12 series (top row of question marks) or at A in the hypothetical 21 series (bottom row of question marks). The Strong Baseball Principle asserts that whatever series of R's and G's at A the question marks in the bottom row might stand for in the 21 run, that same series of R's and G's would also have happened at A in that series of runs had

---

[8]At this moment in my talk at the conference, there were cries of protest from the philosophers in the house. I was told that "If I were hungry, I would eat a candy bar" does not imply the proposition "There exists a candy bar which is the one I would eat were I hungry" (the Candy Bar Principle). I affirmed my commitment to the Candy Bar Principle. I said I wanted to make a rather different point, but I think they all stopped listening then and there. I hope you will not stop reading here and now. If you insist on talking candy, I would suggest that a more accurately analogous proposition is "Either there exists a candy bar which is the one I would eat were I hungry or there does not."

the switch at B been set to 2 instead of 1—i.e., had the runs been 22 runs instead. By the same token, whatever sequence of R's and G's the question marks in the top row represented for the results at B in a series of 12 runs, that same sequence would also have described the results at B had the runs been 22 runs.

This last application of the Strong Baseball Principle, by comparing hypothetical cases, has a different character than the first two, which compare a hypothetical case with the real one, and here it might more accurately be termed the Very Strong Baseball Principle. Returning to the sporting analogy, the Very Strong Baseball Principle applies when the game is, in fact, canceled because of rain. I nevertheless maintain that had the game been played, it would have taken place in exactly the same way, whether or not I watched it. This last assertion may elicit an even more violent objection from the pedant. Is it really reasonable to insist that something should happen in exactly the same way when conditions change very far away from it, when in actual fact it never happened at all?

But is it really any more reasonable, I hasten to add, to insist that such an assertion is impermissible? I maintain that if last night's game had not been rained out, it would have happened the same way whether or not I had watched it on television. Can you prove me wrong when I say this? Wouldn't most unsuperstitious people regard the proposition as true? Indeed, as uninterestingly true? To be sure, the pedant will translate it into a series of harmless statistical assertions, but is it really *wrong* to apply it to the individual case as well? The hallmark of the Strong Baseball Principle at work is this nagging conviction, to which only a pedant could object. For how can one possibly get into any trouble asserting relations between two things neither of which actually happened?

One can. It is worse than bad form; it is bad physics. Let us try it out. We have to replace the first row with some sequence of R's and G's and the sixth row with some other such sequence in such a way that the first and second rows give the right statistics for 12 runs, the fifth and sixth, for 21 runs, *and* the first and sixth for 22 runs. We do not insist that any particular way of doing this is preferable to or any more deserving of some hypothetical reality than any other, but for the Strong Baseball Principle to survive, *some* among the various possibilities must be consistent with these statistics.

Now in 22 runs the colors disagree 85% of the time, so whatever goes into the first row has to disagree with whatever goes into the sixth in about 85% of the positions.

On the other hand the set of R's and G's in the top row can differ from that in the second row in only about 15% of the positions (since they must have the correlations appropriate to a series of 12 runs). The second row is the same as the third row (by the Strong Baseball Principle). The third row differs from the fourth row in only about 15% of the positions, since they give the data in a 11 run. The fourth row is the same as the fifth row (by the Strong Baseball Principle). And the fifth row can differ from the set of R's and G's appearing in the bottom row in only 15% of the positions (since those rows must have the correlations appropriate to a series of 21 runs).

A moment's reflection on the last paragraph is enough to reveal that whatever sequence of R's and G's is in the top row, it can differ from whatever sequence is in the bottom row, in at most about 15% + 15% + 15% = 45% of the positions. But according to the next to the last paragraph whatever is in the top row must differ from whatever is in the bottom row in about 85% of the positions. You cannot have it both ways. Thus the (Very) Strong Baseball Principle is so restrictive as to rule out *every* possibility for the unrealized switch settings. Far from merely being meaningless nonsense, an application of the Strong Baseball Principle to the *Gedanken* demonstration contradicts the observed facts.

5. *Conclusion*

In this demolition of the Strong Baseball Principle, we did not interpret it as demanding the existence in some cosmic bookkeeping office of a list of data for the unperformed runs. We only took it to require that if the *actual* experiment consists of a long series of 11 runs, then among all the *possible* sets of data that *might* have been collected had the experiment instead consisted of 12, 21, or 22 runs, there should be *some* satisfying the condition that, run by run, what happens at one detector does not depend on how the switch is set at the other. If the Strong Baseball Principle is valid, it should be possible to *imagine* sets of $B2$ and $A2$ data such that the $B2$ data produce the right statistics (85% same and 15% different) when combined with the actual $A1$ data, the $A2$ data produce the right statistics (85% same and 15% different) when combined with the actual $B1$ data, and the two sets of imagined data produce the right statistics (15% same and 85% different) for a 22 experiment.[9]

Since it is impossible to imagine *any* such sets of data, then the Strong Baseball Principle has to be abandoned not because it is bad form, unjustifiable, or frivolous to argue from what might have happened but did not, but

---

[9]In Candy Bar terms, the Strong Baseball Principle does not say that there exists a particular sequence of R's and G's which are the colors that would have flashed had a detector been set differently. It only says that among all the mutually exclusive and exhaustive possibilities for such sequences should be *some* that are consistent with the frequencies of flashings characteristic of the four different pairs of switch settings.

because there are no conceivable sets of data for the cases that might have happened but did not, which are consistent with the numerical constraints imposed by the known behavior of the device, when those constraints are further restricted by the Strong Baseball Principle.

This attack is inherently nonclassical. If, in the best *Gedanken* demonstration I could devise, the 85% and 15% had been replaced by 75% and 25%, then the argument would have collapsed. For instead of the top row being able to differ from the bottom by no more than 15% + 15% + 15% = 45%, which is manifestly less than the required 85%, it would only have been possible to bound the difference by 25% + 25% + 25%, which is just enough to provide the required 75%. Only by exploiting *quantum* correlations can one construct an 85%–15% *Gedanken* demonstration. Any model of the device one might devise based on classical physics would necessarily result in 75%–25% or less extreme statistics, and the Strong Baseball Principle would be immune from this kind of refutation by physicists, no matter how dim a view of it philosophers took. I assert this with confidence because classical physics is local and deterministic and in a deterministic world the Strong Baseball Principle makes perfect sense as a manifestation of locality.

Going in the other direction, it is easy to invent fictitious *Gedanken* demonstrations that produce data that refute the Strong Baseball Principle even more resoundingly than does the device. Consider, for example, a hypothetical device in which 85% and 15% were replaced by 100% and 0%, so that the lights always (not just most of the time) flashed the same colors in 11, 12, and 21 runs, and never (not just infrequently) flashed the same colors in 22 runs. Then the argument refuting the Strong Baseball Principle would be even simpler. An 11 run would necessarily result in the same color (say R) at A and B. Suppose instead the switch at A had been set to 2. The Strong Baseball Principle would then assert that R would still have flashed at B and since the same colors always flash in 12 runs, A would still have flashed R. By the same token B would still have flashed R had its switch been set to 2. Therefore, since the setting of the switch at one detector cannot affect what happens at the other, both would have flashed R if both had been set to 2. But when both are set to 2, both have to flash different colors.

No experiment is known that can provide this more compact refutation. Even quantum miracles can go only so far. The 85%–15% statistics are the most extreme I know how to extract from the quantum theory, and although they are strong enough to demolish the Strong Baseball Principle, the argument we went through is somewhat less direct than that available for the 100%–0% statistics.

It is a characteristic feature of all quantum conundrums that something has to have a nonvanishing probability of happening in two or more mutually exclusive ways for startling behavior to emerge. The viewpoints of quantum and classical physics are distinguished, more than anything else, by the impropriety in quantum physics of reasoning from an exhaustive enumeration of two or more such possibilities in cases that might have happened but did not. We are startled when such reasoning fails because as an analytical tool in classical physics and everyday life, it is not only harmless but often quite fruitful. The most celebrated of all quantum conundrums—how can there be a diffraction pattern when the electron had to go through one slit or the other?—is based on precisely this impropriety. It is just where there is room for some interplay between various unrealized possibilities, that one can look for the quantum world to perform for us the most magical of its tricks.

Therefore it is wrong to apply the principle that what happens at A does not depend on how the switch is set at B to individual runs of the experiment. Many people want to conclude from this that what happens at A *does* depend on how the switch is set at B, which is disquieting in view of the absence of any connections between the detectors. The conclusion can be avoided, if one renounces the Strong Baseball Principle, maintaining that indeed what happens at A does not depend on how the switch is set at B, but that this is only to be understood in its statistical sense, and most emphatically cannot be applied to individual runs of the experiment. To me this alternative conclusion is every bit as wonderful as the assertion of mysterious actions at a distance. I find it quite exquisite that, setting quantum metaphysics entirely aside, one can demonstrate directly from the data and the assumption that there are no mysterious actions at a distance, that there is no conceivable way consistently to apply the Baseball Principle to individual events.

APPENDIX: QUANTUM MYSTERIES FOR ANYONE

> We often discussed his notions on objective reality. I recall that during one walk Einstein suddenly stopped, turned to me and asked whether I really believed that the moon exists only when I look at it.
>
> —A. Pais[10]

> As O. Stern said recently, one should no more rack one's brain about the problem of whether something one cannot know anything about exists all the same, than about the ancient question of how many angels are able to sit on the point of a needle. But it seems to me that Einstein's questions are ultimately always of this kind.
>
> —W. Pauli[11]

---

[10]Pais (1979, 907).
[11]From a 1954 letter to Max Born (Born 1971, 223).

Pauli and Einstein were both wrong. The questions with which Einstein attacked the quantum theory do have answers; but they are not the answers Einstein expected them to have. We now know that the moon is demonstrably not there when nobody looks.

The impact of this discovery on philosophy may have been blunted by the way in which it is conventionally stated, which leaves it fully accessible only to those with a working knowledge of quantum mechanics. I hope to remove that barrier by describing this remarkable aspect of nature in a way that presupposes no background whatever in the quantum theory or, for that matter, in classical physics either. I shall describe a piece of machinery that presents without any distortion one of the most strikingly peculiar features of the atomic world. No formal training in physics or mathematics is needed to grasp and ponder the extraordinary behavior of the device; it is only necessary to follow a simple counting argument on the level of a newspaper brain-twister.

Being a physicist, and not a philosopher, I aim only to bring home some strange and simple facts which might raise issues philosophers would be interested in addressing. I shall try, perhaps without notable success, to avoid raising and addressing such issues myself. What I describe should be regarded as something between a parable and a lecture demonstration. It is less than a lecture demonstration for technical reasons: even if this were a lecture, I lack the time, money, and particular expertise to build the machinery I shall describe. It is more than a parable because the device could in fact be built with an effort almost certainly less than, say, the Manhattan project, and because the conundrum posed by the behavior of the device is no mere analogy, but the atomic world itself, acting at its most perverse.

There are some black boxes within the device whose contents can be described only in highly technical terms. This is of no importance. The wonder of the device lies in what it does, not in how it is put together to do it. One need not understand silicon chips to learn from playing with a pocket calculator that a machine can do arithmetic with superhuman speed and precision; one need not understand electronics or electrodynamics to grasp that a small box can imitate human speech or an orchestra. At the end of the essay I shall give a brief technical description of what is in the black boxes. That description can be skipped. It is there to serve as an existence-proof only because you cannot buy the device at the drugstore. It is no more essential to appreciating the conundrum of the device than a circuit diagram is to using a calculator or a radio.

The device has three unconnected parts. The question of connectedness lies near the heart of the conundrum, but I shall set it aside in favor of a few simple practical assertions. There are neither mechanical connections (pipes, rods, strings, wires) nor electromagnetic connections (radio, radar, telephone, or light signals) nor any other relevant connections. Irrelevant connections may be hard to avoid. All three parts might, for example, sit atop a single table. There is nothing in the design of the parts, however, that takes advantage of such connections to signal from one to another, for example, by inducing and detecting vibrations in the table top.

By insisting so on the absence of connections, I am inevitably suggesting that the wonders to be revealed can be fully appreciated only by experts on connections or their lack. This is not the right attitude to take. Were we together and had I the device at hand, you could pick up the parts, open them up, and poke around as much as you liked. You would find no connections. Neither would an expert on hidden bugs, the Amazing Randi, or any physicists you called in as consultants. The real worry is unknown connections. Who is to say that the parts are not connected by the transmission of unknown Q-rays and their detection by unrecognizable Q-detectors? One can only offer affidavits from the manufacturer testifying to an ignorance of Q-technology and, in any event, no such intent.

Evidently it is impossible to rule out conclusively the possibility of connections. The proper point of view to take, however, is that it is precisely the wonder and glory of the device that it impels one to doubt these assurances from one's own eyes and hands, professional magicians, and technical experts of all kinds. Suffice it to say that there are no connections that suspicious lay people or experts of broad erudition and unimpeachable integrity can discern. If you find yourself questioning this, then you have grasped the mystery of the atomic world.

Two of the three parts of the device (*A* and *B*) function as detectors. Each detector has a switch that can be set in one of three positions (1, 2, and 3) and a red and a green light bulb (figure 1). When a detector is set off it flashes either its red light or its green. It does this no matter how its switch is

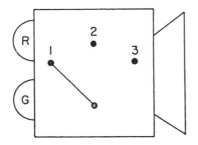

Figure 1.   A detector. Particles enter on the right. The red (R) and green (G)
lights are on the left. The switch is set to 1.

set, though whether it flashes red or green may well depend on the setting. The only purpose of the lights is to communicate information to us; marks on a ribbon of tape would serve as well. I mention this only to emphasize that the unconnectedness of the parts prohibits a mechanism in either detector that might modify its behavior according to the color that may have flashed at the other.

The third and last part of the device is a box (C) placed between the detectors. Whenever a button on the box is pushed, shortly thereafter two particles emerge, moving off in opposite directions toward the two detectors (figure 2). Each detector flashes either red or green whenever a particle reaches it. Thus within a second or two of every push of the button, each detector flashes one or the other of its two colored lights.

Because there are no connections between parts of the device, the link between pressing the button on the box and the subsequent flashing of the detectors can be provided only by the passage of the particles from the box to the detectors. This passage could be confirmed by subsidiary detectors between the box and the main detectors A and B, which can be designed so as not to alter the functioning of the device. Additional instruments or shields could also be used to confirm the lack of other communication between the box and the two detectors or between the detectors themselves (figure 3).

The device is operated repeatedly in the following way. The switch on each detector is set at random to one of its three possible positions, giving nine equally likely settings for the pair of detectors: 11, 12, 13, 21, 22, 23, 31, 32, and 33. The button on the box is then pushed, and somewhat later each detector flashes one of its lights. The flashing of the detectors need not be simultaneous. By changing the distance between the box and the detectors we can arrange that either flashes first. We can also let the switches be given their random settings either before or after the particles leave the box. One could even arrange for the switch on B not to be set until after A had flashed (but, of course, before B flashed).

After both detectors have flashed their lights, the settings of the switches and the colors that flashed are recorded, using the following notation: 31 GR means that detector A was set to 3 and flashed green, while B was set to 1

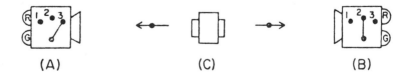

(A)   (C)   (B)

Figure 2.   The complete device. A and B are the two detectors. C is the box from which the two particles emerge.

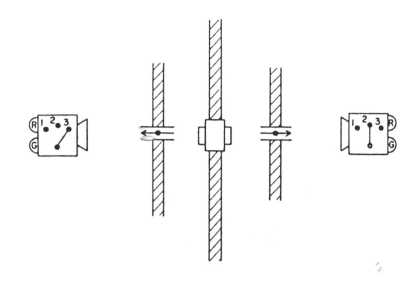

Figure 3   Possible refinement of the device. The box is embedded in a wall that cuts off one detector from the other. Subsidiary detectors confirm the passage of the particles to the main detectors.

and flashed red; 12 RR describes a run in which *A* was at 1, *B* at 2, and both flashed red; 22 RG describes a run in which both detectors were set to 2, *A* flashed red and *B* flashed green; and so on. A typical fragment from a record of many runs is shown in figure 4.

The accumulated data have a random character, but, like data collected in many tossings of a coin, they reveal certain unmistakable features when enormously many runs are examined. The statistical character of the data should not be a source of concern or suspicion. Blaming the behavior of the device on repeated, systematic, and reproducible accidents is to offer an explanation even more astonishing than the conundrum it is invoked to dispel.

The data accumulated over millions (or, if you prefer, billions or trillions) of runs can be summarized by distinguishing two cases:

Case a. In those runs in which each switch ends up with the same setting (11, 22, or 33) both detectors always flash the same color. RR and GG occur in a random pattern with equal frequency; RG and GR never occur.

Case b. In the remaining runs, those in which the switches end up with different settings (12, 13, 21, 23, 31, or 32), both detectors flash the same color only a quarter of the time (RR and GG occurring randomly with equal frequency); the other three quarters of the time the detectors flash different colors (RG and GR occurring randomly with equal frequency).

```
         ___GG  22GG  11
       __2RR 31RG 13RG 22GG 22R
      R 21GR 32RG 11GG 32GR 33GG 2_
    22GG 11RR 11GG 23GG 12RR 32GR 11GG
   G 12RG 13RG 33GG 21RG 13GR 31RR 32GR _
  _GR 13GR 21RG 33RR 13GR 11RR 11GG 13RG 31
 _2GG 32GR 33GG 21GR 21GG 33RR 23RG 21GG 21R
13GR 11GG 32GG 31GR 32RG 33RR 13RR 13RG 12R'
11GG 31RG 33RR 12RG 21GR 11GG 22GG 33GG 23GI
11RR 22RR 12RG 22GG 23GR 12GR 33GG 31GG 13GI
13GR 21RR 33RR 33RR 13RG 23RG 33GG 32RR 12R:
 3RR 32RG 11RR 11RR 11RR 32RG 12RG 21RG 11G
 RG 23RR 21RG 33RR 13GR 12GR 23RG 21RR 32
 R 21GR 12RR 31GR 12RG 13GR 13RG 22RR ]
 23GR 11RR 12PR 33RR 21RG 13GR 21RR
  ^R 12RR 23GG 13RG 21RG 11GG 1?
   ^2RG 32RG 32GR 11GG 22R^
    ^^P 31RG 21^^
```

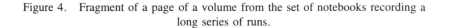

Figure 4. Fragment of a page of a volume from the set of notebooks recording a long series of runs.

These results are subject to the fluctuations accompanying any statistical predictions, but, as in the case of a coin-tossing experiment, the observed ratios will differ less and less from those predicted, as the number of runs becomes larger and larger.

This is all it is necessary to know about how the device operates. The particular fractions $\frac{1}{4}$ and $\frac{3}{4}$ arising in case b are of critical importance. If the smaller of the two were $\frac{1}{3}$ or more (and the larger $\frac{2}{3}$ or less) there would be nothing wonderful about the device. To produce the conundrum it is necessary to run the experiment sufficiently many times to establish with overwhelming probability that the observed frequencies (which will be close to 25% and 75%) are not chance fluctuations away from expected frequencies of $33\frac{1}{3}$% and $66\frac{2}{3}$%. (A million runs is more than enough for this purpose.)

These statistics may seem harmless enough, but some scrutiny reveals them to be as surprising as anything seen in a magic show, and leads to similar suspicions of hidden wires, mirrors, or confederates under the floor. We begin by seeking to explain why the detectors invariably flash the same colors when the switches are in the same positions (case a). There would be any number of ways to arrange this were the detectors connected, but they are not.

Nothing in the construction of either detector is designed to allow its functioning to be affected in any way by the setting of the switch on the other, or by the color of the light flashed by the other.

Given the unconnectedness of the detectors, there is one (and, I would think, only one) extremely simple way to explain the behavior in case *a*. We need only suppose that some property of each particle (such as its speed, size, or shape) determines the color its detector will flash for each of the three switch positions. What that property happens to be is of no consequence; we require only that the various states or conditions of each particle can be divided into eight types; RRR, RRG, RGR, RGG, GRR, GRG, GGR, and GGG. A particle whose state is of type RGG, for example, will always cause its detector to flash red for setting 1 of the switch, green for setting 2, and green for setting 3; a particle in a state of type GGG will cause its detector to flash green for any setting of the switch; and so on. The eight types of states encompass all possible cases. The detector is sensitive to the state of the particle and responds accordingly; putting it another way, a particle can be regarded as carrying a specific set of flashing instructions to its detector, depending on which of the eight states the particle is in.

The absence of RG or GR when the two switches have the same settings can then be simply explained by assuming that the two particles produced in a given run are both produced in the same state; i.e., they carry identical instruction-sets. Thus if both particles in a run are produced in states of type RRG, then both detectors will flash red if both switches are set to 3. The detectors flash the same colors when the switches have the same settings because the particles carry the same instructions.

This hypothesis is the obvious way to account for what happens in case *a*. I cannot prove that it is the only way, but I challenge the reader, given the lack of connections between the detectors, to suggest any other. The apparent inevitability of this explanation for the perfect correlations in case *a* forms the basis for the conundrum posed by the device. For the explanation is quite incompatible with what happens in case *b*.

If the hypothesis of instruction-sets were correct, then both particles in any given run would have to carry identical instruction-sets whether or not the switches on the detectors were set the same. At the moment the particles are produced there is no way to know how the switches are going to be set. For one thing, there is no communication between the detectors and the particle-emitting box, but in any event the switches need not be set to their random positions until after the particles have gone off in opposite directions from the box. To ensure that the detectors invariably flash the same color every time the switches end up with the same settings, the particles leaving the box in each run must carry the same instructions even in those runs (case *b*) in which the switches end up with different settings.

Let us now consider the totality of all case *b* runs. In none of them do we ever learn what the full instruction sets were, since the data reveal only the colors assigned to two of the three settings. (The case *a* runs are even less informative.) Nevertheless, we can draw some nontrivial conclusions by examining the implications of each of the eight possible instruction sets for those runs in which the switches end up with different settings. Suppose, for example, that both particles carry the instruction-set RRG. Then out of the six possible case *b* settings, 12 and 21 will result in both detectors flashing the same color (red), and the remaining four settings, 13, 31, 23, and 32, will result in one red flash and one green. Thus both detectors will flash the same color for two of the six possible case *b* settings. Since the switch settings are completely random, the various case *b* settings occur with equal frequency. Both detectors will therefore flash the same color in a third of those case *b* runs in which the particles carry the instruction-sets RRG.

The same is true for case *b* runs where the instruction-set is RGR, GRR, GGR, GRG, or RGG, since the conclusion rests only on the fact that one color appears in the instruction-set once and the other color, twice. In a third of the case *b* runs in which the particles carry any of these instruction-sets, the detectors will flash the same color. The only remaining instruction-sets are RRR and GGG; for these sets both detectors will evidently flash the same color in every case *b* run.

Thus, regardless of how the instruction-sets are distributed among the different runs, in the case *b* runs *both detectors must flash the same color at least a third of the time.* (This is a bare minimum; the same color will flash more than a third of the time, unless the instruction sets RRR and GGG never occur.) As emphasized earlier, however, when the device actually operates, the same color is flashed only a quarter of the time in the case *b* runs.

Thus the observed facts in case *b* are incompatible with the only apparent explanation of the observed facts in case *a,* leaving us with the profound problem of how else to account for the behavior in both cases. This is the conundrum posed by the device, for there is no other obvious explanation of why the same colors always flash when the switches are set the same. It would appear that there must, after all, be connections between the detectors— connections of no known description which serve no purpose other than relieving us of the task of accounting for the behavior of the device in their absence.

I shall not pursue this line of thought, since my aim is only to state the conundrum of the device, not to resolve it. The lecture demonstration is over. I shall only add a few remarks on the device as a parable.

One of the historic exchanges between Einstein (Einstein, Podolsky, Rosen 1935) and Bohr (1935), which found its surprising denouement in the work of J. S. Bell (1964) nearly three decades later, can be stated quite clearly in terms of the device. I stress that the transcription into the context of the device is only to simplify the particular physical arrangement used to raise the issues. The device is a direct descendant of the rather more intricate but conceptually similar *Gedanken* experiment proposed in 1935 by Einstein, Podolsky, and Rosen. We are still talking physics, not descending to the level of analogy.

The Einstein, Podolsky, Rosen experiment amounts to running the device under restricted conditions in which both switches are required to have the same setting (case *a*). Einstein would argue (as was argued above) that the perfect correlations in each run (RR or GG but never RG or GR) can be explained only if instruction-sets exist, each particle in a run carrying the same instructions. In the Einstein, Podolsky, Rosen version of the argument, the analogue of case *b* was not evident, and its fatal implications for the hypothesis of instruction-sets went unnoticed until Bell's paper.

The *Gedanken* experiment was designed to challenge the prevailing interpretation of the quantum theory, which emphatically denied the existence of instruction-sets, insisting that certain physical properties (said to be complementary) had no meaning independent of the experimental procedure by which they were measured. Such measurements, far from revealing the value of a preexisting property, had to be regarded as an inseparable part of the very attribute they were designed to measure. Properties of this kind have no independent reality outside the context of a specific experiment arranged to observe them: the moon is *not* there when nobody looks.

In the case of my device, three such properties are involved for each particle. We can call them the 1-color, 2-color, and 3-color of the particle. The *n*-color of a particle is red if a detector with its switch set to *n* flashes red when the particle arrives. The three *n*-colors of a particle are complementary properties. The switch on a detector can be set to only one of the three positions, and the experimental arrangements for measuring the 1-, 2-, or 3-color of a particle are mutually exclusive. (We may assume, to make this point quite firm, that the particle is destroyed by the act of triggering the detector, which is, in fact, the case in many recent experiments probing the principles that underly the device.)

To assume that instruction-sets exist at all is to assume that a particle has a definite 1-, 2-, and 3-color. Whether or not all three colors are known or knowable is not the point; the mere assumption that all three have values violates a fundamental quantum-theoretic dogma.

No basis for challenging this dogma is evident when only a single particle and detector are considered. The ingenuity of Einstein, Podolsky, and Rosen lay in discovering a situation involving a *pair* of particles and detectors, where the quantum dogma continued to deny the existence of 1-, 2-, and 3-colors, while, at the same time, quantum theory predicted correlations (RR and GG but never RG or GR) that seemed to require their existence. Einstein concluded that, if the quantum theory were correct, i.e., if the correlations were, as predicted, perfect, then the dogma on the nonexistence of complementary properties—essentially Bohr's doctrine of complementarity—had to be rejected.

Pauli's attitude toward this in his letter to Born is typical of the position taken by many physicists: since there is no known way to determine all three n-colors of a particle, why waste your time arguing about whether or not they exist? To deny their existence has a certain powerful economy—why encumber the theory with inaccessible entities? More importantly, the denial is supported by the formal structure of the quantum theory which completely fails to allow for any consideration of the simultaneous 1-, 2-, and 3-colors of a particle. Einstein preferred to conclude that all three n-colors did exist, and that the quantum theory was incomplete. I suspect that many physicists, though not challenging the completeness of the quantum theory, managed to live with the Einstein, Podolsky, Rosen argument by observing that though there was no way to establish the existence of all three n-colors, there was also no way to establish their nonexistence. Let the angels sit, even if they cannot be counted.

Bell changed all this by bringing into consideration the case b runs and pointing out that the quantitative numerical predictions of the quantum theory ($\frac{1}{4}$ vs. $\frac{1}{3}$) unambiguously ruled out the existence of all three n-colors. Experiments done since Bell's paper confirm the quantum-theoretic predictions.[12] Einstein's attack, were he to maintain it today, would be more than an attack on the metaphysical underpinnings of the quantum theory—more, even, than an attack on the quantitative numerical predictions of the quantum theory. Einstein's position now appears to be contradicted by nature itself. The device behaves as it behaves, and no mention of wave-functions, reduction hypotheses, measurement theory, superposition principles, wave-particle duality, incompatible observables, complementarity, or the uncertainty principle is needed to bring home its peculiarity. It is not the Copenhagen interpretation of quantum mechanics that is strange, but the world itself.

As far as I can tell, physicists live with the existence of the device by implicitly (or even explicitly) denying the absence of connections between its pieces. References are made to the "wholeness" of nature: particles, detectors, and box can be considered only in their totality; the triggering and flashing of detector A cannot be considered in isolation from the triggering and flashing of detector B—both are part of a single indivisible process. This attitude is sometimes tinged with Eastern mysticism, sometimes with Western know-nothingism, but, common to either point of view, as well as to the less

---

[12]Clauser and Shimony (1978). For a less technical summary, see d'Espagnat (1979).

trivial but considerably more obscure position of Bohr, is the sense that strange connections are there. The connections are strange because they play no explicit role in the theory: they are associated with no particles or fields and cannot be used to send any kinds of signals. They are there for one and only one reason: to relieve the perplexity engendered by the insistence that there are no connections. Whether or not this is a satisfactory state of affairs is, I suspect, a question better addressed by philosophers than by physicists.

I conclude with the recipe for making the device, which, I emphasize again, can be ignored

> The device exploits Bohm's version (1951, 614–619) of the Einstein, Podolsky, Rosen experiment. The two particles emerging from the box are spin $\frac{1}{2}$ particles in the singlet state. The two detectors contain Stern-Gerlach magnets, and the three switch positions determine whether the orientations of the magnets are vertical or at $\pm 120°$ to the vertical in the plane perpendicular to the line of flight of the particles. When the switches have the same settings the magnets have the same orientations. One detector flashes red or green according to whether the measured spin is along or opposite to the field; the other uses the opposite color convention. Thus when the same colors flash the measured spin components are different.

> It is a well-known elementary result that, when the orientations of the magnets differ by an angle Θ, then the probability of spin measurements on each particle yielding opposite values is $\cos^2(\Theta/2)$. This probability is unity when $\Theta = 0°$ (case a) and $\frac{1}{4}$ when $\Theta = \pm 120°$ (case b).

> If the subsidiary detectors verifying the passage of the particles from the box to the magnets are entirely nonmagnetic they will not interfere with this behavior.

BELL'S THEOREM: A GUIDE TO THE IMPLICATIONS

JON P. JARRETT

At a recent public lecture[1], Harvard physicist and University Professor Emeritus Edward Purcell told his audience of his happiness that he has lived to see a philosophical problem settled in the laboratory. There are many (and I count myself among them) who regard science and philosophy as inseparably intertwined in their mutual concern for understanding the structure of the world. Nevertheless, it would be highly misleading, and perhaps a trivialization of the relation between science and philosophy, to maintain that philosophical problems are *routinely* addressed in the laboratory. Professor Purcell's remark that day was prompted by his assessment of the astonishing degree to which recent experiments bear directly on issues of a traditionally philosophical character. I, too, am happy that this situation to which Purcell referred has taken place in *my* lifetime.

The situation in question concerns recent experimental investigations of fundamental questions (call them "metaphysical," if you like) associated with Bell's theorem, the subject of this paper. In a paper published in 1964, physicist John S. Bell first proved a version of the theorem which has since borne his name; but this early work, its subsequent refinements and generalizations, and the associated experimental investigations are profitably viewed as the remarkable, most recent stages of a line of inquiry extending back over the past half-century. The degree to which we have in the present situation achieved a satisfactory culmination of this line of inquiry is a contro-

Earlier versions of this paper were read at the University of California at Berkeley, Harvard University, Cornell University, and the 1987 American Philosophical Society Central Division Meeting in Chicago. I am grateful to each audience for helpful comments and criticisms. Particular thanks go to Don Howard and Allen Stairs for their insightful commentaries at the Chicago APA meeting. I wish also to thank Robert Jarrett for taking time off from his retirement to build the Mermin Contraption used during the above-mentioned presentations and depicted on p. 64.

[1]E. M. Purcell, "The principle of certainty: Philosophical reflections on quantum mechanics," lecture delivered at Harvard University, 20 March 1987.

versial matter, but that Bell's work and related research reveal, with an empirical force absolutely unprecedented in such "metaphysical" domains, a magnificently peculiar feature of the world—*that* (at least in my view) is beyond dispute.

In 1935, a paper claiming Albert Einstein among its authors appeared in the *Physical Review*. The argument presented in this paper, the now-famous Einstein-Podolsky-Rosen (or "EPR") argument, appealed to an ingenious *Gedankenexperiment* to expose what those authors called the "incompleteness" of quantum mechanics. There followed on the heels of the EPR paper a response, also in *Physical Review,* by Niels Bohr, (1935) who offered a defense of quantum mechanics against this charge of "incompleteness." Bohr's reply has been widely regarded, at least within the physics community, as a decisive refutation of the EPR argument.

However, the adequacy of Bohr's defense was never universally acknowledged, not even within the physics community, and a controversy concerning this matter of "completeness" has persisted in some form or other to the present day. Although the intervening years did see important developments on various significantly related fronts, the possibility of experimentally pursuing the issues dividing Einstein and Bohr emerged only with the justly celebrated theorem of John Bell.

The purpose of this paper is to provide an orientation of sorts to the results associated with this "experimental metaphysics." For that purpose, it will be helpful first to provide a bit more in the way of general background which will permit me subsequently to call attention to a wonderful irony in all of this.

Set in the context of experiments of a special class, Bell's theorem establishes the incompatibility of quantum mechanics and various sets of constraints to which the qualifier "local realistic" is often attached. Bell's theorem is a derivation, with a set of these "local realistic" constraints taken as premises, of a so-called "Bell inequality," a constraint on the empirical predictions of any theory satisfying the premises. Lest I inadvertently give the impression that local realism is supposed to be some precisely defined, monolithic philosophical position, I want to emphasize that I use the term "local realism" in a deliberately vague way to refer to a worldview of a highly general sort. Here, in broadest outline, is the local realistic picture of the world:

All physical phenomena are produced by the interactions of various physical entities whose existence and whose characteristics are independent of our knowledge of them. The ontological inventory may include such things as particles and/or fields of the sorts with which we are familiar, but may also (or instead) include things we do not even know how to describe. The physical state of each entity or system of entities is exhaustively characterized by a list of precise values of appropriate physical magnitudes. No particular set of those physical magnitudes with which we are familiar need be in the list or even definable in terms of those which do occur in the list. Needless to say, if there is to be any contact between such a state-description and our experience, there must be *some* connection between the observable properties of macroscopic bodies and the physical magnitudes posited in state-descriptions; but the attitude of the local realist is that we should not uncritically assume that any particular familiar notions (such as the simultaneous position and momentum of a particle with respect to a specified frame of reference) are well defined in domains which extend arbitrarily far beyond our limited experience. Therefore, the local realist assumes only that there is *some* set of appropriate properties of physical entities of each type. The complete state-description for a system consists then in the precise specification of these properties of the physical entities that make up the system. Such a state-description implicitly includes everything we might say "physically is the case" for that system. Finally, local realism requires that the laws which govern physical interactions and the evolution of physical states, while they need not be deterministic, must respect considerations of what is commonly called "locality"; roughly speaking, "action-at-a-distance," in some suitable sense of that phrase, is prohibited.

Regarded as a program in natural philosophy, local realism (again putting it crudely) is the position according to which all phenomena are to be embedded into a unified picture of this general sort. Anyone not already familiar with Bell's theorem might well be at a loss to imagine a phenomenon which is not compatible with local realism, so broad and so innocuous-sounding are its strictures; and yet, as I will attempt to describe more clearly in what follows, we now have strong empirical evidence that indeed there are phenomena which defy embedding in any such local realistic picture of the world.

I previously alluded to an irony. It is this: Einstein's own worldview, whatever is to be included in a properly detailed characterization of it, surely falls within the bounds of local realism;[2] and yet, the class of experiments which provide the context for Bell's theorem consists of modified versions of the Einstein-Podolsky-Rosen *Gedankenexperiment;* and these experiments, whose results not only violate Bell-type inequalities, but do so in excellent accord with the predictions of quantum mechanics, provide the evidence which tells against local realism.

This brings me to my central aim. Approaches to this subject vary considerably. Bell-type inequalities can be derived from different sets of

---

[2]See, e.g., Fine (1984b) and Howard (1985) for careful analyses of Einstein's views.

premises; some versions of Bell's theorem employ one or another premise (some formulation of determinism, for example) which other versions do not; in some treatments certain premises are consolidated and others not made explicit, and, in general, things just get formulated differently in different expositions. I claim no special virtues for the approach I will follow here beyond its utility for drawing attention to matters I take to be of greatest significance.

I intend to proceed with the aid of the mysterious objects shown in the photograph below. They will serve to illustrate the experimental setup. I will state and discuss a number of constraints appearing in Bell-type arguments and I will state (without proof) and comment upon a number of results linking these constraints in various ways to each other and, by way of Bell's theorem, to Bell-type inequalities. Along the way, I will offer my assessment of the implications of Bell's theorem and I will conclude by suggesting a reason for believing that the final chapter of the story begun in 1935 by Einstein, Podolsky, and Rosen remains to be written.

1. The Mermin Contraption

In discussions of Bell's theorem aimed at a general audience, a first obstacle has often been the setting out, in some suitably intelligible way, of the relevant features of the associated experimental context. This involves talk of Stern-Gerlach devices and spin-$\frac{1}{2}$ particles, or photon polarization measurements; and while the relevant features here are really not all that complex, they do add an extra layer of jargon through which the layperson must struggle in order to grasp the principal concerns. As physicist David Mermin has noted, this can also constitute an obstacle of sorts for people who know *too much* quantum mechanics and are unable to set aside their "quantum theoretic prejudices" so as to appreciate the underlying strangeness (1985, 41–42).

In various of his papers on Bell's theorem, Mermin circumvents this obstacle by introducing a representation which cleverly caricatures the actual experimental situation (1981a; 1981b; 1985). I have shamelessly stolen Mermin's representation as my blueprint for the gadgets shown in the photograph, but so as to give credit where credit is due, I hereby dub these devices arranged as depicted "the Mermin Contraption." In his 1981 paper (reprinted in this volume), Mermin describes his contraption in a way which beautifully sets the stage for my subsequent discussion.[3]

Someone reading this description might justifiably wonder what in the

---

[3]See Mermin (1981a), the three paragraphs beginning: "The device has three unconnected parts . . ." reprinted in this volume, pp. 50–51.

The Mermin Contraption provides a highly schematic representation of actual Bell-
type experimental setups. (Photo courtesy of Carol Roberts.)

world this amazing contraption *does*. Think of the devices *A* and *B* as detectors of some similar sort, and of *C* as a source of what I will call "particles,"
although I do this for convenience without intending any firm ontological
commitment. Each detector has a switch which can be set in any one of three
positions, and a pair of lights, one red and one green. Atop the source is a
button, the pressing of which is followed almost immediately by the illumination of one of the lights at *A* and one of the lights at *B*. From one trial to the
next, the switch of each detector is independently and randomly set before the
button is pressed and the lights come on. (I should add that in the most recent
actual experiments of this sort, (Aspect, Dalibard, and Roger 1982) the
operation represented here by the setting of the switches occurs in that brief
interval between the time the button is pressed and the time at which the lights
come on!)

The hypothesis that whatever influences the devices do exert on each other do occur in the manner suggested by way of the passage of particles (or something) from the source to the detectors is in keeping with a variety of straightforward observations of which the following are but two examples: the elapsed time between the pressing of the button and the illumination of a light increases as the distance between the source and the detector increases—this could be checked by positioning the detectors sufficiently far away; and when a barrier is placed between the source and a detector, neither light of that detector comes on after the button is pressed.

Think of the detectors as each performing a measurement (any one of perhaps three different sorts corresponding to the three different switch positions) on the source emissions. The outcome of each such measurement is given by the color of the illuminated light. The experiment consists of a great many such measurement trials. The data recorded for a typical trial might look like this: 13GR. In this trial, for example, the switch at detector A was in position 1 and the switch at detector B was in position 3, the green light came on at A and the red light came on at B. Statistics are compiled and for each of the nine pairs of switch settings, the corresponding fraction of the trials yielding each of the four pairs of color outcomes is computed. Amazingly, these statistical results provide direct evidence that, whatever is in these boxes and whatever it is that takes place when the button is pressed and the lights come on, the operation of the Mermin Contraption cannot be subsumed under any theory compatible with local realism.

2. Minimal characterization of empirical theories governing Bell-type phenomena

Any empirical theory governing the operation of the Mermin Contraption must ascribe states to the measuring devices and the source emissions which determine (presumably by way of appropriate physical laws) functions of the form $p_\lambda^{AB}(x,y|i,j)$, where λ is the state of the source-emissions (the pair of particles); x and y are the measurement outcomes (red or green) at A and B respectively; and i and j are respectively the A and B detector-states, including the switch settings and whatever else may be relevant. In this notation and in parts of the subsequent discussion, only the detector switch settings are explicitly mentioned, but a more general treatment could easily be given. I will call these p_λ^{AB} functions the "elementary joint probability" functions.

For instance, in this notation, $p_\lambda^{AB}(R,R,|1,2)$ is the probability our theory assigns to the illumination of both red lights in those trials for which the A switch is in position 1, the B switch is in position 2, and the two particles are in state λ. To say that the two particles are in state λ is not necessarily to be understood as entailing an ascription of a state to each particle individually (but neither are such cases excluded). Instead, only the weaker claim that λ is the state of the composite two-particle system is intended. Note also that this characterization explicitly accommodates intrinsically indeterministic theories. For present purposes, deterministic theories (or, to permit a harmless ambiguity, theories whose *states* are deterministic) constitute special cases for which the range of these elementary probability functions is $\{0,1\}$.[4]

In order to interpret these p_λ^{AB} functions as probabilities, the following conditions must be satisfied:[5]

$$0 \le p_\lambda^{AB}(x,y|i,j) \le 1$$

and

$$\sum_x \sum_y p_\lambda^{AB}(x,y|i,j) = 1,$$

where the sums are taken over the possible measurement outcomes. In what follows, it will be useful to refer to the marginal and conditional probabilities defined by:

$$p_\lambda^A (x \mid i,j) \equiv \sum_y p_\lambda^{AB} (x,y \mid i,j),$$

$$p_\lambda^B (y \mid i,j) \equiv \sum_x p_\lambda^{AB} (x,y \mid i,j),$$

$$p_\lambda^A (x \mid i,j,y) \equiv \frac{p_\lambda^{AB} (x,y \mid i,j)}{p_\lambda^B (y \mid i,j)} ,$$

$$and \quad p_\lambda^B (y \mid i,j,x) \equiv \frac{p_\lambda^{AB} (x,y \mid i,j)}{p_\lambda^A (x \mid i,j)} .$$

For example, $p_\lambda^A(G|2,2)$ is the probability our theory assigns to the outcome "green" for a measurement at A when both detectors have their switches set to position 2, λ is the two-particle state, and the outcome at B is unspecified; and $p_\lambda^B(R|2,3,G)$, for example, is the conditional probability for getting an illuminated red light at B in trials for which the A and B detector switch settings are 2 and 3 respectively and λ is the two-particle state, *given* that the green light comes on at A.

---

[4]A further comment may help to clarify my use of the term 'determinism' here. One sometimes speaks of states that evolve deterministically in time, and such states are contrasted with those whose temporal evolution is stochastic. Please note that my use of the term 'determinism' has nothing to do with temporal evolution; instead, it has to do with the definiteness of the predicted outcomes of the possible measurements performable on a system while it is *in* a given state.

[5]Here and throughout, such conditions are implicitly universally quantified over all values of the free variables for which the expressions are defined.

Finally, it will also be convenient to build into the notation a way to refer to trials in which it is decided to perform a measurement on just one particle of the pair. In such cases, asterisks will denote both the "switch setting" of the nonoperative detector and the corresponding "outcome" of the nonmeasurement. Thus, $p_\lambda^{AB}(R,*|1,*)$, for example, is the probability assigned to the outcome "red" when device A switch setting is 1, λ is the two-particle state, and no measurement is performed at B. By suitably interpreting the sums over possible measurement outcomes in the formulae already introduced, these cases require no further modification of the notation. From previous definitions, for example,

$$p_\lambda^A (R \mid 1,3) \equiv \sum_y p_\lambda^{AB} (R,y \mid 1,3)$$

$$= p_\lambda^{AB} (R,R \mid 1,3) + p_\lambda^{AB} (R,G \mid 1,3),$$

where the sum is performed over the two possible outcomes of the B measurement. Similarly,

$$p_\lambda^A (R \mid 1,*) \equiv \sum_y p_\lambda^{AB} (R,y \mid 1,*) = p_\lambda^{AB} (R,* \mid 1,*),$$

where the sum reduces to the one term corresponding to the one possible "outcome" of the B nonmeasurement.

It is important to understand that it is not to be assumed that from one trial to the next the states are the same, or even that we are in any position to know what the actual states are or, for that matter, what goes into a state-description (other than the detector switch settings, of course). It is precisely this ignorance on which deterministic accounts of these experiments rely, and which motivates the so-called "deterministic hidden-variables" theories. Such theories attempt to "explain" the observed statistics of measurement outcomes along the following lines: we might get, say, the red light at A and the green light at B in 42% of those trials in which the switch settings are 1 at A and 2 at B just because in precisely 42% of those trials, the two-particle system turned out to be in a state λ such that $p_\lambda^{AB}(R,G|1,2) = 1$. The statistical character of the observed results is then not in the least a sign of any intrinsic indeterminism in nature; we merely suffer from ignorance of the underlying "hidden variables" that make up the "true" states, the "hidden" deterministic states, the λs. This is precisely the way we think about the statistics that occur in classical statistical mechanics.

Of course, we should permit such ignorance of underlying states even in the more general intrinsically indeterministic cases (which is to say, for theories whose elementary probability functions do take on values intermediate between 0 and 1). But we must then further require of any such theory (deterministic or not) that it provide a probability density function, $\varrho(\lambda)$, on the set Λ of hidden states. Roughly speaking, our theory has to tell us how frequently particle-pairs *do* emerge in each of the possible two-particle states. From this information (from the elementary probability functions and the distribution of hidden states), our theory permits us to calculate statistical predictions for the results of many trials by forming the appropriately weighted averages of the elementary probabilities. The predicted relative frequency of outcome x at A and outcome y at B in trials in which the A switch setting is i and the B switch setting is j is then given by

$$ p^{AB} (x,y \mid i,j) = \int_{\lambda \in \Lambda} p_\lambda^{AB} (x,y \mid i,j) \, \rho \, (\lambda) \, d\lambda. $$

Such predictions are to be compared with the observed statistics to assess the empirical adequacy of our theory. In this manner, the posited hidden states make empirical contact with observable phenomena.

Bell's theorem shows that the statistical predictions so generated by any such theory, if that theory satisfies certain other "local realistic" constraints, must conflict with the experimentally well-confirmed predictions of quantum mechanics. The predictions of any such theory satisfy Bell's inequality, which is violated in actual experiments in precise quantitative agreement with the predictions of quantum mechanics.

I cannot emphasize strongly enough that the claim here is *not* that our best theoretical account of these phenomena, namely that of quantum mechanics, is not a local realistic account. That quantum mechanics is not a local realistic theory is, of course, true; but the claim being made here goes considerably beyond this. The claim is that we have strong empirical evidence (I repeat—strong *empirical* evidence) that *no* local realistic theory is true of our world.

Reflecting on the character of these experiments, this ought to sound preposterous to anyone unfamiliar with the subject. How can such a startling conclusion be drawn from such ordinary-looking experimental results? At this point, I simply refer the reader to other sources for actual derivations of Bell-type inequalities. I wish to note only that the level of mathematical sophistication required to follow these derivations is not great; it is quite natural at first to feel that the argument is so simple that the reasoning which leads from it to the claim that the evidence for something so strange and wonderful (something still to be described more fully herein) turns on the value of one rather mundane experimentally measured quantity exceeding another *must* be a swindle. Referring to the experimentally well-confirmed statistical predictions of quantum mechanics—predictions which, as previously noted, do violate the Bell-type inequalities constraining any local realistic account of the experimental data—physicist Richard Feynman said, "I've entertained my-self always by squeezing the difficulty of quantum mechanics into a smaller and smaller place, so as to get more and more worried about this particular item. It seems to be almost ridiculous that you can squeeze it to a numerical question that one thing is bigger than another. But there you are . . . ." (1982, 485).

3. Discussion of results

What follows may be regarded as a logical "road map" of results related to Bell's theorem.[6] It will be convenient to express these results in terms of the constraints listed here:

DETERMINISM $\qquad p_\lambda^{AB} (x,y \mid i,j) \in \{0,1\}$

CONSERVATION $\qquad p_\lambda^{AB} (R,G \mid i,i) = p_\lambda^{AB} (G,R \mid i,i) = 0$

LOCALITY $\qquad \begin{cases} \text{A) } p_\lambda^A (x \mid i,j) = p_\lambda^A (x \mid i,*) \\ \text{B) } p_\lambda^B (y \mid i,j) - p_\lambda^B (y \mid *,j) \end{cases}$

COMPLETENESS $\qquad \begin{cases} \text{A) } p_\lambda^A (x \mid i,j,y) = p_\lambda^A (x \mid i,j \mid *,j) \\ \text{B) } p_\lambda^B (y \mid i,j,x) = p_\lambda^B (y \mid i,j) \end{cases}$

STRONG LOCALITY $\quad p_\lambda^{AB} (x,y \mid i,j) = p_\lambda^A (x \mid i,*) p_\lambda^B (y \mid *,j)$

Consider the following claims:

(I)	DETERMINISM & CONSERVATION & LOCALITY	$\Rightarrow$ BELL-TYPE INEQUALITY

(II) LOCALITY $\Leftrightarrow$ NO SUPERLUMINAL SIGNALS

Determinism, for present purposes, is just the requirement that the theory specify probabilities of 0 or 1 for all possible outcomes of all possible Mermin Contraption measurements. While this is not a general requirement of local

---

[6]Some further details are given in Jarrett (1984) and references therein. Shimony (1986) recasts parts of Clauser and Horne (1974) and Jarrett (1984) into a form which comports neatly with the present work. Several other variants of Bell's theorem and related research can be found in sources listed in the bibliography recently compiled by Ballentine (1987).

realism, it is surely of interest to consider theories in which state-descriptions actually entail the outcomes of possible measurements. Determinism, as formulated here, is satisfied by all such theories.

The conservation condition derives its name from the quantum-theoretic analysis of these experiments, wherein it expresses the conservation of angular momentum. That quantum-theoretic analysis is not, however, to be assumed in any of what follows. Instead, I tentatively offer as warrant for this condition (subject to a qualification to be mentioned later) this simple empirical fact: In the data of Mermin Contraption experiments, in each trial in which the A and B switch settings are the *same*, the lights which come on at A and B agree in color. Hence, for the time being, conservation is put forward as a necessary condition for empirical adequacy.

Locality is an independence condition of a special sort. Consider this half of the locality condition, the A half: $p_\lambda^A(x|i,j) = p_\lambda^A(x|i,*)$. (Similar considerations can be seen to apply to the B half). Once it is recognized that there is a suppressed universal quantification over j, the switch setting of detector B, the dependence of the left-hand side on the switch setting at B is seen to be illusory. This condition simply asserts that the probability for color x coming on at detector A when the outcome at B is not specified is a function of only λ and the switch setting (the "state") of the participating detector. In particular, this probability is independent of the switch setting (the "state") of the distant detector B, or whether B is even prepared to perform a measurement. The probability at A is fully determined "locally," so to speak, by λ and i alone.

The link between locality and relativity theory is expressed in claim (II), which asserts the equivalence of locality and a prohibition of superluminal signals. Relying only on an admittedly imprecise characterization of the "no superluminal signals" condition, it is not difficult to see how one might argue at least for the implication from right to left. Suppose locality is violated; suppose for the sake of illustration, that we have a theory which violates locality in this way: For some two-particle state s, $p_s^A(R|1,2) \neq p_s^A(R|1,3)$. Since our concern here is with "in principle" considerations, let us further suppose (for these purposes) that experimenters can reliably prepare particles in the state s of interest. (This is, of course, not a supposition we could hope to justify in practice; but *this* point is quite irrelevant to present purposes.) Now imagine a large ensemble of Mermin Contraptions, all with two-particle systems prepared in state s and the switches of all the A detectors at setting 1, and imagine the detectors positioned arbitrarily far from the sources. After the particles are all in flight from the ensemble of sources toward their respective detectors, an experimenter E_A positioned near the ensemble of A detectors is preparing to record the relative frequency of the occurrence of illuminated red lights among the A detectors. An experimenter E_B, similarly positioned near the ensemble of B detectors, is still in the process of making up his or her mind about whether to set all of the B detector switches in position 2, or instead to set them all in position 3. Experimenter E_B can wait until the very last instant before arrival of the particles, so that no causal signal— not even light—would have time to reach the other wing of the contraption in time to influence the A measurement outcomes. In such a circumstance, in which experimenter E_A's statistics are sensitive to the B detector switch settings, if there were a suitable prior arrangement between experimenters E_A and E_B, E_A could infer from the relative frequency of illuminated red lights among the A detectors, the outcome of experimenter E_B's decision. (To put it just a bit more carefully, E_A could, in principle, infer E_B's decision with a degree of confidence approaching certainty.) In this manner, experimenters E_A and E_B could communicate in a form which apparently is possible only through mechanisms which violate the relativistic prohibition of super-luminal signals.

Returning now to claim (I), the conjunction of determinism, conservation, and locality entails a Bell-type inequality. This claim mirrors the form of Bell's original 1964 argument. Since the predictions of quantum mechanics satisfy conservation and locality, but violate the Bell-type inequality, it is an immediate consequence that no theory satisfying determinism is compatible with quantum mechanics. Please understand that this is not simply the claim that quantum mechanics violates determinism. That claim is *obviously* true; Bell's theorem is not needed to establish *that*. What is asserted here, rather, is the much stronger claim that *no* deterministic theory, *no* theory which posits deterministic physical states underlying the allegedly "incomplete" state descriptions of quantum mechanics, can reproduce, at the observational level, the statistical predictions of quantum mechanics.

Although these statistical predictions of quantum mechanics are, generally speaking, well confirmed, there are reasons I will mention presently for resisting the conclusion that claim (I) provides an adequate basis for the experimental refutation of deterministic theories of Bell-type phenomena. I will claim, nevertheless, that these reasons do not rescue determinism.

Let us move to claims (III), (IV), and (V):

(III)	COMPLETENESS & CONSERVATION & LOCALITY	$\Rightarrow$ DETERMINISM

(IV)	COMPLETENESS & CONSERVATION & LOCALITY	$\Rightarrow$ BELL-TYPE INEQUALITY

(V)	DETERMINISM	$\Rightarrow$	COMPLETENESS
	COMPLETENESS	$\not\Rightarrow$	DETERMINISM

Note that claim (IV) follows immediately from claims (I) and (III). The new constraint appearing in these claims is the one called "completeness." On the basis of claim (V), we can recognize claim (IV) as another version of Bell's theorem, more general than claim (I). Claim (IV) replaces the constraint of determinism with the weaker constraint of completeness.

Like locality, completeness is an independence condition of a special sort. Consider this half of the completeness condition, the A half: $p_\lambda^A(x|i,j,y) = p_\lambda^A(x|i,j)$. (Similar considerations can be seen to apply to the B half.) This condition asserts that for a specified pair of detector switch settings i and j and a specified two-particle state λ, the probability for getting the light of color x to come on at A is invariant under conditionalization on the outcome y at B.

The aspect of local realism captured by the completeness condition, along with the motivation for giving it this name, is suggested by considerations which are quite familiar to philosophers of science.[7] A humble example will suggest the general idea. Suppose one were to learn that in a given large, randomly selected population, heart attacks are less prevalent among people who own jogging shoes than in the population as a whole. (I venture to say that this would not be thought implausible.) One knows, of course, that the likelihood of avoiding a heart attack is not in the least diminished by purchasing a pair of jogging shoes and tossing them into the closet. Mere ownership of jogging shoes is not a causal factor in the prevention of heart attacks. The negative correlation between heart attacks and owning jogging shoes is explained by some other factor, such as regular exercise, which is a genuine causal factor in the prevention of heart attacks and one which is positively correlated with the ownership of jogging shoes. Consequently, if we restrict our attention to that segment of the population consisting of those who exercise regularly, we should no longer find a correlation between ownership of jogging shoes and the presence or absence of heart attacks. That correlation

---

[7]See, e.g., Salmon (1984) for a recent discussion of common-cause explanation.

gets "screened-off" by this more complete specification of the population under consideration.

In the context of Mermin Contraption experiments, completeness underlies the possibility of providing an analogous story about correlations between the pairs of lights that come on at A and B. Intuitively, "complete" state-descriptions must include all causally relevant factors contributing to the outcomes of measurement-events, the illumination of detector lights. The completeness condition asserts that once the detector-states and the two-particle state are fully specified, the probability determined for a specified outcome at one detector is invariant under conditionalization on the outcome at the other detector; that is to say, the outcome-pairs are uncorrelated. In any local realistic account, any such correlations would have to be explained in terms of the presence of some unknown common causal factor, all of which are taken to be included, at least implicitly, in complete state-descriptions; these complete state-descriptions screen off the correlations.

It is a necessary condition for "completeness" of a theory in the sense employed by Einstein, Podolsky, and Rosen, that "every element of the physical reality . . . have a counterpart in the physical theory" (1935, 777). Without claiming that the notion of completeness employed in the EPR argument implies my completeness condition in any strict sense, I do hope it is clear that the EPR wording suggests an informal link of just that sort.

I mention in passing that locality and completeness, as formulated here, are independence conditions which are themselves independent of each other. To put it in a somewhat oversimplified way: one (locality) asserts that a measurement outcome at one wing of the Mermin Contraption does not depend on the switch setting of the distant detector, and the other (completeness) asserts that the outcomes at the two wings are themselves mutually independent. Neither condition implies the other.

Claim (III) corresponds in a rough way to the EPR argument, although the two formulations are quite different. I want to take advantage of their correspondence, however, to emphasize the point, sometimes obscured in popularizations, that Einstein, Podolsky, and Rosen do *not* argue for the incompleteness of quantum mechanics on the basis of any *assumption* of determinism. On the contrary, they argue (in effect) that in certain experimental contexts, determinism *follows* from their notion of completeness along with other conditions either satisfied by quantum mechanics or required by relativity. They conclude, then, that in such experimental contexts, since quantum mechanics is not deterministic (in the relevant sense), it must be incomplete. I emphasize further that for Einstein, Podolsky, and Rosen, incompleteness does not follow from indeterminism alone. I hope my corresponding claim (V) seems plausible from previous considerations, so that in claim (III), it is no surprise to find that both locality and conservation are essential to the implication.

In fact, the empirical grounds for the conservation condition employed in versions of Bell's theorem represented by claims (I) and (IV) are weaker than my account here suggests. In my presentation of the idealized Mermin Contraption, I have neglected an important feature of the actual experiments, to wit, that real particle detectors are horribly inefficient. As a consequence, conservation is difficult to confirm experimentally. In terms of the Mermin Contraption, there are many trials in which a light comes on at only one wing of the contraption even though both detectors are prepared to perform measurements. This gives rise to nonnegligible experimental uncertainty attaching to the strict correlations in the predicted statistics required by conservation. It is possible that concealed violations of conservation occur. While there seem to be no positive reasons for assuming that such secret violations do occur (and other reasons for believing that they do not occur), we are attempting to draw rather significant conclusions from the experimental data and it would be nice if we could dispense with conservation as a premise in the derivation of Bell inequalities. Fortunately, we can.

(VI)	LOCALITY & COMPLETENESS	$\Leftrightarrow$	STRONG LOCALITY

(VII)	STRONG LOCALITY	$\Rightarrow$	GENERALIZED BELL-TYPE INEQUALITY

(VIII)	DETERMINISM & LOCALITY	$\Rightarrow$	GENERALIZED BELL-TYPE INEQUALITY

Claim (VII), yet another version of Bell's theorem, asserts that a generalized Bell-type inequality follows from a condition called "strong locality," which (according to claim [VI]) is just the conjunction of locality and completeness. (As an aside, I should just mention that these generalized Bell-type inequalities pertain to Mermin Contraptions whose detectors sport four switch positions.[8])

Finally, regarding claim (VIII) (which is an immediate consequence of

---

[8]This is so only if the detectors are to be regarded as identical devices (i.e., having the same range of possible measurement types). Otherwise, since measurements of only two distinct types

claims [V] through [VII]), it is intriguing to ponder the dilemma this would pose for Einstein, were he with us today. Experimental violations of generalized Bell-type inequalities would appear to force him to relinquish determinism or to undertake a major overhaul of relativity theory. These generalized Bell-type inequalities are considered to be very well tested; and if we regard locality as secure in virtue of its link with relativity (because relativity is extraordinarily well confirmed independently), then I think it fair to say that we now do have strong empirical evidence against determinism. Until these experiments of recent years (and these have all been within the past two decades), one might reasonably still have hoped that the indeterminism of quantum mechanics would be revealed to be like that of classical statistical mechanics. No longer.

But giving up determinism is not enough. By claim (VII), the experimental evidence against the generalized Bell-type inequalities is also evidence against strong locality; and taking locality as secure in virtue of its link with relativity, claim (VI) mandates that we give up completeness. Quantum mechanics, over and above its fundamental indeterminism, does violate the completeness condition as formulated here, and now even this more radical feature of the quantum-mechanical worldview has been confirmed by experiment; and with this evidence for incompleteness, local realism ceases to be tenable.

4. Confessions and reflections

I cannot in good conscience conclude my guide to the implications of Bell's theorem until I make a confession. First of all, claim (II), asserting the equivalence of locality and "no superluminal signals," is at best a little misleading. It can be shown, given a rather more careful formulation of the "no action-at-a-distance" condition required by relativity theory, that it is in one way a bit stronger and in another way a bit weaker than the locality condition formulated here. Still, I think it fair to say that for purposes of assessing local realism, they are at least "effectively" equivalent. Consider for example that to give up locality *without* giving up relativity would be to allow that superluminal communication of the sort previously described could occur reliably without the mediation of a signal by the transport of matter-energy. I hardly need point out that the mere logical availability of such a maneuver (giving up locality in *this* way) provides no solace for the local realist.

---

need to be performed by each of the two detectors (with four distinct measurement types altogether), detectors with two switch settings each would suffice.

Second, everyone knows that no experiment or series of experiments can be conclusive in the very strictest sense. Provided we are willing to countenance sufficiently bizarre supplementary hypotheses, we can maintain any cherished belief in the face of any apparent evidence to the contrary. I have neglected to mention various auxiliary assumptions standardly employed in Bell-type arguments. It is generally assumed, for example, that because of the absence of connections linking the devices, the distribution of hidden states is independent of the detector switch settings and that there are no causal links between the act of selecting the A switch setting and the act of selecting the B switch setting. The most recent Bell-type experiment largely circumvents the need for the latter assumption, but the former is still crucial. While the assumption that the distribution of two-particle states is independent of the detector switch settings is an extremely plausible assumption, it should be acknowledged that this assumption *could* be false. Yes, it is logically possible that in each of the several different actual Bell-type experiments, there were unknown hidden connections, perfectly "local realistic" in character, which brought about a correlation in the distribution of initial two-particle states and the premeasurement detector-states in just such a way so as to yield the observed quantum statistics. (Remember that it is not enough for the theory to yield statistical predictions which violate Bell-type inequalities; the predictions must be in good numerical agreement with the well-confirmed predictions of quantum mechanics.) In the absence of any independent justification, the hypothesis of such a diabolical conspiracy of nature is surely so blatantly *ad hoc* that no self-respecting local realist could entertain it seriously. Abner Shimony, Michael Horne, and John Clauser put it this way:

> Skepticism of this sort will essentially dismiss all results of scientific experimentation. Unless we proceed under the assumption that hidden conspiracies of this sort do not occur, we have abandoned in advance the whole enterprise of discovering the laws of nature by experimentation. (Bell et al 1985, 101)

Conspiratorial mechanisms of various sorts have been proposed (under other descriptions, of course), and these approaches deserve consideration, but I fail to see any suitably independent justification for taking them seriously at the present time.[9]

Third, I must admit that my own Mermin Contraption, the one in the photograph above, does have hidden connections. The connections are produced not by mysterious Q-rays, but by garden-variety electrical wires con-

---

[9]I by no means want to suggest that all such approaches should be dismissed outright. For all anyone really knows, future developments *could* force us to view these matters in a quite different light. See, e.g., Bohm and Hiley (1981) and Fine (1982d) for some clever proposals for reconciling local realism and experimental results.

cealed in the platform. They can be discovered quite easily by examining the contraption firsthand. The corresponding devices in the actual experiments are, of course, not secretly connected in this or in any other deliberate fashion.

Finally, I probably should also confess to maintaining a convenient silence with respect to certain complications associated with the application of Bell's theorem to theories that admit time-dependent states. But because these complications seem not to lead to any interesting loopholes through which local realism may escape, and moreover because a somewhat tedious analysis shows that they do not alter any of the general implications I want to draw, it is perhaps just as well not to digress any further to discuss the matter here (see Jarrett 1986).

I can now proceed with a clear conscience to some concluding remarks. Einstein, Podolsky, and Rosen were right: quantum mechanics *is* incomplete in an important sense. But Bohr was also right: this feature of quantum mechanics (whatever we decide to call it) is not to be construed as a defect in the theory. The experiments associated with Bell's theorem provide strong evidence that *any* empirically adequate theory must incorporate this feature so flagrantly antithetical to local realism. We cannot reasonably hope that completeness will be reinstated in some future theory which supersedes quantum mechanics.

But what *is* incompleteness? Recall that both locality and completeness are independence conditions and both have to do with the "unconnectedness" of the parts of the Mermin Contraption. Locality expresses a form of independence between the outcome at one wing and the premeasurement state of the distant detector. As previously discussed, if there were any dependence of this sort, experimenters could (at least in principle) exploit it for purposes of superluminal communication, a form of action-at-a-distance which appears to be radically at odds with relativity. Completeness, on the other hand, expresses a form of independence between the outcomes at the two wings of the Mermin Contraption. Complete state-descriptions render the outcomes uncorrelated. However, no correlations of the sort associated with violations of completeness can be exploited for superluminal communication because it is a consequence of the failure of determinism that measurement outcomes are not (even in principle) under the control of experimenters.

Incompleteness, then, appears to represent a connectedness of some sort between spatially distant events (more accurately, between spacelike-related events), which nevertheless does not directly contradict relativity. Abner Shimony has introduced the term "passion-at-a-distance" to refer to this subtle form of nonlocality. In principle, it is only the uncontrollability resulting from indeterminism that stands in the way of superluminal communication via passion-at-a-distance. One begins to feel that, as regards relativity theory, even if incompleteness does respect some operationally construed letter of the law, it nevertheless violates it in spirit.

Can anything more be said to make sense of incompleteness? Can any "explanation" be given of the process by which the correlations associated with Bell-type phenomena are maintained? Attitudes toward such questions differ markedly. Many physicists regard them as pseudoquestions. The urge to find a further explanation, some claim, reflects a serious failure to grasp the deeper lessons of quantum mechanics. But on the subject of correlations, Bell himself has written:

> You might shrug your shoulders and say "coincidences happen all the time," or "that's life." Such an attitude is indeed sometimes advocated by otherwise serious people in the context of quantum philosophy. But outside that peculiar context, such an attitude would be dismissed as unscientific. The scientific attitude is that correlations cry out for explanation. (1981, 55)

With local realism out of the running, what would *count* as an "explanation" of Bell-type phenomena? The question is an important one and I regret that I have no direct answer to offer. I believe that the infamous quantum-mechanical "measurement problem" (which I have made no attempt to set forth here) is a genuine problem for quantum mechanics. It is unfortunate that some presentations make it appear to be little more than an irritating lament about the nonclassical character of the quantum world. But when the problem is given a careful formulation,[10] the measurement problem reveals that quantum mechanics (at least as it is presently understood) has not accomplished a task which it sets for itself. No *quantum-mechanically* acceptable account of measurement interactions has yet been given. Although this is not a claim with which I would expect a great many physicists to agree, it is a claim for which (in a suitably sharpened formulation) a solid case is available. (At any rate, that is my opinion.) Moreover, Bell-type phenomena, which appear to reveal "nonlocal" influences of certain measurement events, have the character of dramatic special cases of the measurement problem. Therefore, a genuine resolution of the measurement problem might reasonably be expected to shed some further light on incompleteness; and this, I claim, coupled with the foregoing evaluation of the legitimacy of the measurement problem, provides at least some basis for the view that Bell-type phenomena are yet to be properly understood.

Quantum mechanics has long been known to have a "nonlocal" character. Once again, it does satisfy the condition I have been calling "locality," but it violates the condition I have been calling "strong locality," the condition which results from conjoining locality and completeness. There is now

---

[10]Stein (1972) gives a masterful treatment.

strong experimental evidence that this aspect of quantum mechanics must be shared by any acceptable empirically adequate theory. To *this* extent, I consider it entirely appropriate to say that a philosophical problem has been settled in the laboratory.

Having said *that,* however, I must add that I also believe it is appropriate to say that important questions remain. "Quantum connectedness," "passion-at-a-distance," "incompleteness"—whatever we choose to call it—this feature of the post–"local realistic" worldview challenges any straightforward notion of the individuation of physical objects. In terms of the Bell-type experimental setup, for example, we are barred from representing what we ordinarily would call the "*two* particles" in a given "pair" *as* two distinct entities; although detectable in spacelike measurement events, "they" form a single object, connected in some fundamental way that defies analysis in terms of distinct, separately existing parts. Providing an adequate elucidation of this inseparability, and of the measurement process itself, an elucidation which gives us a more wholehearted reconciliation between relativity theory and the "passion-at-a-distance" associated with this inseparability—this is the grand, unfinished task.

THE WAY THE WORLD ISN'T: WHAT THE BELL
THEOREMS FORCE US TO GIVE UP

LINDA WESSELS

Bell's theorem shows that certain assumptions about the nature and behavior of quantum systems, in specific kinds of experiments, imply that data obtained in those experiments will satisfy Bell's inequality. Cushing describes these experiments, sometimes called *Bell experiments,* earlier in this volume. The reader should refer to Cushing's sketch of the experiment in his figure 3 when specific aspects of the experiment are discussed below. Bell experiments are designed to ensure that any correlations between the measurement outcomes at *A* and *B* could be due only to a correlation between the properties of the electrons at the time they are measured, and that this correlation could only be caused, in turn, by an interaction between the two electrons that took place (or at least began) while the electrons were together in the source, but ended before they reached *A* and *B*.

Quantum mechanics predicts that the data gathered in the Bell experiments will not satisfy Bell's inequality. If Bell's inequality had been experimentally confirmed, one of the most successful theories physics has ever produced would have been overthrown. Since, in fact, the data does not satisfy Bell's inequality, nothing so spectacular as science in disarray has occurred. We do find ourselves in an interesting and potentially exciting situation, nonetheless. A realist might describe the situation this way: Rather than simply providing an opportunity for improving the abstract mathematical laws of quantum theory, the experimental verdict on Bell's inequality offers a tool for discovering descriptive truths about micro-objects. Since quantum systems do not satisfy Bell's inequality, one (or more) of the premises from which the inequality follows must be wrong. By examining those premises, we might be able to uncover the false one and learn something new about the quantum world. An antirealist would describe the situation more circumspectly: The empirical failure of Bell's inequality may allow us to discover some restrictions the quantum world imposes on what kinds of models will be able to capture its observable effects.

Two complications arise immediately. First, since Bell's original work, others have proved similar theorems using slightly different premises and concluding with slightly different inequalities. One reason for examining alternative premises is that Bell's original assumptions seem overly restrictive. In particular, Bell assumed a deterministic link between the objective properties of quantum systems and the outcomes of measurements on them. But perhaps the properties of quantum systems determine only probabilities of measurement outcomes, or perhaps it takes the properties of both a system *and* a measuring device to determine probabilities of the measurement outcomes. It turns out that even weaker assumptions like these imply inequalities like Bell's, and these inequalities also are not satisfied by the experimental data. (Such inequalities are called "Bell inequalities"; the theorems proving them are called "Bell theorems.") Thus, there are several different sets of premises that might be examined, each leading to a different experimentally falsified Bell inequality.

The second complication results from a simple truth of logic: more than one set of premises can imply the same conclusion. Thus, even choosing to focus on a particular Bell inequality does not determine the set of premises to be examined.

The task is not, then, simply to analyze some given set of premises. We must first choose a set that promises to be interesting and informative. To that end, I recently formulated a set of premises which, together with some theorems of probability, jointly imply a Bell inequality (1985).[1] These premises promise to be interesting because they satisfy two heuristic guidelines: (1) they are as weak as possible regarding determinism; and (2) they reflect commonly held presuppositions about a wide range of physical objects and processes. They are presented in the next two sections.[2] In section 3, the role of relativity in the Bell theorems is analyzed. Then, in section 4, the premises

---

[1] In Wessels (1985), it was shown that these premises plus some probability theorems imply factorizability, and said that factorizability plus some additional assumptions would give a Bell inequality. But, in fact, no further assumptions are needed. In particular, no additional premises are needed to deduce the Clauser-Horne inequality. The other assumptions I had in mind really are needed only to go from the raw data collected in a Bell experiment to the relative frequencies that appear in a Bell inequality, e.g., the assumption that the two electrons in each pair have the same velocity, or the assumption that there is no correlation between the objective properties of a system measured and the failure or success of the device used to measure it.

[2] The premises presented here are basically the same as those in Wessels (1985), though the same names are not used (*QS* was not even given a name), and here they are formulated less technically (hence less precisely). The only insignificant difference is that *H1* and *H2* are replaced here by *M2* and *P2;* this eliminates the need for *EL7.*

are examined to see which might be abandoned, and the implications of abandoning them are discussed. A brief conclusion is provided in section 5.

1. Premises about quantum systems

The set of premises divides naturally into four subsets. The first subset is found on table 1. Its first premise, *QS,* is explicitly about quantum systems. It says that quantum systems subjected to Bell experiments can be treated as bodies. A body is an object that is contained in a relatively well-defined, localized, spatial surface, thus has a well-defined spatial location, and in addition, remains distinguishable from other objects even while its physical characteristics (including location and spatial surface) change. We commonly conceive of objects as bodies, and this conception of objects also underlies many scientific theories. Examples of common bodies are tables, trees, persons, and clouds; examples from science include planets, charged spheres, point particles, infinitesimal bits of wire, and (at least before quantum mechanics) water molecules and subatomic particles. As these examples indicate, an object can be treated as a body for some purposes, but as something else for others—say, as a collection of bodies, or as part of some other kind of system. Thus *QS* does not imply that quantum systems must *always* be treated as bodies, only that they *can* be for the purposes of analyzing the Bell experiments.

Next are premises about bodies in general. *B1* says a body has properties that it possesses independently of whether any attempt is made to examine them. The set of objective properties of a body at a given time is called its *state,* or sometimes (to distinguish it from a quantum-mechanical state) its *objective state.* Of course, the kinds of objective properties that are of interest are those characteristics of a body that causally influence the outcome of an observation or measurement of the body.[3] More generally, objective properties play certain roles in our models of physical interaction. Models for interactions (among bodies) are the subject of the remaining *B* premises.

How do we describe an interaction between two spatially contiguous

---

[3]In the case of quantum systems, the objective properties need not be thought of as precise values of quantum-mechanical observables. "Secondary properties" in general might be thought of as propensities, grounded in objective properties. Color, for example, is the propensity to respond to impinging radiation in a certain way. An object has this propensity because it, in turn, has certain objective properties. If in certain circumstances a quantum system "has" a precise value for certain observables, this might well be construed as a secondary quality—its propensity to interact with a measuring device in a way that yields the precise value as the measured result. In other circumstances it may "have" no precise value, i.e., no propensity to yield a unique precise value, though it still has objective properties.

TABLE 1:

Assumptions about Quantum Systems

QS. The quantum systems measured in Bell experiments can be treated as bodies.

B1. A body has objective properties—properties it has independently of whether it is measured, or observed in any other manner.

B2. Interactions among bodies satisfy either the contiguous interaction model or the influence model or some combination of the two.

B3. Evolution of the objective state of a body in the presence of influences, satisfies a probabilistic law. This law gives the conditional probabilities for the objective state of the body at a time t', given the objective state of the body at any previous time t, and the magnitudes of the influences, at the locations of the body, during $(t,t']$.

B4. Probabilities for the evolution of the state of a body over a period $(t,t']$ depend *only* on the objective state of the body at t and the magnitudes of the influences at the position of the body during $(t,t']$; they are statistically independent of states of the body before t, and of any other circumstances (e.g., properties of other objects or influences at other places) before or during $(t,t']$.

B5. An interaction lasting a period $(t,t']$ between two bodies that are spatially contiguous at t and t', in the presence of influences, satisfies a probabilistic law. This law gives conditional probabilities for the objective states of the bodies at t' given their objective states at t and the magnitudes of the influences at the location of the interaction during $(t,t']$.

B6. For two bodies contiguous at times t and t', interacting during $(t,t']$ in the presence of influences, the joint probabilities for the objective states of these bodies at t' depend only on the objective states of the bodies at t, and the magnitudes of the influences at the location of the interaction during $(t,t']$; they are statistically independent of states of the bodies before t, and of any other circumstances (e.g., states of other objects or influences at other places) before or during $(t,t']$.

SC. If neither an interaction between two bodies before or at a time t, nor interactions between each of these bodies and other systems prior to or at t, nor some combination of these has caused the objective states of the two bodies to be correlated at t, then the states of the two bodies are not genuinely correlated at t.

bodies, for example, a collision between two billiard balls? We suppose that each ball has a set of objective properties, including color, mass, shape, position, velocity. Some of these properties change in value over time, some remain constant. This pattern of change and constancy in values is often called the time *evolution* of the body, or of its properties or its objective state. When a solitary ball is isolated from any disturbing influences, this evolution will satisfy certain regularities: color, mass, shape, and velocity remain the same, position (in general) changes. When two such balls (otherwise isolated) collide, the evolution of each will not (in general) satisfy those same regularities. Their velocities will change, and perhaps other properties too—shape, for example. The effect the balls have on each other, because they collide, is represented by the difference their collision makes in their evolution. In addition, it is also the objective properties of the individual balls that determine what this effect will be. How much the velocity of each ball changes depends on the properties of both balls (mass and velocity) just prior to their collision.

The above example used a sort of scientific language to describe an interaction between contiguous bodies. Objective properties play these same roles when we give commonsense-level descriptions of "close encounters"; the descriptions are just not so precise or quantitative. We might attribute to dogs such properties as large or small, intelligent or dull, fast or slow, and say that Fido managed to chase Spot away because Fido is large and fast while Spot is small and dull. The role of the objective properties is basically the same.

Call this the *contiguous interaction model*. It applies when two bodies interact while in contact: their properties just prior to contact determine the effects of the interaction; the effect on each body is just the difference between its evolution after the interaction, and the evolution it would otherwise have exhibited if the interaction had not taken place.

Another model is used when an interaction does not involve contact. It also underlies our commonsense analyses of such interactions, but takes its clearest form in scientific applications. In this case, the way the interaction affects each body is captured by considering the effect of an "influence" present at that body's location, but generated by the other body. An interaction between two distant bodies is analyzed as two interactions, each involving the effect of a local influence on a single body. Call this the *influence model*. Forces acting at a distance are one way such influences have been modeled. Other ways of implementing the influence model are more intuitively plausible—those in which interactions are mediated by the transmission of small particles, for example, or by the propagation of disturbances through an intervening medium (a gas, a fluid, an ether) or a field.

Three features give the influence model its "bite" in science: (1) the influence has a well-determined value at every place in the region surrounding its source, (2) the effect of the influence on a body at one of those places is quantifiable also, and (3) that effect depends only on the characteristics of the influence at that place and the properties of the body at that place. This means that positing an influence to explain the effect of a particular body on another has precise, quantifiable consequences for other similar pairs of bodies in similar situations—consequences that can therefore be tested empirically. Characteristic (3) makes influence interactions *local*. The effect of an interaction between two bodies is depicted as the effect on each body due to conditions only in its immediate vicinity. The effect of the loudspeaker on my ear is determined by the objective properties of my ear plus the magnitude of the vibrational disturbance in the air just surrounding my ear; the disturbance at my ear depends, in turn, on the objective properties of the loudspeaker that generated it and characteristics of the air through which the disturbance propagated. Another example: on classical theory, the gravitational force of the sun on the earth is determined by the gravitational potential at the earth plus certain objective properties of the earth; that gravitational potential depends, in turn, on objective properties of its source, the sun. The influence model is incorporated in an even more subtle way into the relativistic account of the same phenomenon.

When an interaction between two distant bodies significantly affects both, it may be difficult to get precise predictions of the outcome. But the model proves particularly convenient when the effect on one of the interacting bodies is negligible. Then, once the influence due to the unaffected body is determined, its properties (and those of the intervening medium) can be ignored; the total effect of the interaction can be captured (approximately) simply by looking at the effect on the other body due only to conditions in its immediate vicinity. The model is also used for treating interactions involving more than two bodies. Again, such applications can in general pose difficult (if not impossible) computational problems. But if the effect of the interaction on all but one of the bodies is negligible, that effect can be found by treating all the others as (approximately unreacting) sources of influences. The effect of the gravitational influence of the sun and the planets on Halley's comet is determined by the objective properties of Halley's comet plus the gravitational potential at the location of the comet; the latter is determined by the (approximately) unaffected properties of the sun and planets. For situations involving several interacting bodies where two of them are in contact, the two models can be combined. The interactions between all but two of the bodies can be captured by the influence model, while the interaction between the two can be treated as a contiguous interaction model.

B2 says these are the models that apply to interactions among bodies. It is *B3* that makes the implementation of the influence model nondeterministic.

B4 fits the locality of the model into this nondeterministic framework: evolution of a body in a period $(t,t']$ depends *only* on the state of the body at t and the influences at the location of the body during $(t,t']$. This means, in particular, that the evolution of the body does not depend on its own states before t, nor on any other circumstances before or during $(t,t']$, where "other circumstances" include properties of other bodies and of surrounding media (since any contributions of these *during* $(t,t']$ are already incorporated in the influences, and contributions of these *before* t have already had their effect taken into account in the evolution of the body prior to t), and also includes influences before and during $(t,t']$ at places other than the location of the body.

B5 makes the contiguous interaction model nondeterministic, and allows a contiguous interaction to take place in the presence of influences due to yet other bodies. (Contiguous interactions *not* in the presence of influences are covered by *B5* as a special case—when the set of influences is empty.) *B5* allows for the possibility that *during* their interaction, contiguously interacting "bodies" may not have the characteristics of a body. They may not be individualizable, for example, or have individual positions. For this reason, contiguous interaction is characterized as an interaction between bodies that are spatially contiguous just before and just after their interaction, when they *can* be treated as bodies with position. It *is* supposed that the interaction itself has a location, or more intuitively, that the complex body formed during such an interaction has a position.

B6 is similar to *B4*. It makes explicit the locality of the influence model, now as it combines with the contiguous model: the outcome of an interaction between two (initially and finally) contiguous bodies depends *only* on the initial states of these bodies and the influences at the location of the interaction. In particular (again), the interaction outcome does not depend on states of the interacting bodies prior to their interaction, nor on any other circumstances before or during their interaction—circumstances such as properties of other bodies or of surrounding media, or influences at earlier times or at other places.

The last premise about bodies, *SC* (sufficient cause), concerns the explanation of correlations. When there is a lawlike correlation, we suppose that there must be a cause for this correlation. Of course, cases of statistical dependence arise even in series of event-pairs that are not causally related, simply coincidentally. A lawlike correlation between two types of events, however, is a failure of statistical independence that occurs almost always in very long series of these types of events; correlations predicted by natural laws fall into this class. We standardly assume that such correlations are not mere coincidence, but that there is something going on when these types of events occur that is sufficient to cause the correlations—either an interaction between the correlated events or some common cause. The particular kind of correlated events *SC* talks about are the simultaneous occurrences of objective states of two bodies. *SC* is crucial to any derivation of Bell inequalities.

2. The other premises

The remaining premises are listed in table 2. The second subset of premises is about measurement devices and measurement processes. *M1* says measurement devices are bodies and measurement processes are contiguous interactions. *M2* and *M3* concern the relation between the objective properties of measuring devices and some of their important observable characteristics. The initial setting of a device (the way it is set up to do a measurement), and also its observable display of a measurement result, presumably just *are* particular configurations of some of its objective properties. *M2* recognizes that the initial setting of a device may not uniquely determine *all* of its objective properties by supposing that the initial setting determines only a probability distribution on the possible initial objective states of the device.[4] When a measurement ends, the relevant connection goes the other way, from the objective properties of the device to its display of measurement results. Supposing this display just is a configuration of objective properties gives the deterministic connection posited by *M3*. A merely statistical connection seems implausibly loose. Nonetheless, even if *M3* is weakened by allowing the properties of the device to determine only *probabilities* for displayed results, a Bell inequality follows.

The third subset of premises simply characterizes the Bell experiments. Consult figure 3 in the Cushing essay while reading them.

The final subset consists of two premises about regularities in the pairs of electrons emitted from the source. *P1* is an explicit part of the description of the experiment (see the discussion accompanying figure 3 in Cushing's essay). *P2* is also a standard premise of Bell theorems. It supposes that the production of electron pairs by the source is at least statistically regular enough to yield well defined probabilities on the objective states of the electrons at that moment when interaction between them ends.

---

[4] It is more plausible to suppose that the initial setting of a device only determines probabilities on *some* of the objective properties of the device, since it is reasonable to assume that not all of the objective properties affect in a significant way the *measuring* operation of the device. These "idle" properties might be configured any old way when the initial setting is made, hence there may be no lawlike relation between the two. Thus there is a weaker version of *M2* that could be adopted. It too, however, results in a Bell inequality.

TABLE 2:

Assumptions about Measurement Devices and Processes

M1. A measurement device can be treated as a body; a measurement on a quantum system can be treated as an interaction between two (initially and finally) contiguous bodies, the measuring device and the quantum system measured.

M2. For each possible objective state of a measuring device, there is a well-defined conditional probability that the device will have that state at the beginning of a measurement, given the initial setting of the device at that time.

M3. The result displayed by a measuring device at the end of a measurement is determined uniquely by the objective properties of the device at the end of the measurement.

Assumptions about the Experiment

E1. Any interaction before or during a run of the experiment, between either electron or either measuring device, and any systems other than these, can be treated on the influence model. These "other systems" are not (significantly) affected by the interactions, their influences are (to a sufficient approximation) the same for every run of the experiment, and these influences do not cause a correlation between the objective states of the paired electrons, nor between the measuring devices, nor between either electron and either measuring device.

E2. Before the measurements on a pair of electrons, there has been no interaction that significantly affects the measurement outcomes between either measuring device and either electron, or between one device and the other.

E3. For each pair of electrons, there is a time before the measurements when interaction between them ends, and they do not interact again during the rest of the experiment.

E4. During the measurements, the measuring device at A does not interact with the electron measured at B, nor the device at B with the electron measured at A.

Assumptions about the Pairs of Electrons Emitted by the Source

P1. The pairs of electrons emitted by the source are anticorrelated in spin, i.e., measurements for spin along the same direction on both electrons will give plus for one, minus for the other.

P2. There exist well-defined, joint probabilities for the objective states that the emitted pairs of electrons will have (respectively) at the moment, prior to measurement, when they cease interacting.

3.   The role of relativity theory

Discussions of the Bell theorems often cast relativity theory in an important role. Sometimes it is asked: Does the experimental falsification of the Bell inequalities imply that relativity is false? Or, turning the question around: Which premises of a Bell theorem follow from relativity, and hence cannot be rejected? Both questions may appear puzzling, since relativity is not mentioned explicitly in any of the above premises, nor does it seem to be logically related to any of them. Those in the second and fourth subsets, in particular, seem quite independent of relativity. The others require closer scrutiny, however.

The first subset contains statements about interaction models; we might expect relativity to place constraints on the influence model, in particular. For the first signal principle of relativity says that light is the fastest possible signal: no signal, whether a moving particle or the propagation of a disturbance through some medium or field, can travel faster than light. This requires that the influences in the influence model cannot be propagated at speeds greater than that of light. Notice, however, that none of the B premises talk at all about how quickly influences are propagated. *If* experiments had confirmed the Bell inequalities, we might have tried to develop a detailed theory of quantum systems that incorporates the *B* premises, making sure all influences satisfied the first signal principle. Since the Bell inequalities have been falsified, however, our task is simply to examine the premises that imply them; constraints (like the first signal principle) on how these premises might have been worked into a full, coherent theory are irrelevant.

Relativity may seem even less relevant to the third subset. Yet it is here that relativity theory *does* come into play. For even though these premises are not directly implied by relativity, relativity can and has provided epistemological support for some of them.

The *E* premises describe the experiment used for testing a Bell inequality. Hence the question of whether they are true only makes sense given a particular experimental setup: Does the setup guarantee, or more realistically, give good reason to believe, that these premises are satisfied? For *E1* and *E2*, this amounts to asking whether the experiment has been prepared properly. Have the source, paths of electrons, and measuring devices been arranged so that neither the electrons nor the devices will bump against any surrounding objects? Are the surroundings kept stable (to a sufficient degree of approximation)? Has the apparatus been shielded from any external influence that might correlate the properties of the emitted electrons? Does the setup ensure that before the measurements at *A* and *B*, no correlating interactions have taken place between the measuring devices, or between either device and either electron? Has all possibility of some common cause for such correlations been eliminated? A "no" on any of these simply means that the setup is not appropriate for testing the Bell inequality.

More interesting is the question of whether *E3* and *E4* are satisfied in a given experimental setup.[5] How would a setup ensure that the electrons in each pair stop interacting before they reach their respective measuring devices? Or that while one electron is being measured at *A*, the *A* apparatus itself is not at the same time influencing the outcome of the measurement at *B*? One obvious strategy is to use some auxiliary assumptions about what the unwanted interactions might be like. We might try assuming, for example, that an interaction between paired electrons is due to an exchange of small particles. We could then put a particle barrier between the two electrons after they are emitted. By comparing the data obtained with the barrier in place and with the barrier absent, we could tell whether such an interaction is present. It would be the data obtained with the barrier in place, when the interaction is cut off, that should satisfy a Bell inequality, for that would be data obtained in an experiment that satisfied *E3*.

A similar approach could ensure that *E4* is satisfied. We might try supposing, for example, that the strength of the unwanted interaction diminishes with distance. We could then look for a change in the data as the distance between *A* and *B* is increased. Particularly interesting would be the data obtained when the distance was so large that the intensity of the interaction almost disappears, for under these conditions the experiment would satisfy *E4*. Other hypotheses about the nature of an interaction between paired electrons, or between the measuring devices, or between an electron and the device *not* used to measure it, would motivate other kinds of experiments.

Another way to ensure *E3* and *E4* is to take as an auxiliary assumption the first signal principle. For if *A* and *B* are placed so far apart that only a signal faster than light could travel between the electrons as they approach measurement, the first principle seems to *guarantee* that *E3* is satisfied. The same experimental arrangement also seems to guarantee *E4*, since only a signal faster than light could travel between *A* and *B* during the measure-

---

[5]The predictions given by *quantum mechanics* for these types of experiments also depend on assuming *E3*. Thus the fact that, for a given setup, the predictions of quantum mechanics are borne out might be taken as indirect evidence that *E3* has been satisfied. It would only make sense to go on to ask for further assurance on *E3* (that is, to try to generate experimental designs inspired by the kind of auxiliary assumptions discussed below), if one were to conjecture that the quantum-mechanical treatment is deceptive—that even though quantum mechanics treats two systems as noninteracting (the expression for potential in the Schrödinger equation contains no term representing an interaction between the two systems), the systems might nonetheless be interacting. This would be to conjecture that somehow the quantum-mechanical formalism has built into it the ability to take account of such an interaction even when, on the face of it, the systems are being treated as noninteracting. Tests of the kind discussed below would function as tests of this conjecture.

ments. Of course, some currently unrecognized peculiarity in the structure of space-time could wipe out this guarantee. For example, a wormhole in space-time that allows a slower-than-light signal between A and B, even when they are very far apart (Shimony 1978), would allow $E4$ to be false even when relativity is true. Until there is reason to take such an idea seriously, however, it is irrational to suppose that $E4$ is not satisfied when A and B are appropriately far apart.

Clearly, the first signal principle is a particularly good auxiliary assumption to use. The other types of assumptions suggested above posit specific forms of interaction (e.g., particle exchange), or specific characteristics of those interactions (e.g., reduction of intensity with distance). The restriction on propagation speed imposed by the first signal principle applies to all interactions, however; no specific kind of mechanism or characteristic need be postulated. Furthermore, there is no independent evidence for the more specific posits. The first step in using such auxiliary assumptions is to test for the posited interaction or characteristic, i.e., to test the auxiliary assumption itself. Relativity theory is already independently well confirmed. It is no wonder that the experiments proposed for testing Bell inequalities, and those actually carried out, all rely on relativity to guarantee $E3$ and $E4$. Relativity theory is a very convenient auxiliary assumption. It is in this capacity and *only* in this capacity, that it enters into the picture.

4. *Which premise is wrong?*

It has often been claimed that quantum mechanics forces a radical departure from the classical view of nature because it shows that at bottom, nature is indeterministic—the most fundamental physical processes are stochastic. But some proofs of Bell inequalities show that the peculiar behavior of quantum systems in the Bell experiments cannot be accommodated *simply* by giving up determinism. A glance at the above premises makes this clear. Except for $M3$, all the premises take the laws governing the evolution of quantum systems to be completely nondeterministic. And, as noted above, $M3$ can also be weakened to a nondeterministic form. The only way to weaken these premises further is to suppose that quantum systems behave so irregularly that they satisfy no laws at all, not even probabilistic ones.[6] This is

---

[6]This is not *quite* true—see note 4. The situation discussed there does not change the basic point being made here, however. If the weaker form of $M2$ were used, a Bell inequality would still follow, and there would be no way of further weakening $M2$ without going to the very unlikely assumption that the connection between initial setting and objective properties is not even governed by a statistical law.

highly unlikely, however, given that quantum mechanics itself provides re-markably accurate statistical laws for the observable behavior of quantum systems. Strengthening the premises by supposing the laws are *more* deterministic will not help either. This just gives premises that are special cases of the original ones, hence premises that also imply a Bell inequality. One important lesson that work on the Bell theorems has taught is this: We cannot model the strange behavior of quantum systems simply by relinquish-ing determinism, while leaving the rest of our traditional view of nature unchanged. Until we find out how our view should be altered, it is not even clear whether determinism must be given up.[7]

What changes in our traditional view *will* suffice to capture the peculiar behavior of quantum systems? To answer this question, we ask: Which of the premises listed above should be rejected? It is easier to begin with the reverse question: Which premises are not candidates for rejection?

Consider first *B3*, *B4*, *B5*, and *B6*. Notice that it is *B2* that says that our two interaction models apply to bodies. *B3*, *B4*, *B5*, and *B6* simply ensure that these models are implemented nondeterministically, by positing that the appropriate kinds of probabilistic laws govern the interactions of bodies. To reject any of these four is to suppose that they are wrong about how this implementation should go. There seem to be only two ways they might have gone wrong. Either the kinds of laws they posit are too weak or they are too strong—either more determinism should be assumed, or (since there seems no way to make the laws less deterministic) we must suppose that interactions among quantum systems are not governed by any laws at all. But I have just argued that strengthening the premises by postulating more determinism only results in assumptions that still jointly imply a Bell inequality. And, it was also argued, it is unlikely that quantum behavior is so irregular that it cannot be captured by even statistical laws. Hence, the trouble is not with how the interaction models are implemented. If we have doubts about *B3*, *B4*, *B5*, or *B6*, the trouble is either with *B2*, which says that the interaction models do apply to bodies, or with *QS*, which says that quantum systems are in that category. *B3*, *B4*, *B5*, and *B6* are not likely candidates for rejection.

Similar arguments take *M2*, *M3*, and *P2* out of the running. All three postulate certain regularities in the behavior of bodies—measuring devices in

---

[7]In discussion, van Fraassen questioned the significance of this point, given Fine's proof (1982b) that if factorizability is assumed, there is for any correlation experiment (like a Bell experiment) an adequate stochastic hidden-variables theory if and only if there is an adequate deterministic theory. First, of course, Fine's proof supports the claim that work on the Bell theorems has shown the irrelevance of the determinism issue. Second, however, note that Fine has shown equivalence only on the assumption of factorizability. The question then is: Does factorizability presuppose any measure of determinism, or is it also insensitive to the determinism issue? This is the question Wessels (1985) tries to answer.

M2 and *M3,* pairs of electrons in *P2*. To strengthen any of them by assuming more determinism, or to weaken *M3* by postulating only statistical regularities, will do no good—a Bell inequality still follows. To weaken *M2* or *P2* is to assume there is no statistical regularity in a situation where independent reasons suggest there is. Consider first *M2*. Since the initial setting of a device just is a configuration of its observable characteristics, and observable characteristics just are configurations of objective properties, it is hard to explain how measuring devices could consistently give reliable results unless there are at least the sort of laws posited by *M2*. (See notes 4 and 5 for a correction of this argument.) Now consider *P2*. Since the objective properties of bodies are causally responsible for the results of measurements on them, it would be hard to understand how quantum mechanics could capture the regularities in the measurement results of Bell experiments unless there were at least the statistical regularities posited by *P2*. These are not decisive arguments, of course. One might be able to find some ingenious way to blame the Bell inequalities on *M2* or *P2*. These considerations do show, however, that none of *M2, M3,* or *P2* are very likely candidates for rejection. And *P1* seems an even less likely candidate for rejection, since it is a prediction of quantum mechanics that is, apparently, well confirmed.[8]

What about *M1?* It is hardly doubtful that large-size objects like measuring devices are bodies, so it seems reasonable to classify a measurement on a body as an interaction between bodies. We might worry that measurements on *quantum systems* are peculiar, however, since the measured object is so strange, i.e., we might question whether a measurement on an electron will satisfy one of our two interaction models. It is not *M1* that says how a measurement interaction is to be modeled, however. If we have doubts about how to model measurement, then it is *B2* we must question; if we doubt that electrons being measured should be treated as bodies, then *QS* should be questioned. *M1* is not the problem.

In a properly arranged experiment the premises in the second subset are also well supported. In such an experiment, care has been taken to ensure satisfaction of *E1* and *E2*. What Shimony calls "conspiracies in the backward light cone" are always possible, of course, but mere possibility gives little reason for seriously questioning these premises. And, I have argued, all available evidence indicates that relativity is a reliable auxiliary assumption. Hence, in experiments designed to rely on this auxiliary assumption, *E3* and

---

[8]Judgment on *P1* may not really be quite this straightforward. Don Robinson, currently completing his Ph.D. in the History and Philosophy of Science Department at Indiana University, is exploring the possibility that limitations placed by quantum mechanics itself on the precision of spin measurement may imply that the predicted strict correlation can never be found experimentally.

E4 are satisfied. Since the experiments that have actually falsified Bell inequalities were properly set up and did rely on relativity to guarantee *E3* and
E4, we have good reason to think that all the *E* assumptions were satisfied.

What is left? *QS, B1, B2,* and *SC*. And even this list is too long. For, I
will argue, if *B1* is rejected, we might just as well reject *QS*. Consider first
QS. If we were to reject *QS*, how would we conceive of quantum systems?
We might suppose they are more like fields, with no localized surface boundary, but spread out over a large region, or that a quantum system is really
some pairing up of body and field. It appears, however (though it has not been
proved in detail), that if the proposed quantum field or body-field had the
standard field or body-field properties, and the traditional interaction models
for these kinds of entities were assumed, the field or body-field interactions
would satisfy premises very similar to the *B* premises, and that would also
jointly imply a Bell inequality. In short, moving to some other traditional way
of modeling physical objects would also fail to capture the behavior of quantum systems when they are subjected to a Bell experiment. Rejection of *QS*
(and its field and body-field counterpart) would leave us with no notion of
how to hang objective properties on quantum systems or how to model their
interactions.

Suppose that instead we reject *B1*. Some bodies might still have objective properties, but we might decide that quantum bodies do not. But if we
give up the idea that quantum bodies (or fields or body-fields) have objective
properties, then what is the point of retaining *QS?* The objective properties of
a body are the key to how we describe the behavior of bodies and their
interactions. Maintaining *QS* while rejecting *B1* would affect our ability to
understand the quantum world in the same way as rejecting *QS* itself.

Thus the question of how our view of nature must change in the face of
the Bell theorems boils down to the question of whether we give up *QS, B2,*
or *SC*. We have finally arrived at the hard question, for it is not even clear
how to go about getting the answer. What is clear is that rejecting any one of
the three will result in a radical change in our traditional view of nature.
Rejecting *QS* (and its field and body-field counterparts) leaves us with no
intuitive model of quantum systems, and no idea of how to model their
interactions—it leaves us only with a mathematical apparatus for predicting
probabilities of experimental outcomes. Those who believe that a scientific
theory *is* merely an instrument for predicting and theoretically organizing
observable phenomena may not find this consequence particularly shocking.
An instrumentalist might even see *QS* as the obvious choice for rejection. It is
not necessarily the instrumentalist's best choice, however, for even those who
argue against a literal reading of theories recognize that intuitive models are
often valuable parts of the prediction machinery. Thus even an instrumentalist
might be inclined to save *QS* (or some field or body-field counterpart), and

look for some way of adapting one of our traditional models for physical objects by rejecting either *B2* or *CS*.

B2 might be rejected with the idea that quantum systems influence each other in some strange new way. It has often been suggested that even when two quantum systems are not interacting in ordinary ways, i.e., not in ways captured by either the contiguous or interaction models, their properties or objective states are somehow "entangled" or "non-separable" or in some sense "superimposed." For the most part, such suggestions look like just another way of saying *B2* should be rejected; no positive account is provided of what the new interaction would be like. The one exception is Teller's proposal in this volume that there may be objective relations among quantum systems, relations that are not supervenient on the objective properties of each of the individual systems themselves. Hence (my gloss on Teller), the influence model does not capture the way distant quantum systems influence each other's behavior, because that model assumes that such influences depend only on the objective properties of each system individually. Whether Teller's proposal should be accepted is a separate issue. It serves here as an example of a genuine alternative to *B2,* and further, as an illustration of how radically our view of nature would have to change if *B2* is rejected.

Rejecting *SC* is equally drastic. In fact, people who are unacquainted with the details of the Bell theorems often react with disbelief when I tell them that *SC* might have to be given up. "That is crazy! Clearly you people just don't understand yet what is going on in those experiments." Nonetheless, several philosophically astute and otherwise mentally healthy people have argued that perhaps *the* lesson of the Bell theorems is that *SC* is wrong. (See van Fraassen, 1982a, and Fine, in this volume.)

5. *Concluding remarks*

The empirical falsification of the Bell inequalities requires that one of the premises presented in sections 1 and 2 above be abandoned. I have argued that *QS, B2,* and *SC* are the most likely candidates and that rejection of any one of the three will result in a significant and radical change in our view of nature. Of course, many people have claimed many times before that quantum mechanics forces a radical change in our view of nature. It is only now, however, with the Bell theorems, that we finally have good arguments to support this claim, and to support, further, an analysis of what kinds of change might be required.

To put this talk of radical change in perspective, however, a more conservative observation should be added. While the results of experimental and philosophical analyses of the Bell inequalities do require a significant departure from the way we standardly model physical objects, they have only minimal consequences for our conception of everyday objects and of most objects studied by science. For the Bell theorems give no reason to doubt that *these* objects can be treated as bodies (or fields, or body-field combinations) with objective properties, or that their interactions satisfy one of the traditional interaction models, or that genuine correlations in their objective properties have some cause. The Bell theorems only show that our traditional models are not satisfied by *all* objects in nature—in particular, they fail for objects and processes at the *micro*level. This does not mean that we cannot by and large continue to employ the traditional views with confidence in the traditional domain of *medium-* and *large*-sized objects. For nonrealists, of course, this mismatch in models is not too hard to live with. It does, however, leave anyone attempting a realistic interpretation of quantum mechanics with a hard, though perhaps interesting, problem—that of explaining how quantum systems can be so radically different, given that they are the building blocks for all the other objects in the physical world.

THE CHARYBDIS OF REALISM: EPISTEMOLOGICAL
IMPLICATIONS OF BELL'S INEQUALITY

BAS C. VAN FRAASSEN

In skepticism and realism, empiricist epistemology has its Scylla and Charybdis. The main role of skepticism today is *in reductio:* If a position is shown to lead to skepticism, it is thereby refuted. But fleeing from that danger, we are hard put to steer clear of the metaphysical rocks and shoals of realism. I shall leave the first danger aside for now.[1] Concerning epistemic realism I shall argue that, given one plausible way to make it precise, it is refuted by Bell's inequality argument. Realists will presumably wish to formulate their views on epistemology so as to avoid this refutation, and I shall end with some helpful suggestions.

1. Epistemic realism

The medieval nominalist-realist debate was to a large extent about what we would today call causal properties and dispositions. These figured in the

[This paper was published in *Synthese* 52 (1982): 25–38; some material, enclosed in square brackets, has been added to relate it more clearly to other papers in this volume.] This research was supported by National Science Foundation grant SES 80-005827, hereby gratefully acknowledged. The paper began as a commentary on Putnam (1982); although it took independent shape, the reader will discern my debt in the way I see and state the problems addressed. This paper is also deeply indebted to Richmond H. Thomason, "Prescience and Statistical Laws" (ms. circulated November 1980) which argues cogently for theses about knowledge similar to the ones I defend for reasonable expectation, in relation to causal dependence.

[1]There seem to me two main types of solutions to the problem of skepticism; they are the solutions of idealism and of voluntarism. I would classify Putnam's response to the question, whether we could be brains in a vat, as belonging to the idealist type. The voluntarist tradition, associated with Augustine, may be briefly, if perhaps cryptically, conveyed by the slogan that skepticism is the ass's side of the Buridan's Ass Problem.

explanation of regularities in nature; for they determined how a given sort of thing could be or behave, how it would develop if left alone, and how react if acted upon. When the nominalist critique led to skepticism about the reality of these properties, the realist response was not simply to declare that the observed regularities would be unintelligible without them, though that was part of it. They also argued, unnervingly: If there is nothing to explain the regularity, no reason for it, there is also no reason for it to continue, and hence we can have no reason to expect its persistence. The nominalist position in philosophy of nature would, in other words, lead to skepticism, to the impossibility of reasonable expectations about the future.

This argument can easily be found also in later philosophy, connected with either metaphysical or scientific realism. Thus Peirce's critique of Mill, and his arguments for "Thirdness" (law or physical necessity) are prime examples; not surprisingly perhaps given his avowed debt to Scotus.[2] Similar arguments in contemporary philosophy of science I have discussed elsewhere.[3] A recent book that straddles the two concerns, by Harré and Madden, provides a further striking example.[4]

Let me attempt a preliminary statement of the doctrine I shall call epistemic realism. Consider the question: *How is reasonable expectation about future events possible?* ("Future" may be replaced by "unobserved" for generality.) The recurrent idea that there is some rational form of simple extrapolation from the past, something like rules of induction, may be especially appealing to empiricists because it holds out hope for a presuppositionless, nonmetaphysical answer. But it is an idea that goes into bankruptcy with every new philosophical generation.[5] The answer that I shall call "realist" is: *Reasonable expectation of future events is possible only on the basis of some understanding of (or, reasonable certainty about) causal mechanisms that produce those events.*[6]

Support for this answer springs easily to mind. If there were no causal mechanism that makes litmus paper turn red in acid, then the past regularity (it always has) is a mere accident or coincidence, and there is no reason to

---

[2]See especially Charles S. Peirce, *Essays in the Philosophy of Science*, ed. V. Thomas (New York: Liberal Arts Press, 1957), pp. 157–158, 166, and John F. Bolen, *Charles Peirce and Scholastic Realism* (Seattle, Wash.: University of Washington Press, 1963).

[3]Van Fraassen, *The Scientific Image* (Oxford: Oxford University Press, 1980), chap. 2.

[4]R. Harré and E. H. Madden, *Causal Powers* (Oxford: Blackwell, 1975), see especially pp. 70–71.

[5]As Putnam points out, and I agree, the logical problems brought to light by Goodman's analysis cannot be circumvented, a fact sometimes obscured by their arid logical form.

[6]There is no implication here that the understanding need go very deep; perhaps it is quite possible to have reasonable certainty that there is some causal mechanism or other, producing sequences of some vaguely described type or structure, that is producing the observed sequence, without any certainty at all about more intimate features of the mechanism.

think it will be in the future. What is more, if there is granted to be a cause, everything hinges on what it is like. This is brought out graphically by Bertrand Russell's refutation of simple induction. If there were a simple inductive rule, it would have to go from premises of the form "proportion m/n of past X's were A's" to some conclusion about future X's. Being a rule, it must lead from structurally similar premises to structurally similar conclusions. But now consider two persons applying this rule. The first looks back over past days and, arriving at the premise that the sun has always risen, concludes that it will continue to rise. The second is Russell's example of a man falling down the Empire State building. As he passes an open window on the twenty-seventh floor, he is heard to say "Well, so far so good."

The difference between the two cases lies clearly in the underlying mechanism that produces the two sequences. Their initial segments are structurally alike—but that means nothing at all. To think that it does is exactly like making one of two assumptions: either that all structurally similar observed sequences are produced by similar causal mechanisms, or else that the producing mechanism is irrelevant to the continuation. Since both those assumptions are manifestly unreasonable, to think or reason as if one had made them is unreasonable too.

This is the case for epistemic realism. To discuss it properly, we must make it more precise, and this means mainly that we must make the notion of causal mechanism more precise.

2. *Causality*

Since I shall attack the position, I need to explicate the notion of causal mechanism in a way that is, while nontrivial, as weak as possible. This was also the problem that Hans Reichenbach faced when he wished to reconcile the ideas of causality that had played such a central role in the development of relativity, with the indeterminism of quantum mechanics. He thus developed the concept of *common cause* which has more recently been explored by Salmon and Suppes.[7]

Suppose there is a correlation between two (sorts of) events, such as lung cancer and heavy smoking. That is a correlation in the simultaneous presence of two factors: having lung cancer *now* and being a heavy smoker *now*. An explanation that has at least the form to satisfy us traces both back to a common cause (in this case, a history of smoking which both produced the smoking habit and irritated the lungs). Characteristic of such a common cause is that, relative to it, the two events are independent. Thus present smoking is

---

[7]For fuller presentation and references, see *The Scientific Image*.

a good indication of lung cancer in the population as a whole; but it carries no information of that sort for people whose past smoking history is already known. The second characteristic is that the cause lies in the past (if the two events are spatially separated, in the common part of their absolute past cones). Obviously A and B are independent relative to (A & B); so if the first characteristic alone were taken into account, the notion would be trivial.

The common cause picture is one that can certainly fit an indeterministic world; correlations, rather than events as such, require an explanation, and this is given by tracing (stochastic) processes back to their intersections.

[Let us make this precise, using P to stand for probability; thus $P(A|B)$ is the probability of A, on the supposition of B.] There is a positive correlation between A and B exactly if $P(A|B) > P(A)$, or equivalently if $P(A \& B) > P(A)P(B)$, a symmetric relationship. Similarly we call quantities X and Y uncorrelated only if $P(X = a|Y = b) = P(X = a)$ for all values a and b of these quantities (assuming throughout that the events in question have positive probability). Negative correlation of A and B is obviously equivalent to positive correlation of $\bar{A}$ and B, hence if all positive correlations are explained, so are all negative ones. A common cause C for the correlation of A and B must have, by Reichenbach's definition, the property that A and B are independent (not correlated) conditional on C, and similarly conditional on $\bar{C}$.

Generalized to the case in which the cause itself is a variable factor Z— and not just a yes-no event—this becomes

$$P(A \& B|Z = x) = P(A|Z = x)P(B|Z = x)$$

for all values x of that quantity.

In realistic examples, the events A and B are often outcomes of experiments. That the experiment is going to be done at all is, of course, an independent point; what we are meant to explain causally is that the outcome is thus and so if the experiment is done. Hence the statement of correlation takes the form: $P(A \& B|A* \& B*) = P(A|A* \& B*)P(B|A* \& B*)$. [Here A* stands for the proposition that a measurement has been made to determine A.] If there is a spacelike separation between the two experiments, we suppose that either could be stopped at will before termination, and therefore that $P(A|A* \& B*) = P(A|A*)$. This supposition may be false, for it is conceivable that there is a preestablished harmony, and the experimenters are caused to perform the B*-experiment in just those cases in which experiment A* is performed and has outcome A. A little common sense should help us here when we are discussing a specific, realizable experimental arrangement, though we must keep the preestablished harmony possibility in mind if we contemplate general conclusions.

I am now going to describe a conceivable phenomenon (the one described by Bell) in which there is a correlation for which there can exist no
common cause. The argument presupposes no physics at all. But we can
remark that quantum mechanics allows for phenomena of this sort, and predicts the correlation when they occur. Since quantum mechanics is a well-
supported theory, it is reasonable to have expectations in accord with its
predictions. Therefore it is possible to have reasonable expectations of future
events not based on any understanding of, or certainty about, causal mechanisms that produce those events. My analysis of Bell's argument will give this
conclusion a large degree of philosophical autonomy and generality.[8]

3. Surface description of a phenomenon

There are two generals, Alfredo and Armand, who wish to strike a
common enemy simultaneously, unexpectedly, and very far apart. To guarantee spy-proof surprise, they ask a physicist to construct a device that will give
them a simultaneous signal, whose exact time of occurrence is not predict
able. This is a science-fiction story—their physics is like ours but their
technology much advanced (it happened in a galaxy long ago and far, far
away . . .). The physicist gives each a receiver with three settings, and
constructs a source which produces pairs of particles travelling toward those
receivers. In each receiver is a barrier; if a received particle passes the barrier,
a bell rings. The probability of this depends on the setting chosen. But when
the two generals choose the same setting, one member of the pair of particles
passes if and only if the other does not. Alfredo and Armand agree to choose a
common setting at predetermined time t, and then Alfredo will strike the first
time his bell does not ring while Armand will strike as soon as his bell does
ring.

The story makes clear that no theory is presupposed in the description of
what happens. Before looking at possible theories that might explain this
curious correlation (which in itself is perfectly possible so far, even from a
classical point of view) I shall make this description precise and general.

3.1. The experimental situation: Two experiments will be made, one on
each of a pair of particles produced by a common source, referred to as the left
(L) and the right (R). Each experiment can be of three sorts, or be said to have

---

[8]I circulated this analysis in ditto form under the name of "Baby Bell" in January 1981. The
literature on the subject is voluminous; see especially J. S. Bell (1964, 1966), and J. F. Clauser
and A. Shimony (1978).

one of three *settings*. The proposition that the *first* kind of experiment is done on the *left* particle will be symbolized $L1$; and so forth.

Each experiment has two possible outcomes, *zero* or *one*. The proposition that the *second* kind of experiment is done on the *right* particle and has outcome *zero* will be symbolized $R10$; and so forth. Note that $L1$ is equivalent to the disjunction of $L11$ and $L10$. To allow general descriptions, I shall use indices i, j, k to range over $\{1,2,3\}$ and a, b over $\{0,1\}$. In addition let $\bar{x} = 1 - x$, so that $\bar{a}$ is the opposite outcome of a. [Professor Mermin's "uniform notation" prescribes the use of A and B instead of my L and R (which stood for "aLfredo" and "aRmand", as well as for "left" and "right"). As in his notation, i and j stand for the chosen settings of the apparatus, but I use a and b (instead of x and y) for outcomes.]

A situation in which the two experiments are going to be done on a single particle-pair can be described in terms of a small field of propositions, generated by the logical partition:

$$PR\text{surface} = \{Lia\ \&\ Rjb : i,j = 1,2,3 \text{ and } a,b = 0,1\}$$

which has thirty-six distinct members. [The proposition $Lia\ \&\ Rjb$ describes completely what experiments are done, and what the outcomes are. The generated field of propositions includes also the less informative propositions Li (setting i at left apparatus), Rjb (setting j and outcome b at right apparatus), and so forth. Thus we have here the resources for any sort of factual description at the surface level.]

3.2. Surface probabilities: Probabilities for these propositions come from two sources. First, we may have some information about how the two settings will be chosen (possibly, to ensure randomness, by tossing dice). This gives us probabilities for the propositions in a coarser partition:

$$PR\text{choice} = \{Li\ \&\ Rj : i,j = 1,2,3\}$$

Secondly we may have a hypothesis or theory which gives information about how likely the outcomes are for different sorts of experiments. Because there may be correlation, the information optimally takes the form of a function

$$P(Lia\ \&\ Rjb|Li\ \&\ Rj) = p$$

giving the probability of the (a,b) outcome for the (Li,Rj) experimental setup. Let us call this function P a *surface state*. Note that it is not a probability function on our field; but it can be extended to one by combining it with a probability assignment to PRchoice (which may be called a *choice weighting*). Such a probability function on the whole field [which incorporates the information of one given choice weighting and one given surface state] may be called a *total state.* We can also derive marginal probabilities for the individual experiments, such as $P(Lia|Li) = \Sigma \{P(Lia \text{ \& } Rjb|Li \text{ \& } Rj) : j = 1,2,3 \text{ and } b = 0,1\}$. Note that hypotheses concerning the surface state are directly testable: we simply choose the settings, and start the source working, and do the relevant frequency counts—see how often the bells ring—to follow our story.

3.3. Perfect correlation: The special case I wish to examine satisfies two postulates for the surface states.

    I. *Perfect Correlation* $P(Lia \text{ \& } Ria|Li \text{ \& } Ri) = 0$
   II. *Surface Locality* $P(Lia|Li \text{ \& } Rj) = P(Lia|Li)$
                            $P(Rjb|Li \text{ \& } Rj) = P(Rjb|Rj)$

It should be emphasized again that these probability assertions are directly testable by observed frequencies.[9]

The Perfect Correlation Principle can be stated conveniently as: *parallel experiments have opposite outcomes.* [Similarly, the Surface Locality Principle says that the *outcomes* at either apparatus are statistically independent of the *settings* at the other. Thus neither general can signal to the other by twirling the settings on his own receiver! In these principles no mention is yet made, of course, of any hidden variables, but only of the detectable phenomena.] My formulation, symmetric in *L* and *R,* is a simple condition on the surface states. If both principles hold, we obviously have

$$P(Lia|Li) = P(Lia \text{ \& } Ri\bar{a}|Li \text{ \& } Ri)$$

but the reader is asked to resist counterfactual (and dubious) inferences such as that if the *L*1 experiment has outcome *one* then if the *R*1 experiment had been done, it would have had outcome zero.[10]

---

[9]It will be clear, because II holds (note also that II follows from IV) that there is no question here at all of the two generals being able to signal each other faster than light. It is true that by their arrangement, they reap some benefits which, as their enemy may surmise, they would reap from having signals faster than light if there were such. Imagined conflicts between the described situation and relativity theory lie solely in controversial assumptions about what a relativistic indeterministic theory would have to be like. [Correction: the parenthetical remark is mistaken. This is a case of Simpson's Paradox; because we are not given a probability distribution on the hidden factor *q* in IV, it is indeed possible to have stochastic independence on the hidden level, but correlations on the surface level. The point stands, however: because of surface locality, which is independent of theory, no possible emendation will allow superluminal signals.]

[10]Certain elegant simplifications of Bell's argument [e.g., P. H. Eberhard 1977, and N. Herbert and J. Karush 1978] rest on assumed principles about counterfactual conditionals that have been controversial or definitely rejected in the general theory of such conditionals developed by Stalnaker, Lewis, and others. This is discussed in forthcoming papers by Brian Skyrms and Geoffrey Hellman. [See B. Skyrms 1982, and G. Hellman 1982.]

4. Common causes as hidden variables

When principles I and II hold, there is a clear correlation between outcomes in the two experiments. What would a causal theory of this phenomenon be like? It would either postulate or exhibit a factor, associated with the particle source, that acts as common cause of the two separate outcomes in the examined probabilistic sense. I shall refer to this as "the hidden factor"; not because I assume that we cannot have experimental or observational access to it, but because it does not appear in the surface description (i.e., in the statement of the problem).

Symbolizing the proposition that this hidden factor has value q [which is usually written as $\lambda = q$] as Aq, the space of possibilities now has the still finer partition:

$$PR\text{total} = \{Lia \ \& \ Rjb \ \& \ Aq: i,j = 1,2,3;$$
$$a,b = 0,1; \text{ and } q \in I\}$$

where I is the set of possible values of that factor. [Thus the proposition Lia & Rjb & Aq says that the experiment was carried out with setting i and outcome a at the left, setting j and outcome b on the right, and while the value of the hidden factor was q. Of course, this proposition cannot be checked directly; it concerns the postulated reality behind the phenomena, according to the causal theory under consideration.] A total state must be a probability function defined on the (sigma-) field generated by this partition. (Let us not worry about how to restate this in case I is uncountable; as will shortly turn out, that precaution is not needed.)

III. *Causality* $P(Lia \ \& \ Rjb | Li \ \& \ Rj \ \& \ Aq) =$
$\qquad\qquad\qquad P(Lia | Li \ \& \ Rj \ \& \ Aq) P(Rjb | Li \ \& \ Rj \ \& \ Aq)$

[Principle III is also called *factorizability* (A. Fine), *completeness* (J. Jarrett), and *outcome independence* (A. Shimony). For note that III is made up from a theorem about probability, by deleting "& *Lia*" from the second multiplicand.]

IV. *Hidden Locality* $P(Lia | Li \ \& \ Rj \ \& \ Aq) = P(Lia | Li \ \& \ Aq)$
$\qquad\qquad\qquad\quad P(Rjb | Li \ \& \ Rj \ \& \ Aq) = P(Rjb | Rj \ \& \ Aq)$

[Principle IV is also called *locality* (J. Jarrett) and *parameter independence* (A. Shimony).]

V. *Hidden Autonomy* $P(Aq | Li \ \& \ Rj) = P(Aq)$

[This requirement V is usually specified informally, as ruling out certain kinds of cosmic conspiracy or preestablished harmony, which would make the hidden factor at the outset depend on the apparatus setting to be chosen whether then or later. A drawback of this kind of notation, which writes the hidden factor as a subscript to the probability function, is that it cannot express this.]

I shall break up the ensuing argument into three subarguments, in which these postulates are separately exploited. But I shall say a few words here to defend the idea that a proper causal theory must satisfy all three.[11]

Causality is just the probabilistic part of the Common Cause principle stated before. The other two, *Hidden Locality* and *Hidden Autonomy,* are meant to spin out implications of the idea that it is the common cause alone, and not special arrangements or relationships between the two separate experimental setups, that accounts for the correlation. If we had only III to reckon with, it could be satisfied simply by setting.

$$Aq = (Li_i a_t \ \& \ Rj_t b_t)$$

where the actual settings chosen are (Li_t, Rj_t) and the actual outcomes are $(a_t b_t)$. But the common cause is meant to be located at the particle source, in the absolute past of the two events, which have spacelike separation. Now the choices of the experimental settings, and of the particular type of source used, can all be made beforehand, or else in any temporal order, and by means of any chance mechanisms or experimenters' whims you care to specify.

To put it conversely, if the probability of a given outcome at L is dependent not merely on the putative common cause, but also on what happens at R, or if the character of that putative common cause itself depends on which experimental arrangement is chosen (even if after the source has been constructed) then I say that the two outcome-events have not been traced back to a common cause which explains their correlation. Of course I am not saying that nature must be such as to obey these postulates—quite the opposite. These postulates describe causal models, in the "common cause" sense of "cause," and the question before us is whether all correlation phenomena can be embedded in such models.

4.1. Causality alone: A deduction of partial determinism: Principles I and III alone already imply that when parallel settings are chosen, the process is deterministic, the common cause determines the outcome of the experiments with certainty. For [remembering that I rules out having the same outcome when the same settings are chosen, and] abbreviating "*Li & Rj & Aq*" to "*Bijq*" we derive from those two principles:

$$0 = P(Lia \ \& \ Ria|Li \ \& \ Ri)$$

---

[11]Since II follows from the postulate that IV holds (for all values q), II will not play an overt role in the argument. [Correction: see end of note 9. But again the point stands: only IV, and not II, plays a role in the deduction of Bell's inequalities.]

$$= P(Lia \ \& \ Ria | Li \ \& \ Ri \ \& \ Aq)$$
$$= P(Lia | Biiq) P(Ria | Biiq)$$

But since the product is *zero,* one of the two multiplicands must be *zero.* The other will be *one.* For example, if $P(Li1|Biiq) = 0$ then $P(Li0|Biiq) = 1$. But setting $a = 0$ in the above deduction we conclude that if $P(Li0|Biiq) \neq 0$ then $P(Ri0|Biiq) = 0$, and hence $P(Ri1|Biiq) = 1$. So we see that, conditional on *Biiq* [and therefore, *whenever* parallel settings are chosen, if such a common cause is present], all experimental outcomes have probability *zero* or *one.*

I doubt very much that Reichenbach can have perceived this consequence of his principle, because he had explicitly designed it so as not to require determinism for causal explanation. Had Einstein read Reichenbach and perceived this consequence in time, he could have added a little codicil to the Einstein-Podolsky-Rosen paradox: According to the Common Cause principle, conditional certainties of the sort found in that paradox can exist only if they are the result of a hidden deterministic mechanism, so quantum mechanics is incomplete. See how much we have got—and we have hardly begun!

4.2. Hidden Locality: A deduction of complete determinism: We have just deduced that conditional on the antecedent [proposition that the same setting *i* is chosen for each apparatus, and common cause factor *q* is present, i.e.] (*Li & Ri & Aq*), all probabilities for outcomes are *zero* or *one.* But *Hidden Locality* [IV] says that this antecedent contains irrelevant information as far as the outcome at either side, separately, is concerned. [Adding this principle to our previous result, therefore, we arrive at]:

$$P(Lia | Li \ \& \ Aq) = P(Lia | Li \ \& \ Ri \ \& \ Aq) = zero \ or \ one$$

This follows from I, III, and IV together. It says that, given the value of the hidden variable that acts as common cause, the outcome of any performable experiment on either side is determined with certainty.

4.3. Hidden autonomy: The testable consequences: It is a tenet of modern philosophy, owed perhaps to Kant, that the mere assertion of causality, or even determinism, has no empirical consequences. Any phenomena at all can be embedded in a causal story: only specific causal hypotheses have testable consequences. Nothing we have seen so far refutes that tenet, for all the consequences drawn have been about the hidden variable and not about the surface phenomena themselves. But we come now to the peculiar twist that Bell discerned.

We can begin with Wigner's observation [1970] that, given the preceding, there are only eight relevant classes of values for the hidden variable. (And accordingly, no generality in the causal theory will be lost if we say that the variable has only eight possible values.) For these values can be classified by their answers to the questions:

(a) Suppose Li. Is it the case that $Li1$?

(b) Suppose Rj. Is it the case that $Rj1$?

Given Aq, [as we have now seen,] each of these questions receives a definite *yes* or *no* answer (with probability *one*). And indeed, the answers to the second type of question are determined by those to the first:

$$P(Rj1|Rj) = P(Rj1|Lj \ \& \ Rj)$$
$$= P(Lj0 \ \& \ Rj1|Lj \ \& \ Rj)$$
$$= P(Lj0|Lj \ \& \ Rj)$$
$$= P(Lj0|Lj)$$
$$= 1 - P(Lj1|Lj)$$

which deduction uses principles I, III, Iv.

Since there are three questions of form (a), each with two possible answers, these answers divide the hidden-variable values into $2^3 = 8$ types. Let us say that q is of type (a_1, a_2, a_3) when this value q predicts outcomes a_1, a_2, a_3 for arrangements $L1$, $L2$, $L3$ respectively. And let us abbreviate the assertion that the actual value is of this type to $Ca_1a_2a_3$. Precisely:

$$Ca_1a_2a_3 = v\{Aq:P(L1a_1|L1 \ \& \ Aq) =$$
$$P(L2a_2|L2 \ \& \ Aq) =$$
$$P(L3a_3|L3 \ \& \ Aq) = 1\}$$

This is an ordinary finite disjunction of form $(Aq_1 \ v \ . \ . \ . \ v \ Aq_m)$ if the set of values of the hidden variables is finite (and we know now that we can assume that without loss of generality). Thus we have not introduced new propositions; we are still working within the field generated by PRtotal.

Suppose now that we have chosen settings $L1$ and $R2$. What is the probability that we shall get outcomes $L11$ and $R21$? Let us put it a different way. Supposing Aq, what must the value q be like if we are to get outcomes $L11$ and $R21$? It must clearly be of type $(1,0,b)$ for some value b or other. In other words, this outcome will happen only if $(C101 \ v \ C100)$ is the case. But that proposition has a probability of its own—and that is our answer.

The argument I have just given tacitly presupposes Principle V of *Hidden Autonomy* for it assumes that the choice of $L1$ and $R2$ as settings does not affect the probabilities for value q. Let us state the argument precisely. To begin [and using V at once to get to the second line]:

$$p(1;2) = P(L11 \ \& \ R21|L1 \ \& \ R2)$$
$$= \Sigma \ \{P(Aq)P(L11 \& \ R21|L1 \ \& \ R2 \ \& \ Aq):q \in I\}$$

The notation $p(1;2)$ is our abbreviation for future reference. We notice now that in the summation, the conditional probability equals *zero* except in cases where q is of type $(1,0,1)$ or of type $(1,0,0)$. Hence we have:

$$p(1;2) = \Sigma \{P(Aq):q \text{ is of type } (1,0,1) \text{ or } (1,0,0)\}$$
$$= P(C101.\text{v}.C100)$$
$$= P(C101) + P(C100)$$

In just the same way we deduce

$$p(2;3) = P(C110) + P(C010)$$
$$p(1;3) = P(C110) + P(C100)$$

Adding up the first two equations we get the sum of four probabilities, two of which appear again in the equation of $p(1;3)$. Hence

$$p(1;2) + p(2;3) \geq p(1;3)$$

And this is Bell's famous inequality. [The other inequalities, formed by permutations of the numbers 1,2,3, are deduced in the same way.]

It hardly needs pointing out that the numbers $p(i;j)$ are surface probabilities by their definition (in which the hidden variable does not occur). So this inequality is testable directly by means of observable frequencies. So our quite metaphysical-looking principles have led us to an empirical prediction! What is more, the Einstein-Podolsky-Rosen paradox has variants which satisfy our surface description and for which quantum mechanics predicts the violation of this inequality. And finally, experimentation so far has produced overwhelming support for the quantum-mechanical predictions. The conclusion is surely inevitable: there are well-attested phenomena which cannot be embedded in any common-cause model.

5. Epistemological conclusion

I have made an effort to present the deduction of Bell's inequality shorn of all superfluous mathematical technicalities and woolly interpretative commentary. (A reader as yet unfamiliar with the literature will be astounded to see the incredible metaphysical extravaganzas to which this subject has led.) In addition I began the deduction with theoretical postulates (III–V) that follow directly from the idea that correlation phenomena must have common-cause explanations.[12]

• Returning now to epistemology, let us again ask when it is possible to have reasonable expectations about future events. Assuming (as we surely all agree) that it is reasonable to base one's expectations on well-supported

---

[12][See van Fraassen (1985). These issues will be explored further in a book presently in progress.]

scientific theories, we are reasonable to expect the persistence, whenever the relevant conditions obtain, of the correlations predicted by such theories. And this point is quite independent of whether we are provided with a causal explanation—or even with the possibility thereof.

In response to the situation highlighted by Bell's inequality, I suggest a picture of theories, of the enterprise of theorizing, and of how we justify our expectations that has nothing to do with causation or determination (however partial) of the future by the past. When I gave a surface description, I was doing (in a modest way) what Suppes calls constructing a model of the data. When I followed this with postulates about a hidden factor acting as common cause, I was constructing a family of theoretical models in which such phenomenal structures were to be embeddable. And when the inequality was deduced it became clear that only a proper subclass of the data models could be embedded in any of the theoretical models. Empirical adequacy of a theory consists in its having a model that all the (models of) *actual* phenomena will fit into. In some cases, the methodological tactic of developing a causal theory will achieve this aim of empirical adequacy, in other cases it will not, and that is just the way the world is. The causal terminology is descriptive, in any case, not of the (models of the) phenomena, but of the proffered theoretical models.[13] So pervasive has been the success of causal models in the past, especially in a rather schematic way at a folk-scientific level, that a mythical picture of causal processes got a grip on our imagination. But to say that is itself as metaphysical as any other causal talk, is eloquent testimony, perhaps, that the grip is firm.

APPENDIX (1988)
On Explanation in Physics

I would like to add some reflections on the nature of explanation in view of the Notre Dame Conference. Do the violations of Bell's inequalities require an explanation? In one sense that is merely the question whether we—or science—need a good theory adequate to those phenomena. To that question the answer is obviously *yes*. But in another sense the question amounts to: Do we—or science—need a theory or account which does not merely fit the phenomena, but yields an explanation (of the correlations involved) which tells us "the reason why"? The contrast between the two questions is that

---

[13]A burning question at this point is clearly: What account can be given of testing and confirmation, the process that leads to reasonable theory choice or acceptance? I have given a preliminary statement of an account of theory choice as a case of decision making in the light of conflicting criteria [van Fraassen 1984].

between mere description and real explanation, a venerable distinction if not a supremely clear one.

That second question may have the answer *no* even if the first is answered *yes*. Arthur Fine argued for that point of view, to my mind entirely convincingly. The other point of view seemed the more popular however. As I see it, that leads to the following recipe or program for research:

1. List a number of "traditional" principles, which together imply Bell's inequalities.
2. Classify these principles as more or less "deep" or "fundamental."
3. Strike some of the less fundamental ones (so that Bell's inequalities are no longer derivable) to arrive at a neo-classical philosophy of nature that can accommodate quantum mechanics.
4. Offer the remaining principles, and their logical independence from the discarded ones, as explanation of the possibility of correlations without Bell's inequalities.

But what sort of project is this? Consider (1) first. "Traditional" is meant to suggest that these principles were part of the metaphysics (!) underlying classical physics. Would any reputable historian of science make this claim? Is it backed by historical research? Or does it represent the author's conviction that classical physicists really believed them implicitly? Or that they can be logically deduced from classical physics? Or were they arrived at anecdotally, e.g., by a quote from Newton that refers to space as God's sensorium?

My suspicion is that the set of principles described in (1) will be such that one can assert at most that classical physicists could have held them if they had been sufficiently schooled in recent philosophy to understand them—and that the author is convinced he or she would have held them in their place. The main support for this will be that they imply Bell's inequalities, and that everyone was surprised when those were violated. But this could have supported many alternatives.

Similarly, what are the grounds for (2) and (3)? The author likes the remaining part and struck the ones he or she disliked? Obviously, if the set of principles is not redundant for the deduction, Bell's inequalities will cease to follow if any one of them is discarded. Finally, what precise criterion of explanation is invoked in (4)? I am a little at a loss: How can you explain a phenomenon by dropping assumptions? However that may be, is the invoked criterion for explanation, if any, a good one—and is it demonstrated that we now have an explanation which meets that very criterion?

There appear, in the history of science and philosophy, six types of explanation which attempt to render correlations unmysterious. They are *chance, coincidence, coordination, preestablished harmony, logical identity*, and *common cause*. The first two were described and distinguished by Aristotle (*Physics* II, 4–6). The Aristotelian tradition admitted a significant amount of indeterminism—individual events, though generally produced by some cause, could also just happen *by chance.* An observed correlation could be like that. In the play *Rosencrantz and Guildenstern Are Dead,* the same coin tossed by one particular person always comes up heads. To say it is a fair coin, and the person honest, is to attribute this to chance—any sequence $A_1, \ldots, A_n$ of outcomes, after all, has the same probability $(1/2^n)$ as any other. To offer this diagnosis, however, is to assert also that the correlation is extremely unlikely to persist—it was a mere accident. The diagnosis is that there was no "real" correlation.

Coincidence is not the same. When you and I meet in the market, it is by coincidence—if my going to the market is for reasons that have nothing to do with yours. (This was Aristotle's example.) But it is not chance, for our trips to the market do not just happen, we each could explain them "causally." As with chance, we stretch the word a bit if we call this explanation. But "coincidence" is certainly a good response to "Why did you two meet in the market?" This same pattern of response remains apt even if we meet there often, but still for reasons having nothing to do with each other's movements. Indeed, it seems to me the coincidence could be universal and still just coincidence. We have one apparent example of such a case in relatively recent physical hypotheses. While the example may be false, it does not seem to raise philosophical puzzles.

The example is this. In E. A. Milne's cosmology, atomic clocks and astronomical clocks induce two different time metrics. But the two are related by a logarithmic function. Of course it is mere convention which process you adopt as clock. The correlation is still a fact about the two processes, one microphysical and the other astronomical, which are otherwise unrelated as far as the theory goes. To elaborate on this, suppose we identify a third type of process which has two stages, so we can write it as ABC. Suppose that its two stages are always equal by the atomic clock: $t'(C) - t'(B) = t'(B) - t'(A)$. Physics might be able to explain this equality, and entail that it will always be found in such a process. Then, of course, we also find that in sidereal time, we have an equality of ratios: $t(C)/t(B) = t(B)/t(A)$. (As we know, according to Milne, $t'(X) = \log t(X)$, for all dateable events X.) Our third process might be sufficiently macroscopic that physics also has an explanation of this regularity, which is a correlation with astronomical processes. These two regularities are indeed two faces of one regularity, but only contingently so, just as a matter of fact—read "coincidence"—because the sidereal year is not logically related to the period of the atomic clock.

By *coordination* I mean a correspondence effected by signals (in a wide sense): some energy or matter traveling from one location to another, and acting as partial producing factor for the corresponding event. The situation need not be deterministic—there can be indeterministic signaling if the signal is not certain to arrive and/or not certain to have the required effect. But the word 'travel' must be taken seriously. Hence this explanation cannot work for corresponding events with spacelike separation. To speak of instantaneous travel from X to Y is a mixed or incoherent metaphor, for the entity in question is implied to be simultaneously at X and at Y—in which case there is no need for travel, as it is at its destination already. Correlation between distant events happening to parts of the same extended entity, however, are not *ipso facto* less mysterious.

Recent experiments by Aspect and others leave little or no hope for coordination to explain the quantum mysteries. We have three types left, and the first, *preestablished harmony* is instructive exactly because of the reasons for its proposal (by Malebranche and Leibniz). Consider the correlation between such mental events as my decision to raise my arm, and the bodily rising of my arm. We cannot attribute this to chance; and if there is free will, it is not a coincidence in the sense of Aristotle. For we predict that the correlation will persist even if, for the sake of experiment, I make these decisions in some new way or even randomly. Now these philosophers saw no causal mechanism that could coordinate these events through "signals" from mind to body. To call the phenomenon a preestablished harmony, of course, can only have one of two functions—to postulate an entity which has either predetermined my apparently free actions, or coordinates the two series of events "from outside"—or else, *to admit that we have no explanation but to refuse to consider the correlation mysterious nevertheless.*

Explanations through what I shall call *logical identity* are the most beautiful and satisfying, and also the most treacherous. In modern terms: the correlation between corresponding values of variables A and B is explained if each is definable as a function of a third variable. Then we say, A and B just *are* $f(C)$ and $g(C)$. An example of a statement of form "A just *is* $f(C)$" is the familiar "temperature just *is* the mean kinetic energy of the molecules." Clark Glymour has given a clear and instructive astronomical example: Ptolemy and Copernicus on oppositions (Glymour 1985).

The correlation is wholly unmysterious when it merely reflects an identity. This was, of course, the paradigm of a scientific account in Aristotle's *Posterior Analytics:* a scientific account states exactly what a (kind of) thing *is,* and what this implies is exactly what the account can explain about the thing. (Examples are of the same form as above; e.g., "thunder just is the quenching of fire in the clouds.") A contemporary example is furnished by hidden variable models; when successful, they give this sort of explanation. But this sort of explanation by definition is treacherous, for the new definitions have certain empirical presuppositions whose correctness is *a priori* uncertain. In addition, this procedure has demonstrable limits that showed up exactly in the "no hidden-variable" theorems. Those theorems defeat any attempt to explain EPR-style correlations in this fashion. I will come back to this in a moment.

The final pattern of explanation is causal. Correlation of simultaneous separated events can be explained if we can find a *common cause,* that is, if both can be traced back to some events in their common past. As pointed out originally by Reichenbach who formulated it, this is a concept of causation that still makes sense in an indeterministic universe.

This insight of Reichenbach's described by him in the 1920s and 1930s—and well known to Einstein, though I do not know how early—was one of great importance. For it showed exactly how, and to what extent, one could sensibly speak of a causal order in the context of indeterminism. It should be emphasized that Reichenbach has only outlined a pattern, which is not by itself sufficient for explanation. This pattern provides only necessary conditions for causal order in an indeterministic world. However, these necessary conditions entail Bell's inequalities.

The common cause pattern of explanation I have explored in "The Charybdis of Realism"; it fails for certain quantum-mechanical phenomena. Likewise, the logical identity pattern fails there. The latter I have explored in more detail in my "EPR: When is a correlation not a mystery?" (1985). Both its appeal and its defeat are quite telling.

The moral is, it seems to me, that the scientists who constructed quantum mechanics did exactly as much as we—or science—needs. It is possible to gain greater insight into the structure of that theory—and all the papers at the conference gave us such additional insight. It is not possible to reach independent insight into the "fundamental" structure of nature and to illuminate quantum theory by viewing it *sub specie* that structure. But then, that is not needed. There does not have to be a reason for everything.

A SPACE-TIME APPROACH TO THE BELL INEQUALITY

JEREMY BUTTERFIELD

Does the violation of Bell's inequality involve causation between the two wings of the experiment? If so, then quantum mechanics, which violates the inequality, allows superluminal causation; for the relevant events in the two wings can be spacelike-related. Relativity is often taken to prohibit superluminal causation; and I shall so take it—so that causation between the wings means that quantum mechanics conflicts with relativity. (If you believe that relativity does not prohibit such causation, no matter. I will not need the details of relativity, so that for present purposes, our differences are verbal; when I say 'relativity', just read 'the prohibition of superluminal causation'.) On the other hand, if violation of Bell's inequality does not involve causation between the wings, then quantum mechanics and relativity can "coexist peacefully" (Shimony 1978). I believe in peaceful coexistence. And the first aim of this paper is to make a point in favor of such coexistence for stochastic hidden variables models of the Bell experiment (reviewed in section 1).

The point arises from Jarrett's (1984) analysis of the assumption of factorizability that is used in proofs of the Bell inequality for stochastic hidden variables. Jarrett shows that factorizability is the conjunction of two conditions, which he calls completeness and locality. He argues that relativity implies locality, but not completeness; and so he urges that the violation of the inequality impugns completeness, not locality. In section 3, I shall argue that relativity does not imply locality; so that the prospects for peaceful coexistence are rather better than Jarrett allows. (In fact, I shall argue that relativity plays identical roles in justifying completeness and locality: in both cases, the role is partial—another assumption is needed. I shall, however, agree with

For conversations and comments, I would like to thank the participants at Notre Dame, audiences at Western Ontario and Cambridge, and especially Harvey Brown, Nancy Cartwright, Henry Krips, and the editors.

114

Jarrett that completeness is less plausible than locality; and so it, rather than locality, should be given up.)

The second and main aim of the paper is to extend Jarrett's analysis by taking the hidden variable and the states of the apparatus as the total physical state of a region of space-time: hence my title. Thus we get various versions of these assumptions, corresponding to different choices of region. Section 4 will undertake a survey of what regions can be chosen, while making completeness, locality, and the other assumptions of the Bell inequality as plausible as possible. This survey will show how weak the assumptions can be, and thus how puzzling is the violation of the Bell inequality. I regard this exercise as complementing Howard's and Wessels's attempts to "get behind" the assumptions of the Bell inequality.[1]

My two aims are linked. In both cases, I will be concerned with the idea of an event z "screening off" two other events, x and y; that is to say, conditionalization on z renders x and y stochastically independent, $p(x\&y|z) = p(x|z).p(y|z)$. I develop this idea in section 2. The idea and the terminology of "screening-off" was established in the philosophical literature by Reichenbach as one of his conditions for an event z to be a common cause of two others, x and y.[2] However, I shall not need the other conditions for a common cause.[3] In section 2, I will show that each of Jarrett's conditions, completeness and locality, is an example of screening-off. Indeed, each is motivated by Reichenbach's Principle of the Common Cause; and even by a weaker statement which follows from it. The principle states that whenever x and y are correlated—i.e., stochastically dependent: $p(x\&y) \neq p(x).p(y)$— and one does not cause the other, then there is in their common past an event z, which is a common cause in Reichenbach's sense. This implies a weaker statement which is sufficient for our purposes: that if x and y are correlated without direct causation, then there is in their common past an event z that screens them off. And the plausibility of this, together with the belief that in the Bell experiment there is no direct causation between some correlated events (because the events are spacelike-related), makes screening-off, as embodied in completeness and locality, plausible. And it is this similar motivation for each of Jarrett's conditions that yields my first aim. Relativity (i.e., the prohibition on superluminal causation) does not imply locality, any more than it implies completeness: in both cases, another assumption, such as Reichenbach's Principle or the weaker statement above, is needed as well.

Since Reichenbach's Principle and relativity together imply Jarrett's

[1]See Howard and Wessels, this volume; see also Wessels (1985); Howard (1985).

[2]See Reichenbach (1956), p. 163, p. 189.

[3]See Reichenbach (1956), pp. 157–163; Salmon (1984), pp. 159–163, is a recent discussion.

conditions, the violation of the Bell inequality and the desire to retain relativity impugn Reichenbach's Principle. Similarly for the weaker statement that if x and y are correlated without direct causation, then their common past contains a z which screens them off. And so one might hope to avoid being puzzled by the Bell experiment by finding independent reasons for rejecting this statement: for saying that screeners-off need not exist.[4] Van Fraassen has urged such reasons. He gives straightforward and unpuzzling examples, having nothing to do with quantum theory, in which the common past of x and y seems to contain no screener-off. Salmon and Bell have also given examples.

However, I find the violation of the Bell inequality puzzling; and so I want the statement about the existence of screeners-off to be plausible—so that its failing is strange. And this motivates my use of space-time regions. For I will urge (in section 2) that we can save the statement, even for the examples of van Fraassen et al., by taking the screener-off to be the total physical state of a suitable region of space-time. Thus my aim here is not to show that there must be screeners-off; but that in all the straightforward examples, like van Fraassen's, the state of a suitable space-time region is one. The statement that there is always such a region and such a state will then be plausible. And accordingly, we can motivate completeness and locality by taking the hidden variable and the states of the apparatus to be the total physical state of suitable regions (section 3). The use of space-time regions prompts the question: Can we choose regions which make all the assumptions of the Bell inequality plausible? Section 4 argues that the answer is yes. The violation of the Bell inequality will then mean that the statement that there is always a screener-off is false. But its falsity will not be a consequence of straightforward examples like van Fraassen's; it will be a consequence of the Bell experiments' results. And accordingly, I submit that those results are mysterious, precisely because they refute this plausible statement.

1. Stochastic hidden variables

Suppose that pairs of particles are emitted by a source, the members of a pair flying in opposite directions, each going to an apparatus that can measure spin in various directions. We suppose that for each direction, there are just two possible results, $+1$, and -1. Let the wings be labeled A (left) and B (right); let i be the direction chosen for measurement in A and let j be the

---

[4]This strategy is common to Fine and van Fraassen (this volume); but the motives differ. Fine has defended a local realist interpretation of quantum theory (e.g., 1973, 1982d); so he is pleased to find assumptions of Bell inequalities which go beyond local realism, and which he can reject. But van Fraassen wishes to rebut scientific realism, and he sees Reichenbach's Principle as the basis of an argument for it.

direction chosen for measurement in B. By measuring many pairs for a given choice of i and j, we can collect statistics about the correlations between the particles.

Let us first recall deterministic hidden-variable models. ('Deterministic' is a misnomer, since determinism—the state at one time fixing the state at another—is not the issue; 'determinate value' would be better, as several authors point out.)[5] Such a model makes two assumptions. (1) It assumes that for each direction and each particle, there is some result, $+1$ or -1, that the particle is bound to give. Some kind of locality is built into this assumption: for the model assumes that the result for one particle in a given direction is independent of the choice of which direction is to be measured in the other wing. This assumption means that each particle can be thought of as having definite values for every direction of spin. One can think of the set of all these values as the particle's state. And so one can think of all the possible states of the particle as forming a set, with each spin-direction giving a map on this set into $(+1,-1)$. Similarly, we can form the set of all possible states of the pair. Let this set of states be S, and let λ (the "hidden variable") be a member of S. So the spin-directions i and j that are measured on the A and B wings determine maps from S to $(+1,-1)$. (2) It assumes that a distribution over λ, $\rho(\lambda)$, is characteristic of the source. Again, some kind of locality is built in, for this distribution is assumed to be independent of the settings i and j.

Let us write x and y, for some particular results (each ± 1) in the A and B wings respectively. Each pair $\langle i,x \rangle$ and $\langle j,y \rangle$ determines a map from S into $\{0,1\}$; let us write these maps as $[i,x]$, $[j,y]$, so that $[i,x](\lambda) = 1$ iff the value of spin in direction i for state λ is x; that is, iff the result of an i-measurement on state λ would be x ($= \pm 1$); and similarly for $[j,y](\lambda)$. Then the probability of getting both the results x and y, with the settings i and j, is got by averaging over the states in S, using ρ:

$$\text{prob}_{i,j}(x,y) = \int [i,x](\lambda).[j,y](\lambda).\rho(\lambda)d\lambda$$

We need to make two points about this equation. First, we are in effect using the maps from S to $\{0,1\}$ to make the results form a Boolean algebra, and thus a suitable domain for a probability function. Second, by having the left-hand side contain i and j in the subscript, rather than after a conditionalization stroke, as in $\text{prob}(x,y|i,j)$, we are assuming that there are many probability spaces (each with the four events $\langle +1,+1 \rangle$, $\langle +1,-1 \rangle$, $\langle -1,+1 \rangle$, $\langle -1,-1 \rangle$), each labeled by its settings $\langle i,j \rangle$, rather than one big probability space, whose events would be all the various quadruples $\langle x,y,i,j \rangle$. Mathematically, there is little difference between these approaches. The big space is, of course, partitioned into subspaces, labeled by $\langle i,j \rangle$, each obtained by

[5]See Earman (1986a), p. 485 fn. 38.

conditionalizing on its value $\langle i,j \rangle$. And given the many spaces, labeled by the settings, it is a mathematical triviality to build a single space whose conditional probabilities prob$(x,y|i,j)$, prob$(x|i,j)$, etc., are the given prob$_{i,j}(x,y)$, prob$_{i,j}(x)$, etc. (The big space we build is not unique: we can assign each label $\langle i,j \rangle$ whatever probability we like, provided they sum to one.) But there is a conceptual difference between the approaches. With many spaces, one is not committed to apparatus settings, i and j, having a probability; with the big space, one is. I do not think that every proposition or event has a probability; and since one thinks of apparatus settings as decided by the experimenter, they seem quite good candidates for not having a probability. In any case, it seems to be worth signaling where one can avoid the assumption that some proposition has a probability. Accordingly, I prefer to use many spaces and the above notation. Nevertheless, for the sake of uniformity with the other papers in this volume, I am happy to use 'prob$(x,y|i,j)$' etc.: we shall never have to calculate suspicious probabilities like prob(i,j), or prob$(i,j|x,y)$. So we write:

$$\text{prob}(x,y|i,j) = \int [i,x](\lambda).[j,y](\lambda).\varrho(\lambda)d\lambda$$

Calculating probabilities in this way makes the model subject to a Bell inequality—whatever the distribution ρ over λ. There is, of course, controversy over the status of the assumptions above, which incorporate some kind of locality.[6] I shall later discuss one point which arises here: that locality as it occurs above is not equivalent to "no action at a distance" in the kind of sense implied by relativity.

I turn to stochastic hidden-variable models. Such a model avoids assumption (1) above. It supposes that for each pair there is some physical state (the "hidden variable"), λ, that prescribes probabilities for the various possible results of measuring spin in the various possible directions in the two wings of the experiment. Thus λ does not determine values for spin; these evolve stochastically in flight. Thus each choice of λ, i, and j determines a probability for results, x and y, in the two wings. I shall call it $p_\lambda^{AB}(x,y|i,j)$. Here the superscript AB signals that the probability is for a joint measurement, and in accordance with the discussion above, the conditionalization on i,j could be replaced by absorbing i,j into the subscript.

We need to make two further points about this probability. First, it is meant to be an objective probability or chance. However, we will not need a detailed theory of chance. We need only note that the basic idea is of irreducible indeterminism, with weights attached to the various possibilities. (I think of this in terms of possible worlds: indeterminism is a matter of worlds that match exactly up to a time, not matching thereafter; and chance is a matter of weights attaching to the various possible worlds that match up to a time. But

---

[6]See Fine's prism models (1982d) and Earman (1986a).

this construal will not materially affect the argument to follow: in particular, we can take possible worlds as sequences of possible states of the total system-plus-apparatus of the Bell experiment.)

Secondly, I have so far presented i and j as coding only the spin-direction chosen by the experimenter. But some treatments (including Jarrett's) allow for "apparatus hidden variables," that code further features of the apparatus that might influence measurement results. I shall also allow this, but to avoid introducing further notation, I shall incorporate all such features in i and j. Thus i will stand not only for the spin-direction chosen on the A-wing, but also for such other features of the A-apparatus as may influence its results; similarly for j. For convenience, I shall still call i and j 'settings', even though only parts of them (in particular, the spin-direction measured) can be set by the experimenter.

Although assumption (1) is gone, there is a locality assumption similar to that built into (1): roughly, an assumption that the probabilities of results are independent of the setting in the other wing. This is usually made precise as: λ prescribes probabilities for each wing irrespective of what is measured in the other wing, and for each λ the probabilities prescribed in the two wings are stochastically independent. That is, one assumes that for all λ, i, and j, and all results x and y, there are "one-wing" probabilities, $p_\lambda^A(x|i)$ and $p_\lambda^B(y|j)$; and that the following equation, expressing stochastic independence, holds:

$$p_\lambda^{AB}(x,y|i,j) = p_\lambda^A(x|i).p_\lambda^B(y|j).$$

This form of the locality assumption is called factorizability; it goes back to Bell (1971) and Clauser and Horne (1974).

A stochastic hidden-variable model also makes assumption (3) above: there is a distribution over λ, $\rho(\lambda)$. And again, there is an assumption of locality—this time, the same as in the deterministic case: $viz.$, this distribution is assumed to be independent of the settings i and j.

Using both factorizability and this last assumption, we can write:

$$p^{AB}(x,y|i,j) = \int p_\lambda^{AB}(x,y|i,j).\rho(\lambda)d\lambda$$
$$= \int p_\lambda^A(x|i).p_\lambda^B(y|j).\rho(\lambda)d\lambda$$

Probabilities calculated in this way are subject to a Bell inequality, whatever the distribution ρ. To be precise, the probabilities on the left hand side are not testable, because the settings i and j are allowed to incorporate apparatus hidden variables which will vary from one trial to another. But testable probabilities obtained by averaging over these variables are also subject to a Bell inequality, whatever the probability distributions on these variables—provided each is independent of the distant setting (see Bell 1971). For simplicity I shall ignore these distributions throughout the paper; all my arguments are unaffected.

Since the inequality is violated, one must give up some assumption made by the model. (And since quantum mechanics violates the inequality, it follows that the more one considers these assumptions natural, the less can one construe the puzzling features of quantum mechanics as deriving solely from its being irreducibly indeterministic—for the refuted model is that.) Analysis of the assumptions has focused on the new locality assumption— factorizability. I shall now review Jarrett's analysis.

Jarrett (1984) shows that factorizability is a conjunction of two conditions. One is about the results in the two wings being independent of each other; Jarrett calls this condition "completeness." The other condition is about the probabilities of results in one wing being independent of the setting in the other; Jarrett calls this condition "locality."

To be precise, a model is *complete* iff each λ, i, and j obey the condition: for all corresponding results x and y, the two probabilities prescribed by λ, i, and j for x, and for y, are stochastically independent. That is, the model is complete iff for all λ, and all i, j, and results x, y:

$$p_\lambda^{AB}(x,y|i,j) = p_\lambda^{AB}(x|i,j).p_\lambda^{AB}(y|i,j).$$

If the single probabilities (on the right) are non-zero, this implies:

$$p_\lambda^{AB}(x|i,j,y) = p_\lambda^{AB}(x|i,j) \quad \text{and} \quad p_\lambda^{AB}(y|i,j,x) = p_\lambda^{AB}(y|i,j)$$

which in Reichenbach's terminology says that for given λ: i and j, taken together, "screen off" x and y from each other. I shall assume that the single probabilities are indeed non-zero, so that completeness expresses Reichenbach's basic idea—roughly, that λ, i, and j specify a probability for x, and are sufficiently informative (logically strong) that the further information that y holds does not affect the probability; and similarly, vice versa. (More details below.)

And a model is *local* iff for all λ, and all i, j, and results x, y: $p_\lambda^{AB}(x|ij)$ does not depend on j; and $p_\lambda^{AB}(y|i,j)$ does not depend on i. ('p does not depend on j' means here simply 'p does not vary, as j varies'.) This independence means we can drop the extra argument behind the conditionalization stroke, and the corresponding superscript, and write $p_\lambda^A(x|i)$, $p_\lambda^B(y|j)$. I shall argue below that this condition, like completeness, is a condition of screening-off.

Jarrett shows that the conjunction of these two conditions is equivalent to factorizability. The argument is summarized in this derivation of factorizability from the conjunction:[7]

$$
\begin{aligned}
p_\lambda^{AB}(x,y|i,j) &= p_\lambda^{AB}(x|i,j,y).p_\lambda^{AB}(y|i,j) \\
&= p_\lambda^{AB}(x|i,j).p_\lambda^{AB}(y|i,j) \quad \text{by completeness} \\
&= p_\lambda^{AB}(x|i).p_\lambda^B(y|j) \quad \text{by locality.}
\end{aligned}
$$

[7]See Jarrett (1984), pp. 582–584; Ballentine and Jarrett (1987), p. 699.

So Jarrett's analysis prompts the question: Which should be given up—completeness or locality? Jarrett himself holds that locality is implied by relativity; and his verdict is that the violation of the Bell inequality shows that completeness fails, while locality and relativity do not. I shall argue in section 3 that relativity does not imply locality; but I shall agree with Jarrett that completeness, rather than locality, should be given up. As a preliminary to that discussion, I need to examine Reichenbach's idea that if events are correlated, there is in their common past a third event that screens them off. I shall maintain that this idea is plausible if we take the third event to be the total physical state of a space-time region.

2. Screening-off

Reichenbach starts from the idea is that if events are correlated (stochastically dependent), but one does not cause the other, there must be some common explanation of both of them. Here, events are localized in space-time and bear causal relations to one another, both deterministic and probabilistic; they form a Boolean algebra, so that they can have probabilities. One can take them to be propositions about regions of space-time. Everyday examples make this idea plausible. For example, suppose we find that the answers given by spouses to questionnaires about politics are correlated: conservative wives tend to have conservative husbands, and so on. It is plausible that for any pair of wife and husband, their answers and the correlation that they manifest have a common explanation, relating to their past. Reichenbach then makes the idea of a common explanation precise, in terms of his Principle of the Common Cause.[8] We do not need the full statement of this, but only this consequence: if two events x and y are correlated, but one does not cause the other, then there is in their common past another event that screens them off. To be precise, if $p(x\&y) \neq p(x).p(y)$, then there is an event z in the common past of x and y that induces conditional stochastic independence:

$$p(x\&y|z) = p(x|z).p(y|z).$$

If $p(x|z)$ and $p(y|z)$ are non-zero, this implies:

$$p(x|y\&z) = p(x|z) \quad \text{and} \quad p(y|x\&z) = p(y|z),$$

which says that z screens off x and y. Again, everyday examples make this plausible. To put the point in epistemic terms: surely if we knew about certain events in the common past of husband and wife, e.g., their mealtime conver-

---

[8]See Reichenbach (1956), p. 163.

sations or their daily paper, we could predict the wife's answers so well that information about the husband's answers would give us no further help. Such events would then screen off from one another the wife's and husband's answers.

Note that x and y must not be included one in the other. The details of the individuation of events do not matter for this point. We need only assume that if z is in the common past of x and y, it is distinct from both; and that if x and z are distinct, then in general $p(x|z) \neq 1$. (These assumptions are hardly contentious; in particular, they are plausible if we take events as having a space-time region, and say that one event includes another iff its region includes the other's region.) Then suppose that x includes y; so specifying x involves specifying y, so that for all z $p(y|x\&z) = 1$; while for a third event z in the common past $p(y|z) \neq 1$—violating the screening off condition. Similarly one can argue that x and y must include no third event; for if this third event is not included in z, then it is specified by $x\&z$ though not by z, and so one expects that $p(y|x\&z) \neq p(y|z)$. Thus if we think of events as having space-time regions, the regions of x and y must be disjoint. In fact, in most of our applications, x and y will be spacelike-related.

In this section, I will give two arguments for taking z to be, not just an event in the common past of x and y, but rather the total physical state of that common past. My first argument turns on the idea of explanation: a single event in the past may screen off without explaining the correlation of x and y. This will lead into the discussion of which space-time region should count as the common past of x and y; there are various candidates, but I shall suggest reasons in favor of some against others. Then I shall give my second, and main, argument: I shall argue that examples where there does not seem to be a screener-off, proposed by van Fraassen, Salmon, and Bell, can be met by taking the screener-off to be the total physical state on one of the favored common pasts. I shall also relate this argument to recent discussions by Wessels (this volume) and Cartwright (forthcoming).

The first argument turns on the idea that a screener-off may not provide the explanation of the correlation between two given events. Indeed, even a screener-off that also satisfies Reichenbach's other conditions for a common cause[9] may not provide the explanation of the correlation. Thus suppose that the wife and husband in the example agreed in some conversation long ago to be conservative, but that later they turned radical, only recently reverting to conservatism. Then we can suppose that their conversation long ago screens off their present conservative answers from each other and even that this conversation satisfies Reichenbach's other conditions for a common cause.

---

[9]See Reichenbach (1956), p. 159 for conditions (5–8); Salmon 1984: p. 159 for conditions (1–4).

But the correlation between their present answers is surely explained by recent influences on their views. (This is not a rebuttal of Reichenbach's Principle of the Common Cause; the principle only says that the common cause that explains the correlation satisfies certain conditions, including screening-off—not that any event meeting the conditions is the common cause.)

This suggests that the correlation is to be explained by conditionalizing on *all* the past events that causally affect x and y; this conditionalization will surely render them independent. Thus in the example, conditionalizing on all the past events that affect the husband's and wife's answers will surely render their answers stochastically independent. But there are two apparent problems with this idea as so far stated. It uses the notion of one event causally affecting another and so it prevents us analyzing this notion in terms of conditional stochastic independence. Also, setting aside analysis, it seems that it will be difficult to apply this idea: How could one know that all causal factors are included? Fortunately, there is a way to overcome these problems—it makes assumptions, but they are natural.

Namely, take the state of the entire common past of x and y to screen them off. That is, we assume the basic idea above: conditionalizing on all affecting events yields stochastic independence. But we also assume: (1) all the events affecting x and y belong to a certain space-time region, their common past; (2) conditionalizing on all the other events in this region, in addition to those affecting x and y, does not disrupt the stochastic independence induced by conditionalizing on the affecting events; and (3) conditionalizing on all the events in this common past is conditionalizing on the total state of the common past. (1) and (3) are certainly natural, especially if we make an appropriate choice of common past (see below). (2) is more contentious; but analogous ideas about probabilities remaining constant under conditionalization have been defended (see Skyrms 1980). Furthermore, assumptions (2) and (3) can be weakened somewhat, as regards their probability structure. As they stand, they increase the size of the probability space: we had a probability distribution with events in its domain, and now the domain is to include also the total physical states of some (perhaps not all) regions of space-time. However, recall the discussion above about the choice between one big probability space, and many, each labeled by its settings $\langle i,j \rangle$; the latter choice avoids some suspicious probabilities. Similarly, here: we can take the total physical state of certain regions to prescribe a probability distribution over some (perhaps not all) events. There is no need to have one big space, with each of these distributions obtained by conditionalizing some single distribution on the total physical state of its region.

Accordingly, we now have the following principle, which I call 'PPSI' (for 'Past Prescribes Stochastic Independence'):

PPSI: If two events are correlated, but one does not cause the other,

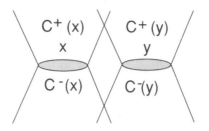

Figure 1.

then the probability distribution prescribed by the total physical state of their common past makes them stochastically independent (and so—with appropriate non-zeros—screens them off).

Given assumptions (1) to (3), PPSI seems plausible in our everyday example; and, of course, in countless others.

But how should we interpret the phrase 'the common past of events x and y'? Because of our interest in peaceful coexistence in the Bell experiment, I shall suppose that x and y are spacelike-related, so that relativity implies that one does not cause the other. If we think of x and y as pointlike, one obvious candidate is the intersection of the backward light cones of the two events; for any event that influences both measurements by a signal that travels no faster than light must be in this intersection. Similarly, if we think of the measurement-events as extended events: the common past can be taken as the intersection of their causal pasts, defined as the union of the backward light cones of their points. Compare figure 1, where $C^-(x)$ is the causal past of x, etc. Indeed, statements similar to PPSI have been considered in the literature on Bell's inequality, with this candidate for the common past being understood throughout.[10]

But there are also other candidates for the common past of x and y. There is all of space-time to the past of a spacelike hyperplane (or more general hypersurface) that lies in the past of x and y, but in the future of the intersection, $C^-(x) \cap C^-(y)$. And there are more restricted candidates: there

---

[10]For examples, this candidate for the common past is implicitly chosen in Hellman's discussion of "stochastic Einstein Locality" (1982), p. 466 (4) and (5); and in Jarrett (1984), p. 587 fn. 15; and it is explicitly chosen in Bell's discussion of "local causality" (1987), p. 54 (2) (see also the Shimony reply in Bell et al. (1985)), and in Earman (1986a, §8).

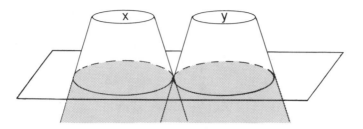

Figure 2.

is the intersection of such a past with $C^-(x) \cup C^-(y)$, the union of the causal pasts. This intersection is shaded in figure 2. At first sight, these seem less natural candidates than $C^-(x) \cap C^-(y)$. They are not uniquely specified: there are many such spacelike hyperplanes, and therefore many different candidates (whether or not we take the intersection). And each candidate contains events that cannot influence at least one of x and y, as well as those events that can influence both. However, these candidates have two compensating advantages.

Each has an advantage over $C^-(x) \cap C^-(y)$. Namely, it is a larger region of spacetime than $C^-(x) \cap C^-(y)$, so that the total physical state of it is logically stronger. PPSI is then more plausible; for more events that might affect x or y are taken into account, and given assumption (2) above, taking other events into account will not disrupt the induced stochastic independence. (And we will see in section 3 that a similar choice makes Jarrett's completeness condition weaker—which will be an advantage.)

And each candidate obeys a natural assumption governing the prescription of probabilities, which $C^-(x) \cap C^-(y)$ violates. Thus recall that the future domain of dependence, $D^+(R)$, of a region R is the set of points p such that every past-directed causal (= piecewise timelike or lightlike) curve through p intersects R; that the past domain of dependence is defined likewise; and that the domain of dependence, $D(R)$, is the union, $D(R) = D^+(R) \cup D^-(R)$. See figure 3. So any event lying outside $D(R)$ can be subject to causal influences (subluminal signals) that do not intersect R; and can therefore have certain causally influenced properties, rather than others, without the causal influence affecting the state of R. It is therefore natural to assume that the state of R does not prescribe probabilities for this event. (At least, not for its

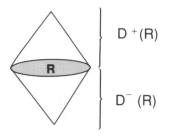

Figure 3.

causally influenceable features; and I will not need to consider probabilities for other features—such as being F-or-not-F.) For the only way that the state of R could prescribe such probabilities would be by averaging over the various possible causal influences that might affect the event. This involves assigning probabilities to these influences (to use as weights in the averaging) and where are these to come from? So it is natural to assume that prescribed probabilities are restricted to the domain of dependence:

> (PrDom)   The total physical state of a space-time region R
> prescribes probabilities only for events lying within the
> domain of dependence, $D(R)$.

Five points of clarification. (1) The 'only' in (PrDom) is important. (PrDom) does not assert that R's state prescribes for all events in $D(R)$, nor that every region R prescribes for at least one event in $D(R)$.

(2) Some theories of chance[11] say that past events have only trivial chances—0 or 1. (PrDom) can agree; the state on region R can prescribe only 0 or 1 for events in its past domain of dependence.

(3) I said (PrDom) was natural. But it may also be unduly cautious. Consider a fair die, and let R be the volume of the die at some time t, when it starts to roll. $D(R)$ will be small; if d is the width of the die, $D(R)$ extends in time only about d/c. We are tempted to say that the state on R—in particular, the symmetry of the die—prescribes a probability of $\frac{1}{6}$ for the event of a 6 turning up, though this event is outside $D(R)$. If we say this, we need to justify some averaging over the other influences, e.g., a gust of wind which at t is outside R; and averaging which does not alter the uniform probability assignment suggested by the symmetry in R. Whether we can justify this is a hard question in the foundations of probability. Thus there are two sources of the

---

[11]For example, that of Lewis (1980), p. 93.

idea of objective probability: indeterminism with weights attached to alternative future events; and averaging over influences that are not fixed—which does not require determinism. The second source is more controversial, since many think that determinism precludes objective probability. Thus (PrDom) is cautious, reflecting only the first source.

(4) (PrDom) assumes, in spirit if not in letter, that there is no superluminal causation. For such causation would mean that events in R's domain of dependence, like those outside it, were subject to causal influences that do not intersect R. Our previous argument applies: How could the state on R prescribe probabilities, without averaging over these influences—and how should that be done? This is not to say that superluminal causation vitiates the whole idea of states of regions prescribing probabilities for events outside the region; perhaps there can be a sensible averaging of influences, at least if superluminal causation is limited enough. But whether such causation is limited or not, it robs the domain of dependence of its distinctiveness: probabilities for events in R's domain of dependence, just like events outside it, will require a weighting of influences that do not intersect R. And in this sense, (PrDom) presupposes that there is no superluminal causation. In this regard, (PrDom) is similar to another condition: Hellman's "stochastic Einstein locality", which says that the probability of an event x should be fixed by the physical state of a spacelike slice stretching across $C^-(x)$—and is therefore not dependent on the state beyond $C^-(x)$. Indeed, Hellman (1982) proposes his condition precisely in order to capture the idea of an indeterministic world, in which there is no superluminal causation affecting events' chances.

(5) I shall not need to assume (PrDom), and thus to reject averaging over influences (cf. (3)); nor to presuppose that there is no superluminal causation (cf. (4))—which is fortunate, since I want to argue for peaceful coexistence. In section 4, I shall use (PrDom), and a similar assumption about determination, to facilitate a survey of the various spacetime regions that λ, i and j could be the state of. But the main point of section 4—that one can choose regions so as to make completeness, and locality plausible—will need neither assumption.

Returning to the PPSI and the question 'What region should we choose as the common past of x and y?' it is now clear what the second advantage is of choosing a region behind a space-like hyperplane. Namely, the intersection $C^-(x) \cap C^-(y)$ has a trivial domain of dependence; i.e., its domain of dependence is itself. But the domain of dependence of these regions includes the events x and y (indeed, the domain of dependence of the entire past of a space-like hyperplane is all of Minkowski space-time). So (PrDom) allows x and y to have probabilities prescribed by the state of these regions; but not by $C^-(x) \cap C^-(y)$. To sum up: given (PrDom) and all the above candidates for the common past of x and y, PPSI must take one of the second group; and given assumption (2) above, PPSI will be more plausible, the larger the region that it takes as the common past. So I favor the second group.

So much for the candidate common pasts of x and y. I turn to my second, and main, argument for taking x and y to be screened-off, not by some single previous event z, but by the total state of their common past. Van Fraassen, Salmon, and Bell have all given examples where there does not seem to be a screener-off. I shall argue that we can take the screener-off to be the total physical state on one of the common pasts of the second group above. (And I shall briefly relate this argument to recent discussions by Wessels and Cartwright.) My overall aim here is not to show that PPSI must be true, but that we can interpret "common past" so that PPSI is true in all everyday examples, and even in physical examples like van Fraassen's. The idea will then be that, so understood, PPSI is plausible, and that it leads to the assumptions of a Bell inequality (sections 3 and 4). The violation of the Bell inequality will then mean that PPSI must be given up. But its falsity will not be a straightforward consequence of examples like van Fraassen's; it will be a consequence of the Bell experiments' results. And accordingly, I submit that those results are mysterious, precisely because they refute PPSI.

Van Fraassen urges that PPSI is straightforwardly and unpuzzlingly false, once we countenance indeterminism (1980, 26–31; 1982b). He points out that if the event of the two particles being emitted is really indeterministic, then the particles can have correlated properties that are not screened off by the total physical state of the past of the emission-event. For example, the particles might have equal and opposite momentum, if they are decay products of a particle that was originally stationary, and momentum is conserved in the decay; but which spatial path the particles fly along (in opposite directions) may be wholly undetermined. This will mean that the particles' momenta are not screened-off from each other by the total physical state of the original particle—or even by the total physical state of the past of the emission-event. For example, if x is the event of particle A entering a certain hemisphere of directions, and y is the event of particle B doing so, and z is the total physical state of the past of the emission-event, then we can have $p(x,y|z) = 0$ while $p(x|z) = p(y|z) = \frac{1}{2}$.

There are countless other examples where screening-off seems to fail. Salmon (1984, 168–177; 1978) considers a collision of two billiard balls (= z), such that if the first goes into a certain pocket (= x), the second goes into another pocket (= y); so $p(y|z,x) = 1$. But the collision is produced by an amateur shot, so that $p(x|z) = p(y|z) = \frac{1}{2}$. So $p(x,y|z) = \frac{1}{2}$ while $p(x|z).p(y|z) = \frac{1}{4}$. Bell (1987, 55) considers a radioactive decay, very similar to van Fraassen's example. Indeed, the existence of such examples is now quite well known in the Bell literature. Hellman (1982, 466) cites Bell's example. And Earman (1986, 470, 478) is no doubt aware of these examples, when he says that screening-off is not implied by the "stochastic Einstein locality" condition that the probability of an event x should be fixed by the physical state of a spacelike slice stretching across $C^-(x)$. Earman's idea seems to be to distinguish a probability being fixed by the state on such a slice, and by the state on the intersection of two causal pasts. That is, he apparently takes these examples to show how stochastic Einstein locality can be satisfied, even while there is some event y outside $C^-(x)$, such that the state of the intersection, $C^-(x) \cap C^-(y)$, does not screen off x and y.

I make the same reply to all these examples. I agree that the event in question (the decay of the original particle, the collision of the balls, etc.) does not screen off (and it is natural to call it a common cause, albeit one that does not produce its effects independently). But there is another event, in fact the state of a larger region of space-time, that does screen off. So the PPSI is saved by taking this state as the screener-off; and thus by taking the screener-off to be the state of the common past, where 'common past' means this region of space-time. I shall show this for van Fraassen's example; the point will carry over to the other examples.

In van Fraassen's example, we must distinguish between correlations that are not screened-off by the emission-event or by the total physical state of its past (agreed: puzzling only to the extent that indeterminism is puzzling); and correlations that are not screened-off even by the total physical state of a larger region of space-time that includes events *later than* the emission-event (truly strange). Thus call the emission event e, so that its causal past is $C^-(e)$. If the emitted particles travel slower than light, then $C^-(e)$ is a proper subset of the intersection, $C^-(x) \cap C^-(y)$. And the total physical state of $C^-(x) \cap C^-(y)$ includes the physical states of the particles during initial segments of their journeys. (Even if the particles travel at the speed of light, $C^-(e)$ is a proper subset of any of our other [larger region] candidates for the common past of x and y. And by choosing a spacelike hyperplane that lies to the future of e, rather than just touching it, we can make the common past include initial segments of the particles' journeys; see figure 2.) But a state that includes the particles' states during these initial segments surely *will* screen off the particles' correlated properties! Certainly it does so in van Fraassen's example and in the other examples above. And similarly in all the everyday and classical physical examples that I know of. In all these cases, PPSI can be preserved by being taken to assert that the state on a region of space-time rather larger than the past of the emission-event/collision screens off. In effect, PPSI is preserved by "pushing forward" to include states that code the two effects of the "common cause". Furthermore, we shall see in sections 3 and 4 that in the deduction of the Bell inequality, there is nothing to prevent our taking PPSI in this weak sense. And so the failure of PPSI is indeed strange.

This rebuttal of the examples is in accord with the treatments of Wessels

(1985, and this volume) and Cartwright (forthcoming). In fact, it summarizes a consensus they share, despite other disagreements. Wessels' aim is similar to mine: to "get behind" Jarrett's completeness and locality conditions. She derives completeness from a number of assumptions articulating a concept she calls "effect locality," which governs how states evolve. The basic idea is that "the evolution of the characteristics of a body B 'in a moment', say $(t, t+dt)$, depends *only* on the characteristics associated with B at t and the external influences felt by B during the moment" (Wessels 1985, 489). Here 'body' is a term of art—an object is a body iff its characteristics evolve in accordance with effect locality. Wessels then obtains completeness by supposing that after emission, each of the two particles is a body. And this means that (in a precise sense which Wessels lays out), the A-particle has a state that, together with the settings, influences the A-result, x, independently of the B-result, y (and vice versa). She also deals with Jarrett's locality, i.e., the A-result's independence of the B-setting; she represents it using an assumption (SC), that correlations must be caused, and an assumption that the apparatuses have no causal history in common.

I regard Wessels as in effect laying out an argument for my rebuttal of the examples above. I said that in everyday and classical physical examples, a state that includes the particles' states during initial segments of their journeys screens off the correlated properties. And Wessels has offered an attractive, uniform account of why there is this screening-off in all these examples: *viz.*, in all of them, objects and their states obey effect locality.

Similarly, Cartwright's recent analysis of common causes in effect lays out an argument for my rebuttal. She adopts a framework for combining probabilities and causation that is used in econometrics. Using this, she proposes a condition for z to be the common cause of x and y, which—*contra* Reichenbach—does not require screening-off; and according to which, the emission-event/collision etc. in the examples above is a common cause of the later events. But she goes on to show that if we add another condition on z, we can derive screening-off. The idea of the extra condition is that the common cause acts independently to produce its effects. She points out that this condition fails in the examples above, when one takes as the common cause the emission-event/collision, etc. But the condition is upheld for the Bell experiment by Wessels's effect locality, since this gives each of the separating particles an individual state through which the emission event influences the two measurement-results. Cartwright suggests that the violation of the Bell inequality impugns the extra condition, not her common cause condition: indeed, in her framework, the quantum state itself is a common cause of the measurement-results.

Naturally, I regard Cartwright, like Wessels, as offering an account of why we can obtain screening-off in the examples above, by "pushing ahead" from the emission-event/collision etc.; namely, pushing ahead makes z satisfy her extra condition. There are plenty of unanswered questions, of course, about how Wessels's and Cartwright's accounts are related: in particular, are they compatible, or are they rivals? But for the argument of this paper, I do not need to broach these questions. All I need is that PPSI can be made plausible by "pushing ahead"; that there are two promising accounts of why this should be so is a bonus.

3. Completeness and locality

I turn to Jarrett's completeness condition. This is clearly related to PPSI: where PPSI refers to the common past screening-off, completeness refers to the trio of λ, i, and j (taken conjointly, not individually) screening off. The differences are: (1) PPSI is a conditional statement, whose antecedent refers to no direct causation, while completeness is a categorical statement that λ, i, and j taken together screen off x and y. (2) In completeness, λ, i, and j are schematic; they are not explicitly associated with a region of spacetime, in particular the common past of x and y.

(1) means, in effect, that PPSI and the assumption of no direct causation imply completeness. So given relativity's prohibition of spacelike causation, completeness is supported by all the everyday examples that support PPSI. As we shall see, Jarrett believes that there is no causation between the spacelike-related measurement-events, so that PPSI's antecedent holds, and completeness is supported.

As to (2), the schematic nature of λ, i, and j means that for Jarrett, completeness can be understood in various ways, depending on how much information they include. In accordance with section 2, completeness is logically weaker, the more that λ, i, and j include, i.e., the logically stronger they are. Thus it is weak if they include the total physical state of the common past of the measurement-events; and weaker, the bigger the region we identify as the common past; see figure 2. Conversely, completeness is stronger, the less they include. For example, suppose that λ includes only information that is in some sense intrinsic to the pair of particles; that i and j include only information intrinsic to the two apparatuses; and that measurement-results are influenced by some factors that are not fixed (determined) by giving λ, i, and j. Then there might well be some correlations between the measurement-results for given λ, i, and j, so that completeness fails. Thus completeness with this λ, i, and j is strong. Note that such correlations do not mean that there is causation between the measurement-events; the correlations could be established in the intersection, $C^-(x) \cap C^-(y)$. Nor do they mean that there is a mysterious "conspiracy of nature" established in this intersection; if λ, i, and j include only intrinsic information (in some sense), completeness can fail through common-or-garden forces influencing the results—just have the "intrinsic-only" λ, i, and j omit information about the forces.

So Jarrett's completeness is ambiguous. In a sense, this is a strength: Jarrett derives factorizability from locality and completeness, while leaving λ, i, and j schematic throughout. But the ambiguity also leaves us with unfinished business. We need to verify that there are specific choices for the information to be included in λ, i, and j that make plausible not only completeness and locality, but the other assumptions required for the derivation of the Bell inequality; e.g., assumption (2) of section 1, that the distribution $\rho(\lambda)$ is independent of i and j. In section 4, I shall verify this, assuming that each piece of information is to be the total physical state on a space-time region. In particular, I shall show that in the context of the other assumptions, completeness can be taken very weakly, with λ, i, and j jointly coding the total physical state of the common past of the measurement-events x and y, in one inclusive, and therefore strong, sense of the phrase 'common past'.

Given such a verification, the violation of the Bell inequality means that if locality and the other assumptions are upheld, completeness fails, even in this very weak sense. And if one adds the assumption that there is no causation between x and y, then this failure of completeness spells a failure of PPSI. This of course is Jarrett's main point. He believes that relativity implies locality, and, relatedly, that there is no causation between the measurement-events. So he upholds locality and concludes that completeness fails. (And although he does not mention the PPSI, it is clear that he also takes this to fail. For his admission that completeness is plausible and that its failure is strange make it clear that he implicitly takes λ, i, and j as coding the total state of the common past of x and y. It is unclear whether he accepts (PrDom), and one of the above candidates for the common past.)

So let us examine Jarrett's argument that relativity implies locality. I shall disagree: his argument needs assumptions besides relativity. As a result, the prospects for peaceful coexistence are rather better than he allows: instead of blaming the violation of the Bell inequality on completeness, one can blame it on locality, without impugning relativity—by blaming one of the other assumptions. But I shall agree with Jarrett that a failure of these assumptions would be stranger than a failure of completeness, so that blaming completeness is best.

Both the disagreement and the agreement arise from the same point: locality is a screening-off condition, just as completeness is. It says that λ and the A-setting i screen off the A-result x and the B-setting j from each other and vice versa. Thus recall what it says: for all λ, settings i, j, and j', and A-results x: $p_\lambda^{AB}(x|i,j) = p_\lambda^{AB}(x|i,j')$ and vice versa. This is clearly a statement of screening-off; although it is unusual in one respect. Namely, we are used to the screener-off being an event in the common past or the state of the entire common past and so we naturally wonder whether λ and i code such an event or state. But we can postpone this until the survey of choices of region in section 4.

Since locality is a screening-off condition, one might expect Jarrett to argue for it, in the way we argued for completeness in (1) above; that is, to argue that relativity and PPSI imply locality. Thus he might assume that neither x nor j causes the other because they are spacelike-related, so that by PPSI, the state of their common past screens off any correlation between them. And he would then have to go on to argue that λ and i taken together code the physical state of their common past. But Jarrett does not proceed like this. He assumes that x and j can be spacelike-related, an assumption I will vindicate in the survey of choices of regions in section 4. And he implicitly assumes something like PPSI in that he infers direct causation from un-screened-off correlations. But he does not explicitly state PPSI, and in partic-ular, he does not argue or assume that λ and i code the state of the common past of x and j. Instead, he argues from a new assumption: roughly, that a correlation between x and j could *not* be established in their common past. And from this he infers, in accordance with something like PPSI, that such a correlation demands direct causation across the spacelike interval between x and j.

In more detail, Jarrett's (1984, 573–576) argument is this. He argues that if locality fails, then there is information-transmission between a result and a setting, assumed to be spacelike-related. Thus suppose that locality fails. Then there is a λ, settings i, j, and j', and an A-result x, such that:

$$p_\lambda^{AB}(x|i,j) \neq p_\lambda^{AB}(x|i,j');$$

(or similarly with A and B wings reversed; but let us take this case.) So for pairs with this value of λ, in experiments with i as the A-setting, probabilities of the A-result carry information about the B-setting. Jarrett is, of course, aware that to argue from this to information-transmission, violating relativity, requires an extra premise ruling out correlations due to a common cause. So he distinguishes correlations that can plausibly be regarded as established in the events' common past from those that cannot. (As we have seen, he finds PPSI plausible, so he includes in the former correlations between results from the two wings for a given pair of settings i and j.) And he assumes that the latter group includes correlations between results in one wing and settings in the other (p. 587 n. 15). That is, he rejects as an incredible conspiracy the postulation of some hidden feature of the common past of the A-measurement-event and the fixing of the B-setting that influences both the probability of an A-result and the B-setting, establishing a correlation between them. Given this new assumption, he infers from the failure of locality that information is transmitted. Thus Jarrett's argument for locality uses more than relativity. It also assumes something like PPSI and it makes the new assumption—that there is no such feature correlating results and settings.

I agree that the new assumption is very plausible. Two points in its defense. (i) It can be weakened somewhat: it need not reject all features correlating A-results and B-settings. It need only reject all such features that are not themselves fixed by λ and i (taken jointly); that is enough to make the violation of locality require information-transmission, with the assistance of something like PPSI. (ii) It does not require that the B-experimenter be free to choose the setting. Nor that the causal history that fixes the B-setting be independent of the causal history of the A-result, where 'j is independent of x' means 'j can vary without x varying'. Nor even that no influence can sct up a correlation of the B-setting and A-result. As remarked in (i), it requires only that the causal histories be disentangled enough that any such influence is fixed by λ and i.

Agreed, Jarrett's presentation[12] looks rather different from my summary. His notation distinguishes hidden variables of the apparatus from the spin-direction measured, and he writes as if experimenters *are* free to choose the settings. Indeed, he introduces a null-setting corresponding to "no measure-ment-made" and argues that a violation of locality would allow the A-experi-menter to infer whether the B-experimenter had chosen to make a measure-ment. However, I see no conflict. Some of Jarrett's remarks make it clear that experimenters' freedom to choose whether to measure, and if so in which direction, is a picturesque way of stating that causal histories are sufficiently disentangled (Jarrett 1984, 573 n 13). I admit that the distinction made by Jarrett's notation and his null-setting have an advantage. We are more confident that (1) the causal history of whether there is a B-measurement, and of which direction, is disentangled from that of the A-result, than that (2) the causal history of a feature of the B-apparatus that is unknown but influences results is disentangled from that of the A-result. Thus at the cost of complex notation, Jarrett makes his new assumption more plausible than I did. But my main point remains: his argument depends on more than the prohibition of spacelike causation by relativity.

Let me sum up the story so far. Completeness and locality are both screening-off conditions and justifying them needs more than relativity. If λ, i, and j jointly code the state of the common past of x and y, in some decent sense of 'common past', then completeness can be justified by the PPSI, using that sense. Locality might be similarly justified but Jarrett, in fact,

---

[12]See Jarrett (1984), p. 573–576; also Ballentine and Jarrett (1987).

argues for it using a new assumption about the causal histories of results and settings being disentangled, and implicitly, some analogue of the PPSI. This assumption is plausible. And we will see in section 4 that λ, i, and j can jointly code the common past of x and y, in a decent sense of 'common past'. I also hold (as others do) that violation of the Bell inequality prompts us to give up completeness or locality; denying other assumptions of the inequality would be *ad hoc*. So if we believe in peaceful coexistence, we must give up PPSI or its analogue implicit in Jarrett, or both.

So which should we give up: completeness or locality? PPSI or its analogue? I think we are hard pressed to choose. These seem to be equally plausible: (1) two results should be screened-off by their common past (λij: completeness); (2) a result and a distant setting, x and j, should be screened-off by λ and i (locality), either because a correlation between x and j would be an incredible conspiracy (Jarrett's new assumption), or because λ and i comprise the common past of x and j. However, on balance, I would agree with Jarrett and give up completeness. Not because giving up locality would impugn relativity, but because result-setting correlations *are* incredible; indeed, they are incredible even for a probability function that does not conditionalize on (or get prescribed by) λ and i. And although the failure of completeness (and PPSI) is certainly strange, we can glimpse why it should be so. In the "entangled" quantum states that violate the Bell inequality, the separating particles cannot really be assigned an individual state; in the terminology of d'Espagnat (1976), each is in an improper mixture.

4. *States of regions*

I will now extend the idea, introduced in section 3, of taking the items that prescribe a probability, λ, i, and j, to be the states of certain space-time regions. The discussion there of completeness took all three items together as a state of a region, whose domain of dependence included the measurement-events. But I shall now survey the ways in which one could associate regions with each of the items separately. The aim is to verify that there are choices that make plausible, not only completeness and locality, but the other assumptions of the Bell inequality, e.g., (2) of section 1, that the distribution $\rho(\lambda)$ is independent of i and j. We shall see, in particular, that in the context of the other assumptions, completeness can be taken very weakly, with λ, i, and j jointly coding the state of the common past of the measurement-events x and y.

The survey will be made easier by three closely related assumptions. One is (PrDom), which restricts probabilities to domains of dependence (section 2). Another is that λ, i, and j contribute to probabilities by influences along causal curves. The third (DetDom), is analogous to (PrDom); it restricts determination to domains of dependence (see below). These assumptions are natural, but they in effect presuppose that there is no superluminal causation. So I note that my overall argument does not need these assumptions. They facilitate a survey of choices, a survey which will yield a choice that prompts us to give up completeness or locality. But making the choice does not require the assumptions.

I think that there are ten principles to be borne in mind in choosing these regions. Some articulate the assumptions above; others are more specific to completeness and locality. Some are precise; others, like the first two principles below, are vague. But I submit that they are all worth imposing on one's choice—and that there is a choice that satisfies them all. I begin with two principles that arise immediately from section 3.

(*λijStrong*) Each of λ, i, and j should be strong; thus their regions should be large. We saw in section 3 that completeness can be strong or weak, according as the information coded in all of λ, i, and j is weak or strong. The same point holds for these items, taken individually: for the stronger they are individually, the stronger they are jointly; so the weaker completeness is. Thus we need to take λ, i, and j as strongly as possible, compatibly with the derivation of the Bell inequality. If we do not, we are liable to underestimate what we need to give up as a result of the violation of the inequality. In particular, if we decide to give up completeness, we are liable to underestimate how weak a form of it must be given up.

Locality gives no similar motivation that each of λ, i, and j be strong. Agreed, locality's claim that λ and i screen off x and j from each other is more plausible, the stronger λ and i: just as completeness is more plausible, the stronger λ, i, and j. But this claim is less plausible, the stronger j (or x). And locality implies also that λ and j screen off i and y from each other. So locality pushes each of i and j in both directions: it pushes each of them, as screener-off, to be strong; and it pushes each of them, as screened-off, to be weak. However, λ occurs in locality only as screener-off; so locality, like completeness, prompts us to take λ as strong as possible.

(*Past*) We saw in section 3 that PPSI and lack of causation between x and y imply completeness, provided λ, i, and j code the common past of x and y. And similarly: PPSI and lack of causation between x and j, and between y and i, imply locality, provided (1) λ and i code the common past of x and j, and (2) λ and j code the common past of i and y. Since PPSI is plausible, we should try to choose regions that give these codings, so as to render completeness and locality plausible. Agreed, 'common past' is vague. But we saw some good candidates in section 2: the past of a suitable spacelike hypersurface, or the intersection of this with $C^-(x) \cup C^-(y)$; see figure 2. So we can hope to use these, or regions as much like them as possible.

So we have two principles. The third is (PrDom), discussed in section 2:

(*PrDom*)   The total physical state of a space-time region R prescribes probabilities only for events lying within the domain of dependence, $D(R)$.

Since we think of λ, i, and j as jointly prescribing single and joint probabilities for the measurement-events x and y, (PrDom) means that these events must lie in the domain of dependence of the union of the regions for λ, i, and j.

Next are two principles based on the assumption that λ, i, and j contribute to probabilities by influences along causal curves. I shall also assume that none is redundant: if two are fixed, probabilities vary with the third. This is not part of the definition in section 1 of a stochastic hidden-variable model, but it is reasonable and worth satisfying if we can.

(*ijContrib*)   We should assume that the region for i must intersect $C^-(x)$, and similarly for j and y. For we expect i to contribute to the probabilities for x by an influence along a causal curve. Furthermore, it is reasonable to require that some such curves do not pass through λ's region. For we want these probabilities to depend on i, for given λ and j; thus we want there to be at least two values of i, $i1$ and $i2$ say, such that $p_\lambda^{AB}(x|i1j) \neq p_\lambda^{AB}(x|i2j)$. (If this were not so, i would be redundant.) If this dependence on i for given λ is to be transmitted by a causal curve, there needs to be a causal curve from i's region to the event x that does not pass through λ's region. For if all such curves passed through λ's region, variation in i could always be accompanied by variation in λ, and so there could be no dependence on i for given λ. Similarly for j and y.

(*λContrib*)   Similarly, λ should contribute to probabilities in both wings by an influence along a causal curve. Thus we should assume that there is a causal curve from the region for λ to the measurement-events, x and y, which does not pass through the regions for i and j. The argument for this is parallel to that for (ijContrib). We want the probabilities to depend on λ for given i and j; if not, λ would be redundant. And we expect λ to contribute to these probabilities along a causal curve. But if all such curves passed through i's or j's region, variation in λ could always be accompanied by variation in i and j, and so for fixed i and j, probabilities could exhibit no dependence on λ.

The sixth principle is motivated by locality:

(*Spacelike*)   The regions for x and j should be chosen so as to be spacelike-related and so should the regions for i and y. For such a choice will mean that direct causation between x and j (i and y) would offend relativity; thus giving relativity a part to play in the justification of locality. Such a choice will also guarantee that x and j, and i and y, are distinct events, which, as we saw in section 2, is necessary if PPSI is to be used to argue for locality, as envisaged in (Past) above. (This principle is parallel to our initial choice of x and y as spacelike-related.)

The seventh and eighth principles affecting the choice of regions are less obvious. They are both about λ, i, and j determining one another, and both are implied by the requirement that locality should hold.

($\lambda Free$)   Neither of the settings determine λ. We can give an argument for this parallel to that for (λContrib): if the A-setting i determines λ, then given i and j, the specification of λ is redundant. But we can also argue from locality. Thus suppose i does determine λ. So different λ's imply different i's. We can now violate locality. For we require that $p_\lambda^B(y|ij)$ should depend on λ. And if varying λ implies varying i, then there are different B-wing probabilities for experiments that differ only in the A-setting i (and thereby in λ), thus violating locality.

This means that there is a tradeoff to be made in the choice of regions for settings i and j: the regions must be as large as they can be, but not so large that i or j determines λ.

($ijFree$) λ should determine neither of the settings. For suppose that λ determines the A-setting i. Then for any other A-setting, i' say, the B-wing probabilities, $p_\lambda^{BA}(y|i'j)$, are 0 or are undefined. But there is some B-result y such that $p_\lambda^{BA}(y|ij)$ is greater than 0. So locality is violated: B-wing probabilities depend on the A-wing setting.

Thus there is also a tradeoff to be made in the choice of λ's region: it must be as large as possible in order to make both completeness and locality weak, but not so large that λ determines i or j.

To apply these last two principles, (λFree) and (ijFree), to the survey of candidate regions, we need to "translate" the notion of determination into space-time terms. So far, the notion has been the usual philosophical one of "fixing," with certain background factors held constant. For example, it is said that the physical fixes the mental, if the laws of nature are held constant: that is, any two possible worlds (or cases or situations) that obey the laws of nature, and with the very same physical facts, have the same mental facts— "no mental difference without a physical difference." Similarly here, with the constant factor being the "laws" of the stochastic hidden-variable model. For i to "determine" λ means that any two possible worlds (sequences of states of particles and apparatus in the Bell experiment) that obey the model, and with the same setting i, have the same λ.

There is a natural causal constraint on determination. The constraint is not specific to the settings and hidden variables. So I give it for any states r and s of some theory that posits causal influences on states:

> If the state r is to determine the state s, any causal influence on s should also influence the state r.

For suppose that in some possible world obeying the theory, some causal influence on *s* does not influence the state *r*. Then provided that the possible worlds obeying the theory form a sufficiently varied set, there is some other world lacking this causal influence on *s* and accordingly differing in regard to *s*, while matching the first as regards *r*; so that *r* does not determine *s*.

This constraint has a natural expression in Minkowski space-time, once we assume that there is no superluminal causation and that *r* and *s* are states of regions; call the regions *R* and *S*. We again use the notion of a domain of dependence, introduced in the discussion of (PrDom) above:

(*DetDom*)   If *r* is to determine *s*, then the region *S* must lie wholly within the domain of dependence, *D(R)*, of the region *R*.

In surveying candidate regions for λ, *i*, and *j*, I shall assume (DetDom), but as mentioned my overall argument does not need it.

Three points in clarification and defence of (DetDom). (1) In effect, it presupposes that there is no superluminal causation. For if there is superluminal causation, then it will be only trivially true—its only instances being regions *R*, *S* with *S* lying within *R*, so that *s* is a part of *r*. The reason is as above: if there is such causation, then a sufficiently varied set of possible worlds obeying the theory will mean that the state *r* does not determine any state beyond *R*—for any such state could be causally influenced without r being causally influenced.

(2) (DetDom) does not assume that the only way that a causal influence on *s* can also influence *r* is by a single "causal chain" (following a causal curve). There could be a single causal chain intersecting the regions of *s* and *r*, or there could be separate causal chains from the influencing state or event. And in both cases, *S* need not lie in *D(R)*. But we should distinguish influence and determination. Determination requires that every possible influence on *s* register on *r*. And given no superluminal causation and a sufficiently varied set of possible worlds, that is equivalent to *S* lying in *D(R)*.

(3) (DetDom) does not assert any kind of determinism, in particular that *r* determines the state on *D(R)*; it only asserts a necessary condition of *r* determining the state *s* of some region *S*—that *r* does not determine anything beyond *D(R)*.

So much for clarifying (DetDom). For our purposes, the point of it is that it means that we can satisfy (λFree) and (*ij*Free) by choosing regions that do not lie wholly in one another's domains of dependence. That is, λ's region is not wholly in *i*'s domain, nor wholly in *j*'s; and neither *i*'s region nor *j*'s is wholly in λ's domain.

Finally, we need a principle about the "other" locality assumption of the Bell inequality: the assumption that the distribution $\rho(\lambda)$ is independent of *i* and *j*. Again, this is a screening-off condition in the same way that completeness and locality are. If our notation allowed a dependence, so that we write $\rho(\lambda|ij)$ (or if one prefers, $\rho_{ij}(\lambda)$; see section 1), then the statement of independence would be: that for all A-settings i and i', and B-settings j and j', $\rho(\lambda|ij) = \rho(\lambda|i'j')$. Which says that ρ screens off λ from the A-setting i and from the B-setting j. Hitherto, the screener-off has been the state of a region: λ, i, and j taken together (in completeness); or λ and i, or λ and j (locality). Now it is a distribution. Fortunately, we can trade in this distribution for the state of a region and get constraints on the choice of region. For clarity, I proceed in two stages.

First, think of λ not as the state of a region, but in the usual way: as the state of the particle-pair, so that λ is fixed by the history of the process of preparing the pair. The use of $\rho(\lambda)$, understood as an objective probability, means that the preparation process is indeterministic. There is some initial state of this process, somewhere in the past of the emission-event; different λ's can result from a given initial state, and they do so with weights characteristic of the initial state. Thus it is natural to think of the distribution ρ as fixed by this initial state, and this initial state is fixed by the total physical state of a certain region of space-time lying in the past of the emission-event. This total state, r say, prescribes the distribution ρ; we could write ρ as p_r. And ρ's independence of i and j amounts to: the state r screens off λ from i and from j:

(†) $\operatorname{prob}(\lambda|r,i) =_{\mathrm{df.}} \rho(\lambda|i) = \rho(\lambda)$;   and similarly for j.

Now, the second stage: let us revert to thinking of λ as the state of a region. The above argument for taking a state r, rather than the distribution ρ, as the screener-off does not depend on λ being just the state of the particle pair. Let λ be the state of a region: we can still posit that there is another region, whose state r prescribes probabilities for λ, and which screens off λ from i and from j. However, (†) above, and our other principles, imply certain conditions.

(†) implies that the regions for λ and i do not touch each other; if they did, continuity across the boundary would mean that i constrained λ, violating (†). *A fortiori*, they should not overlap. Similarly, for λ and j. And similarly the region for r should not touch or overlap the region for i, nor the region for j. Summarizing, we get:

(1)   Regions for the following pairs of states do not touch or overlap: λ and i; λ and j; r and i; r and j.

(1) means that the gaps between the regions must serve as "cushions": in an ideal experiment, for any given choice of states on the regions, one would organize the state of the gap to dovetail with that choice. Also (PrDom) implies:

(2)   The region for λ is in the domain of dependence of the region for r; while for nontrivial probabilities, the region for λ is not in that for r.

And (Past) and (†) together imply:

(3)   The region for r includes the common past of the regions for λ and i; and similarly for λ and j: in some decent sense of 'common past'. (So PPSI can motivate the claim that ρ is independent of i and j).

To sum up, ρ's independence of i and j gives our last principle constraining the choice of spacetime regions:

(ρ*Indep*)   There is a region for a state r conforming to (1), (2), (3), above.

So much for principles. Can we satisfy them? Yes. I shall show that we can do so, with an initial choice that λ is the state of $C^-(x) \cap C^-(y)$.

Recall first that, by (DetDom), we can satisfy (λFree) and (ijFree) by requiring: λ's region is not wholly in i's domain, nor wholly in j's; and neither i's region nor j's is wholly in λ's domain. So for our choice of λ: λ cannot determine i and j provided their regions are partly outside $C^-(x) \cap C^-(y)$; thus satisfying (ijFree). And there are regions for i and j intersecting the pasts $C^-(x)$ and $C^-(y)$, and allowing suitable causal curves from i, j, and λ to the measurements, thus satisfying (ijContrib) and (λContrib); and with domains of dependence sufficiently small to prevent i and j determining λ, thus satisfying (λFree). Thus we can satisfy (λFree), (ijFree), (PrDom), (ijContrib), and (λContrib). The general situation is as in figure 4:

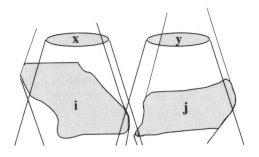

Figure 4.

What about the remaining principles: (λijStrong), (Past), (Spacelike) and (ρIndep)? I shall treat them in turn.

(λijStrong): λ is already strong. The region for λ includes the causal past of the emission-event; and if the particles travel slower than light, it includes initial segments of the particles' trajectories toward the two wings. Also, the regions for i and j are not determined by the principles considered so far. These regions can be extended far beyond what is shown in figure 4. In particular, they do not need to be compact: they can extend infinitely far down the "strips," and to the left and right, outside the causal pasts of x and y. That is: they can extend like this provided that our remaining principles allow it. In this sense, figure 4 is too specific to represent all the choices of region that are, so far, still possible.

(Past): Recall that this requires that certain unions of regions form common pasts of certain pairs of regions—in some sense of common past. More specifically, it requires that: (1) the union of the regions for λ, i, and j form a common past for x and y; (2) the union of the regions for λ and i form a common past for x and j; (3) the union of the regions for λ and j form a common past for i and y. Since 'common past' is vague, it is a matter of judgment what counts as a good precise notion, and we may choose regions that satisfy (1) according to one precise notion of common past, and (2) according to another, and (3) according to yet another—provided that all three are good.

In fact, for the principles considered so far, we are lucky: we can take the common past of two regions to be very large, so that we allow a weak version of PPSI to enter into the justification of completeness and locality. To be precise: in all of (1) (2) and (3), we can take the common past of two regions to be the entire past behind a suitable hypersurface—see figure 2, in section 1. For recall from (λijStrong) that we can extend the regions for i and j down the strips and to the left and right. The only restriction that we must impose on the hypersurface, and that we did not have in the earlier discussion, arises from (λContrib). Thus we want causal curves from λ's region to x and y that do not pass through the regions for i and for j; so the hypersurface which is the future boundary of the union of the regions for λ, i, and j—the hypersurface whose past is the common past—must include the future "nose" of the boundary of λ's region. This hypersurface is thus not wholly spacelike: there are parts of it that must be lightlike, if (λContrib) is to be satisfied. I claim that such a hypersurface nevertheless gives us a good common past.

(Spacelike): Recall that this requires that x and j be spacelike-related; and similarly, i and y. This is easily secured while conforming to all the principles considered so far. Just make i's region not intersect λ's; and similarly, make j's region not intersect λ's. In this sense, figure 4 is too general; intersections which it allows are forbidden by (Spacelike).

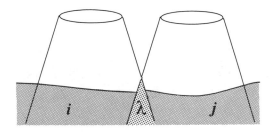

Figure 5.

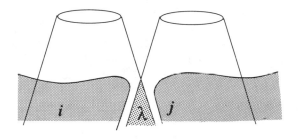

Figure 6.

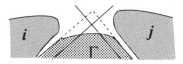

Figure 7.

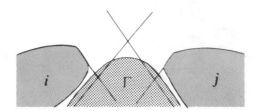

Figure 8.

So these three principles give us figure 5:

(ρIndep): Recall that it has three conjuncts (1) (2) and (3). Its first conjunct (1) requires in part what we have just done to satisfy (Spacelike): that λ and i, and λ and j, have disjoint regions. But (1) also requires that the regions for λ and i, and for λ and j, do not touch. So the future boundaries of λ, i and j do not form a hypersurface. But if we choose regions as in figure 6, we can still claim that λ, i, and j form a common past of x and y. For the gaps between the regions lie outside $C^-(x) \cap C^-(y)$; so there can be no influence along a causal curve from the gap to both x and y.

(1) also requires that the regions for r and i, and for r and j, do not touch. And (2) requires that λ's region lies not in r's region, but in its domain of dependence. This prompts us to choose for r's region, a "base" of λ's region, extending just beyond λ, to get a nontrivial domain of dependence; see figure 7:

Finally, (3) requires that r's region include a common past for λ and i; and for λ and j. We can satisfy this, where 'common past' is understood as 'intersection of the causal pasts', by pushing r's region forward enough to contain the intersections. See figure 8. To sum up, figure 8 shows how we can make plausible all assumptions of the Bell inequality, while taking the screeners-off to be the states of space-time regions. (Other choices of regions are possible too: exercise!)

NONFACTORIZABILITY, STOCHASTIC CAUSALITY, AND PASSION-AT-A-DISTANCE

Michael L. G. Redhead

1. Background

Nonlocality in quantum mechanics (QM) provides a *prima facie* conflict with the foundations of special relativity, which is usually taken to forbid causal processes operating outside the light cone, i.e., at spacelike separation. In my recent work (1987b) various senses of locality have been distinguished and the question of which of these is violated depends crucially on the interpretation of quantum mechanics adopted. In the case of deterministic hidden-variable interpretations, action-at-a-distance, in the sense of instantaneously affecting the possessed value of some attribute belonging to one member of a pair of spatially separated systems by altering the physical arrangement of an apparatus interacting with the other member, appears to be clearly demonstrable. Here a direct conflict with special relativity seems difficult to avoid except by appealing to the problematic topic of tachyonic interactions.

On the other hand, in the case of indeterministic or stochastic hidden-variable interpretations, the issues raised are more subtle.

In a recent paper (Redhead 1986), a discussion was given of the significance of violating the so-called "completeness" condition of Jarrett (see his 1984). Introducing a necessary condition for stochastic causality it was argued that violating Jarrett's condition gave no grounds for inferring a direct stochastic causal link between the outcomes on the two wings of a Bell-type experiment. Instead it was argued that the situation seemed to exhibit, in a precise mathematical way, what Shimony had in mind for introducing the term, "passion-at-a-distance" (1984a). In the present paper I wish to augment and clarify my discussion (1986).

I am grateful to Jeremy Butterfield, Hugh Mellor, David Papineau, and Paul Teller for helpful discussions.

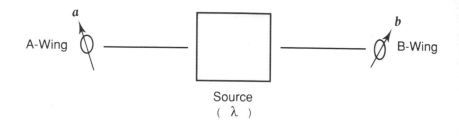

Figure 1.

Stochastic hidden-variable theories enable a proof to be given of the Bell inequality which does not invoke determinism, provided certain conditions are imposed (Redhead 1987b, 98ff). The condition which will concern us here is usually referred to as factorizability. It arises in the following way. Referring to figure 1, the basic mathematical framework for stochastic hidden-variable theories of Bell-type experiments provides a triple joint distribution in the QM state $|\Psi>$, $\text{Prob}^{|\Psi>} (a = \varepsilon_a \ \& \ b = \varepsilon_b \ \& \ \lambda = \varepsilon_\lambda)$, for the physical magnitude a pertaining to the particle on the A-wing of the apparatus to have the value ε_a, while the physical magnitude b for the correlated particle on the B-wing of the apparatus has the value ε_b and the hidden-variable specification of the source λ has the value ε_λ.[1]

The triple joint distribution may also depend on the physical arrangement of apparatus set up to measure a on the A-wing and b on the B-wing, but this dependence, in terms of which Jarrett formulates his (weak) locality condition, will play no role in our own subsequent discussion, and hence will be ignored in our notation.

Now the triple joint distribution can be rewritten in the form

$$\begin{aligned}
\text{Prob}^{|\Psi>} &(a = \varepsilon_a \ \& \ b = \varepsilon_b \ \& \ \lambda = \varepsilon_\lambda) \\
&= \text{Prob} \ (a = \varepsilon_a | b = \varepsilon_b \ \& \ \lambda = \varepsilon_\lambda) \\
&\times \text{Prob} \ (b = \varepsilon_b | \lambda = \varepsilon_\lambda) \\
&\times \text{Prob}^{|\Psi>} \ (\lambda = \varepsilon_\lambda)
\end{aligned} \tag{1}$$

---

[1] It may be helpful to point out that the observables a and b are given, respectively, as $a = \sigma_A \cdot \hat{a}$ and $b = \sigma_B \cdot \hat{b}$. Here σ is the spin operator, $\hat{a}$ is a unit vector specifying a direction in the A-wing, and similarly for $\hat{b}$ in the B-wing.

Note that the dependence on $|\Psi>$ arises only through the third factor, expressing our ignorance of the value of λ in the context of specifying the quantum-mechanical state. The first two factors refer by contrast to objective propensities: the idea is that λ prescribes single and joint probabilities for a and b to take their various values, and does so independently of $|\Psi>$.

(1) is quite general, but in order to derive the Bell inequality it is sufficient to impose the following condition:

$$\text{Prob } (a = \varepsilon_a | b = \varepsilon_b \,\&\, \lambda = \varepsilon_\lambda)$$
$$= \text{Prob } (a = \varepsilon_a | \lambda = \varepsilon_\lambda) \tag{2}$$

so achieving conditional stochastic independence for the magnitudes a and b for a fixed value of λ, commonly referred to as factorizability.

From the violation of the Bell inequality we can infer that (2) must in general fail. (2) is the condition that Jarrett calls "completeness," on the grounds that the state λ of the source fixes *completely* the conditional probability for a to have the value ε_a, specification of the value of b being superfluous.

If Jarrett completeness is violated, how are we to interpret the dependence on the value of b? The usual argument here is that (2) is the condition for λ to *screen off* a from b, i.e., it identifies λ as the common cause of a and b. The failure of (2) means that a common cause explanation of the myste rious Bell correlations cannot be right. So we must invoke an additional direct causal link between a and b, so that a is caused *jointly* by λ and b. This is illustrated schematically in figure 2, where stochastic causal links are indicated by the wavy lines.

But since a and b pertain to particles at spacelike separation, the assumption of a direct *causal* link between a and b raises a *prima facie* conflict with special relativity.

Is there any way we can understand the correlations between a and b which does not lead to such a conflict? We know that a common cause by λ cannot be the whole answer. But suppose it is not even *part* of the answer. Could we model the correlations between a and b entirely in terms of a direct link as illustrated in figure 3?

If the answer were in the affirmative we would be making no progress in resolving the conflict with relativity, but in fact we shall argue for a negative answer which will point us toward an escape route.

In the framework of stochastic hidden-variable theories, this move would mean that the conditional probability $\text{Prob}^{|\Psi>} (a = \varepsilon_a | b = \varepsilon_b)$ given directly by QM would be independent of $|\Psi>$! So we could not possibly reproduce the predictions of quantum mechanics. But the reason for this is clear. In the stochastic hidden-variable approach, as we have seen, $|\Psi>$ indexes our ignorance of the value of λ. If λ plays no role in establishing the

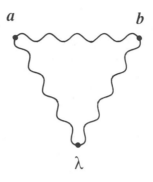

Figure 2.

correlations, they cannot depend on $|\Psi>$. But suppose we give up the stochastic hidden-variable framework entirely and interpret $|\Psi>$ as an objective specification of the state of the particles emerging from the source. Let us then ask the question: Could the conditional probability $\text{Prob}^{|\Psi>}(a = \varepsilon_a | b = \varepsilon_b)$ given directly by QM *under this, more orthodox interpretation of* $|\Psi>$ be the result of a direct causal correlation as illustrated in figure 3?

2. Causal links and robustness

To answer this question we need to introduce a necessary condition for stochastic causality, that sufficiently small disturbances of the cause do not affect the functional form of the causal connection. We will refer to this as the *robustness* criterion for a causal connection.[2]

In the present setup the events that may be cause and effect are the magnitudes a and b taking values ε_a and ε_b, and the functional form of the causal relation is given by the conditional probabilities. It was shown (Redhead 1986) that for two spin-½ particles emerging from the source in the

---

[2]For related notions in discussion of causality, see Lewis (1986, 250) and Skyrms (1980, 12). Lewis maintains that in a causal relation between events it is necessary that the events could have happened in different ways, in particular at different times, while the truth of the causal statement is upheld. Skyrms has the idea of resilience of a conditional probability in a stochastically causal situation, which has its value unaltered by conditionalization on further information, in particular on how the events were brought about. In our discussion we consider just the case where events consist in magnitudes taking values.

Figure 3.

singlet state of their combined spin, $\text{Prob}^{|\Psi>}\,(a = \varepsilon_a|b = \varepsilon_b)$ is not robust under perturbations of either the A-wing or the B-wing of the Bell experiment by the action of arbitrary c-number fields, however weak, where $|\Psi>$ refers to the state of the particles as they emerge from the source. These perturbations change $|\Psi>$ to $|\Psi'>$, and what was shown was that, however weak the perturbation from $|\Psi>$ to $|\Psi'>$, we can always choose magnitudes a and b such that

$$\text{Prob}^{|\Psi'>}\,(a = \varepsilon_a|b = \varepsilon_b)$$
$$\neq \text{Prob}^{|\Psi>}\,(a = \varepsilon_a|b = \varepsilon_b) \tag{3}$$

where the conditional probability on the LHS of (3) is what would obtain if the particles had emerged from the source in spin state $|\Psi'>$.

Furthermore, for fixed a and b, we can always find perturbations, however weak, such that (3) holds. This result shows that a and b cannot be regarded as having their correlations established by a direct causal link, so the situation cannot be as illustrated in figure 3, but rather as in figure 4, where the dotted line indicates a nonrobust link associated with passion-at-a-distance rather than action-at-a-distance.[3]

The conditional probability $\text{Prob}^{|\Psi>}\,(a = \varepsilon_a|b = \varepsilon_b)$ is a property of the a and b magnitudes which depends crucially on the QM state $|\Psi>$—it is as much a property of $|\Psi>$ as it is of a and b and exhibits a harmony or passion of the two magnitudes a and b in the particular state $|\Psi>$.

But, if nonrobustness arises simply from the dependence of $\text{Prob}^{|\Psi>}\,(a = \varepsilon_a|b = \varepsilon_b)$ on $|\Psi>$, why not restore robustness by regarding $|\Psi>$ as contributing a common cause of the correlation? But there's the rub. If $|\Psi>$ is a partial common cause, then it must be overlaid by a direct causal link, since for fixed $|\Psi>$, $\text{Prob}^{|\Psi>}\,(a = \varepsilon_a|b = \varepsilon_b)$ has a robust form as a function of the values ε_a and ε_b. But a direct causal link is unacceptable on relativistic grounds. But if $|\Psi>$ is the whole (common) cause of the correla-

---

[3]Since also $\text{Prob}^{|\Psi>}\,(b = \varepsilon_b|a = \varepsilon_a)$ is not robust under perturbations of the A-wing of the experiment, passion is seen to be, unlike causality, a symmetric relation.

$$a \bullet \text{-----------------} \bullet b$$

Figure 4.

tion then the joint probability $\text{Prob}^{|\Psi>}$ $(a = \varepsilon_a \ \& \ b = \varepsilon_b)$ would have to factorize into $\text{Prob}^{|\Psi>}$ $(a = \varepsilon_a) \times \text{Prob}^{|\Psi>}$ $(b = \varepsilon_b)$, so $\text{Prob}^{|\Psi>}$ $(a = \varepsilon_a | b = \varepsilon_b)$ would be just $\text{Prob}^{|\Psi>}$ $(a = \varepsilon_a)$, which is not at all what QM predicts.

So we have a dilemma, either horn of which is unacceptable. Our response must be to resist the temptation to restore robustness by regarding $|\Psi>$ as contributing a common cause, and accept the dependence on $|\Psi>$, not as a causal dependence, but as indicative of the nonrobustness of the link between a and b, the way of passion rather than of action.

The situation should be contrasted sharply with how we deal with non-robust pseudo-causal correlations produced by examples of the rotating searchlight variety. The illuminations produced at different orientations of the searchlight are correlated, for example, a yellow illumination followed by a yellow illumination. But these correlations depend on how the beam is being filtered as a function of time. Yellow might be followed by red rather than yellow. Robustness of the correlation is restored by regarding the filtering of the beam as part of the cause. There is no objection to this maneuver in the searchlight case, where the total cause of the correlation is correctly understood as a composite common cause provided by the luminosity of the source together with the way it is filtered. In the quantum-mechanical case this natural way of proceeding is blocked by the dilemma posed on the one hand by empirical inadequacy of the totally common-cause explanation and on the other by relativistic difficulties posed by a partial common-cause explanation.

We now revert for a moment to the discussion of stochastic hidden-variable theories and raise the question of whether the combination of common cause and direct cause illustrated in figure 2 could escape from the relativistic objection by replacing the direct wavy line by an acceptable dotted line representing an overlay of the common cause by a direct passion as shown in figure 5.

The answer is no, because for fixed λ, $\text{Prob} \ (a = \varepsilon_a | b = \varepsilon_b \ \& \ \lambda = \varepsilon_\lambda)$ has a *robust* form as a function of ε_a and ε_b. The way of passion is inconsistent with the stochastic hidden-variable approach, in which $|\Psi>$ measures epistemically our ignorance of the state of the source. It only makes sense when $|\Psi>$ is afforded ontic significance in terms of the maximal state specifi-

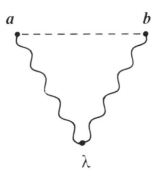

Figure 5.

cation of the particles emerging from the source. This is the essential point where the present discussion corrects my previous analysis (1986).

I want to conclude by making some comments on why robustness should be regarded as a necessary condition for causality. Robustness says that if a and b are two physical magnitudes which are causally related then the cause b screens off a from sufficiently weak disturbances d acting on b. But if this is violated, why not incorporate d as part of the cause of a? But this resets the problem. We now require that b and d jointly screen off a from whatever brings about d. Call this d'. Continuing in this way, I claim that if incorporating d, d', d'' . . . etc. as part of the cause never succeeds in restoring robustness, then we would have no justification in talking about a lawlike causal connection at all. At some point the regress must be halted and a robust functional dependence exhibited. This discussion shows that robustness can be a more general requirement than any "Markovian" condition that the value of a is "fixed" (possibly in a stochastic sense) by the value of b, and not by how that value was brought about.

Furthermore, it should be noted that robustness has nothing to do with stability in the technical sense, that small changes in the value of b produce small changes in the value of a. The stability that is involved is one of functional form.

3.   Robustness and signaling

Finally it should be remarked that in the deterministic case, with a suitable sense of determinism to be spelled out below, robustness implies signaling capability, so we can demonstrate nonrobustness from nonsignaling. To see this, note that

$$\text{Prob}(a = \varepsilon_a) = \sum_{\varepsilon_b} \text{Prob}(a = \varepsilon_a | b = \varepsilon_b) \cdot \text{Prob}(b = \varepsilon_b) \qquad (4)$$

Take, for the deterministic case, the value ε_b to fix the value ε_a via the functional relationship $\varepsilon_a = f(\varepsilon_b)$, where we assume f to be $1:1$, so that both *historical* and *futuristic* determinism is maintained in the terminology of Earman (1986a, 13). Then write $\text{Prob}(b = \varepsilon_b) = \delta(b, \varepsilon_b)$, where δ is the Kronecker delta function, so fixing the value of ε_b at the numerical value b, and $\text{Prob}(a = \varepsilon_a | b = \varepsilon_b) = \delta(\varepsilon_a, f(\varepsilon_b))$ as the robust functional form. Then from (4) we obtain

$$\text{Prob}(a = \varepsilon_a) = \delta(\varepsilon_a, f(b)) \qquad (5)$$

or, in alternative notation, if a is the value of a, when b has the value b, we recover simply

$$a = f(b) \qquad (6)$$

Signaling means that if b is changed to b′, distinct from b, then a is changed to a′, distinct from a. This follows from the fact that f is $1:1$. In the indeterministic case, signaling says that $\text{Prob}(a = \varepsilon_a)$ is changed in functional form, i.e., is not robust under disturbance acting on the B-particle. If the conditional probability is robust under such disturbance, but $\text{Prob}(b = \varepsilon_b)$ is not, it does not follow from (4) that in general $\text{Prob}(a = \varepsilon_a)$ is not robust. In the case of the spin $-\frac{1}{2}$ correlations in the singlet state, where ε_b is a dichotomic variable, it is true that robustness of the conditional probability does imply signaling, provided the marginal $\text{Prob}(b = \varepsilon_b)$ can be changed in a controllable way.[4] But this condition is not satisfied in the Bell experiments with perturbation by c-number fields. In this case we have the situation that $\text{Prob}(b = \varepsilon_b)$ is

---

[4]This follows from the fact that, with $\varepsilon_b = \pm 1$, (4) becomes

$$\text{Prob}(a = \varepsilon_a) = \text{Prob}(a = \varepsilon_a | b = +1) \cdot \text{Prob}(b = +1)$$
$$+ \text{Prob}(a = \varepsilon_a | b = -1) \cdot \text{Prob}(b = -1)$$

Hence

$$\triangle \text{Prob}(a = \varepsilon_a) = \text{Prob}(a = \varepsilon_a | b = +1) \cdot \triangle \text{Prob}(b = +1)$$
$$+ \text{Prob}(a = \varepsilon_a | b = -1) \cdot \triangle \text{Prob}(b = -1)$$
$$= (\text{Prob}(a = \varepsilon_a | b = +1) - \text{Prob}(a = \varepsilon_a | b = -1))$$
$$\cdot \triangle \text{Prob}(b = +1)$$

so, provided $\triangle \text{Prob}(b = +1) \neq 0$ (and also $\text{Prob}(a = \varepsilon_a | b = +1) \neq \text{Prob}(a = \varepsilon_a | b = -1)$), it follows that $\triangle \text{Prob}(a = \varepsilon_a) \neq 0$.

robust, Prob $(a = \varepsilon_a | b = \varepsilon_b)$ is not robust, but Prob $(a = \varepsilon_a)$ is also robust (see Redhead 1986).

In the stochastic case then, robustness and signaling are in general logically quite separate. It so happens that quantum-mechanical correlations cannot be used to signal, but that circumstance cannot be used to demonstrate that quantum-mechanical correlations are not robust.

QUANTUM NONLOCALITY AND THE DESCRIPTION OF NATURE

HENRY P. STAPP

1. Quantum ontologies

One task of philosophers, and perhaps the basic one, is to illuminate for us the kind of world in which we live and our connection to it. Physicists have, however, been proclaiming for more than fifty years that the empirical evidence provided by quantum phenomena demands a radical revision of our ideas about physical reality: that the new phenomena entail a peculiar kind of macroscopic wholeness—a strange sort of nonseparability of macroscopically separated parts of the universe (Bohr 1934, 1935; Stapp 1972). These new theoretical ideas appear to open new perspectives on the place of man in the universe, and to have an impact on the question of values (Wheeler and Zurek 1983; Stapp 1987, 1985b).

In view of these developments, one might expect to find philosophers massively engaged in an effort to determine what these quantum phenomena reveal about the character of the physical world. Any such endeavor must cope, however, with the fact that orthodox physicists, while insisting on the need for radical revision and recognizing certain holistic features, refuse to offer a new ontology: in their view the new theoretical entities upon which physicists base their understanding of quantum phenomena must be interpreted not as theoretical counterparts of entities existing in nature itself, but merely as elements of a computational procedure that allows scientists to form expectations pertaining to observations that appear under certain kinds of conditions.

This reluctance of orthodox quantum physicists to offer an ontology stems in part from their acute awareness of the uncertainties involved in

This work was supported by the Director, Office of Energy Research, Office of High Energy and Nuclear Physics, Division of High Energy Physics of the US Department of Energy under Contract DE-AC0376SF00098.

passing from phenomena to ontology: their theories abound with examples in which identical or nearly identical predictions about phenomena stem from very different assumptions concerning the basic quantum entities. But beyond mere prudence there lies a compelling reason: no known ontology seems acceptable. In particular, every known ontology compatible with the phenomena, as codified by quantum theory, is either a many-worlds ontology (Everett 1957; Dewitt and Graham 1973) or involves explicit faster-than-light influences. In the many-worlds ontology the universe is continuously splitting into an amorphous superposition of noninteracting worlds of the kind we perceive, and the consciousness of each of us is continuously splitting into an amorphous bundle of diverging noninteracting streams of consciousness. On the other hand, every known adequate alternative to this many-worlds ontology involves explicit influences that allow an effect to precede its cause in some frame of reference.

Rather than accepting either of these alternatives the orthodox quantum physicist construes science in a way that relegates ontology to philosophy. Philosophers are thereby seemingly placed in the position of either: (1) apologizing for their failure to produce an ontology—i.e., trying to make us feel comfortable about having no ontology; or (2) embracing an ontology that orthodox physicists reject; or (3) constructing an adequate ontology where physicists have failed. There is, however, another possibility, which is to deduce by careful analysis exactly what constraints are imposed upon ontology by the structure of quantum phenomena. The present work is part of this endeavor, which forms a central focus of this conference.

To set the problem let me briefly describe the situation presented by physicists. Consider for simplicity a classical system consisting of n particles. The state of this system at time t is specified by giving, for each particle i, its position $\vec{x}_i = (x_{i1}, x_{i2}, x_{i3})$ and its velocity $\vec{v}_i = (v_{i1}, v_{i2}, v_{i3})$. Thus the state of the system is specified by a set of $6n$ real numbers: $(x,v) = (\vec{x}_i, \ldots, \vec{x}_n; \vec{v}_1, \ldots \vec{v}_n)$. Usually we do not know the exact position and velocity of every particle. But it may be supposed that any body of information A about the system at time t can be represented by a probability distribution over the various possible values of (x,v). Let this distribution be called $P_A(x,v,t)$. Similarly, for each setup B that can detect the particles of the system there is a quantity $E_B(x,v,t)$ that describes the efficiency of detection at the various points (x,v) at time t. The probability $P(A;B)$ that a system about which we have information A will be detected by setup B is obtained by multiplying $P_A(x,v,t)$ by $E_B(x,v,t)$ and summing (i.e., integrating) over all possible values of (x,v).

Quantum theory is formulated in terms of analogs of the distributions $P_A(x,v,t)$ and $E_B(x,v,t)$. The quantum analogs have properties that differ from those of their classical counterparts. In particular: (1) the classical quantities are nonnegative, whereas the quantum quantities can be negative; (2) there is, in principle, no lower limit to the size of a region in (x,v) space outside of which the classical quantities can vanish, whereas there is a lower "uncertainty principle" limit in the quantum case; and (3) the classical quantities evolve according to Newton's equations, whereas their quantum analogs evolve according to the Heisenberg equations.

The complete quantum mechanical description of the system is given by specifying, for every possible A and every possible B, the corresponding distributions $P_A(x,v,t)$ and $E_B(x,v,t)$, and by specifying also the Heisenberg equation that governs the temporal evolution of these quantities. The orthodox attitude is that these quantities $P_A(x,v,t)$ and $E_B(x,v,t)$ are to be given no ontological status: they are to be interpreted as nothing other than parts of the computational machinery.

Not every scientist is content to have no ontology, and various ontologies have been proposed. To obtain a relativistic theory one usually passes from the simple n-particle system to a field-theory version. But for simplicity I continue to use the n-particle language, where n can now be the number of particles in the universe.

The simplest of the proposed ontologies is that in which the universe is represented by a single function $P(x,v,t)$ that evolves according to the Heisenberg equation (Everett 1957; Dewitt and Graham 1973). This gives the many-worlds ontology mentioned earlier. The peculiarities of this ontology stem from the fact that the quantity $P(x,v,t)$, being the quantum analog of a classical distribution of probability, has, at the macroscopic level, the structural features that would be appropriate to a probability distribution. In particular, *every* possibility is represented, and assigned a certain statistical weight: no real or actual world of the kind we perceive is singled out from the collection of the possible ones.

At the microscopic level, this superposition of all possibilites poses no problem. But the ontology extends in an unbroken way to the macroscopic level. Consequently, we obtain a universe in which can occur a superposition of a live cat and a dead one. No such superposition is ever observed. However, the theory adequately explains this fact: if an observer were to look at the cat, then the Heisenberg equation would entail a splitting of the universe into a superposition of two parts, one having an observer who has perceived a dead cat, the other having an observer who has perceived a live cat. By virtue of the superposition principle, these two macroscopic parts will subsequently evolve essentially independently: for all practical purposes each part will evolve as if the other part did not exist. Thus the brain processes of the observers in the two parts of the superposition would evolve independently, and it is natural to suppose that the two realms of conscious experience would likewise separate.

To accommodate the existing data, it is not necessary to accept such a radical ontology. One can retain the idea that the universe is represented by a single function $P(x,v,t)$, but modify the Heisenberg equation in a way such that virtually nothing is changed at the microscopic level, yet only a single world of the kind we experience emerges at the macroscopic level (Ghirardi et al. 1986; Pearle 1986a, 1986b). The modified Heisenberg equation is designed to produce, dynamically, continual "collapses" of the probability distributions in a way that causes any macroscopic object, such as a pointer on a measuring device (or a cat) to be limited always to a reasonably well-defined position. Thus a superposition consisting of one part in which the pointer has swung to the right and another part in which the pointer has swung to the left (or one part in which the cat has fallen down dead and another part in which it is standing up alive) is excluded; the dynamics forces the system into one state or the other. Suitable collapse mechanisms can be introduced in an infinitude of ways. The ones so far proposed could, on the one hand, be viewed as *ad hoc* mutilations designed to force ontology to kneel to prejudice. On the other hand, these proposals show that one can certainly erect a coherent quantum ontology that generally conforms to ordinary ideas at the macroscopic level.

These collapse ontologies have, however, one feature that conflicts with normal ideas of physicists: they exhibit explicit faster-than-light influences. The simplest way to see this is to consider the analogy to the classical case. When a measurement is performed that gives us information about a system in one region, our information pertaining to other regions may also change. The simplest case is that in which a particle originally known to be either in a first region or a second region is found to be in the first. Then the probability that the particle is in the second region suddenly drops to zero. In the classical case, this sudden change of the probability function in the second region, as a consequence of information obtained by a measurement performed in the first region, does not qualify as an "influence" upon anything in the second region: nothing in the second region has changed; only our information about it has changed. On the other hand, if we were to elevate the classical probability distribution $P(x,v,t)$ to a description of reality itself then the *reality* in the second region would change as a consequence of the measurement performed in the first region, and a faster-than-light influence would appear.

Because of the similarities between the quantum functions $P(x,v,t)$ and their classical counterparts and the obvious nonsense at the classical level of thinking that the collapse of the probability distribution involves any sort of action-at-a-distance, orthodox physicists view with skepticism any realistic interpretation of the quantum mechanical probability distribution $P(x,v,t)$ that entails real faster-than-light influences in connection with what would normally be regarded as a mere change in knowledge.

A third proposed ontology is the pilot-wave ontology (Bohm 1952;

Bohm, Hiley, and Kaloyerou 1987). Here the proposal is to represent the universe by the combination of a single function $P(x,v,t)$ that evolves according to the Heisenberg equation, and a single trajectory in the $3n$ dimensional space of the variable x. This trajectory represents a real world of classical type evolving in the course of time.

The function $P(x,v,t)$ is regarded as a kind of "objective information": it "informs" (controls) the motion of the classical-type world. More specifically, the motion of the moving "world-point" $x(t)$ is controlled by the quantity $P(x,v,t)$ near $x(t)$. Consequently this model also involves explicit faster-than-light influences. They occur because the force at time t upon particle 1 depends upon the locations of all other particles in the universe at time t. Thus if one introduces an external magnetic field in the region where particle 2 is moving, in order to represent a measurement that we perform on particle 2, then variations in this field in this region where particle 2 is moving can instantly affect, in a very large way, the motion of a very far-away particle 1.

Although pilot-wave models involve strong, instantaneous, action-at-a-distance, they will, if built upon a relativistic quantum field theory (assumed to exist), automatically coincide with that relativistic theory at the level of predictions about observations; the instantaneous action-at-a-distance present at the ontological level is not directly detectable at the level of observation. This occurrence at the ontological level of faster-than-light influences that cannot be directly detected at the observational level suggests to orthodox physicists that the ontology is faulty.

Every known comprehensive quantum ontology is, I believe, essentially of one of the three kinds described above, and is thus either of the many-worlds type or involves faster-than-light influences. On the other hand, each of these ontologies ascribes ontological status to $P(x,v,t)$, or to some similar structure. It is thus natural to suspect that the difficulties with these ontologies stems precisely from the fact that they ascribe reality to a quantity like $P(x,v,t)$ that is more akin to knowledge about reality than to reality itself.

This suspicion generates a basic question: Is the need for either many-worlds features or faster-than-light-influences deducible directly from the structure of the *predictions* of quantum theory itself, without reference to the machinery from which the predictions are derived? This question is completely in line with the orthodox view that only the probabilities $P(A;B)$ correspond to reality; that the machinery is nothing but a tool. The question thus posed is whether one can deduce just from an examination of the probabilities $P(A;B)$ provided by quantum theory, with no reference to the machinery from which they are derived, the conclusion that any ontology compatible with these probabilities must exhibit either many-world features or faster-than-light influences.

In connection with this program, a question was raised at the Notre Dame conference that I had no opportunity to answer: Why do I insist on talking about ''influences'' rather than using less contentious concepts such as *nonseparability* and *wholeness,* which are more explicitly exhibited by the quantum formalism. One answer is that I wish specifically to avoid all concepts linked to the formalism of quantum theory. A deeper answer is that we are faced with basic questions about the nature of the physical world. The most critical aspect of our present lack of understanding lies precisely in the general area of nonlocal connections. If we are going to discover nature's form, it is probably imperative to drive as hard as possible at the critical points of mystery. This means formulating and establishing the *most* incisive possible rigorous results about nonlocal connections, rather than settling for some observations about the formalism.

My use of the phrase 'faster-than-light' is not intended to suggest or imply the idea that anything *propagates* or *travels* faster than light. The phrase signifies only that the effects under consideration occur at points lying outside the forward light cones of their respective causes. Far more likely, in my opinion, than anything actually *traveling* faster than light is the likelihood of an intrinsic connectedness of Nature that is alien to the classical notion that spatial separation entails intrinsic separation.

One further point: I do not regard these considerations as mere idle philosophical speculations, never to be tested in the realm of science. Quantum theory was designed to deal with questions pertaining to atomic systems, and the information about them obtainable from macroscopic measuring devices. Serious conceptual problems arise if one attempts to extend the theory as formulated for atomic systems to systems of sizes intermediate to atoms and macroscopic devices, or to cosmological systems such as the early universe. Consequently, quantum theory, like all of its predecessors is probably not the final theory of everything: it is more like the tip of an iceberg. The present studies are, in my view, part of the difficult but important task of constructing a physical theory that encompasses quantum theory and goes beyond.

2. *EPR- and Bell-type arguments*

The argument to be given here is a development of an argument due to John Bell, which is itself a development of an argument due to Einstein, Podolsky, and Rosen (EPR). To make clear the important differences between my argument and those of Bell and others, I begin with a review of the logical structure of the earlier works.

Einstein believed quantum theory to be incomplete, and in 1935 he, together with Podolsky and Rosen, constructed an argument designed to prove this incompleteness. Their argument was later simplified by David Bohm (1951; Bohm and Aharonov 1957), who replaced the *gedanken* experiments considered by EPR with an experimentally feasible spin-correlation experiment. When I refer to EPR, I mean always the Bohm version of the EPR argument.

The EPR argument is based on two premises. The first, called *LOC*, asserts that no influence of any kind can propagate faster than light. The second, called *QT*, asserts that the predictions of quantum theory for the spin-correlation measurements under consideration are valid. In 1964 John Bell showed, in essence, that these two assumptions are mutually incompatible. This result effectively destroys the EPR argument. But more importantly it shows that if the two premises can be formulated in a way that allows the EPR-Bell argument to be rigorously carried through then *QT* implies that *LOC* is false: $QT \Rightarrow \sim LOC$.

The EPR argument was designed to show, more precisely, that quantum theory cannot provide a complete description of "physical reality." Accordingly, the EPR formulation of the no-faster-than-light-influence premise was necessarily intertwined with a concept of "physical reality." This intertwining obscures the significance of Bell's result. For, accepting the validity of the predictions of quantum theory, one does not know whether the implied failure of *LOC* is due to an actual occurrence of faster-than-light influences, or merely to a breakdown of the intertwined reality concept. Thus the conclusions that follow from the original EPR-Bell argument are usually referred to as a failure of "local realism": the original form of the argument does not entail the existence of faster-than-light influences. The program of the present work is to formulate an assumption *LOC* that, on the one hand, is a pure expression of the idea that there is no-faster-than-light influence of any kind, and, on the other hand, is rigorously incompatible with *QT*. This result is to be proved in a framework defined by two other assumptions, called *Free Choices* and *Unique Results (UR)*.

The first of these assumptions asserts that the choices of the two experimenters can, in the present context, be treated as two independent free variables. The second assumption asserts that, for any of the possible localized measurements *M* under consideration here, *if M* is performed, *then* there is a single value *x* such that the unique result of *M* is *x*. This latter assumption rules out many-worlds ontologies.

The key operative concept of the EPR-Bell argument is a property I here call *counterfactual definiteness*. It is defined in connection with a possible measurement *M*. The definition is as follows: *CFD(M)*: There exists a value *x* such that if measurement *M* were to be performed, then the result of that measurement would definitely be *x*.

Notice that this property is not equivalent to the UR assumption mentioned above: $CFD(M)$ asserts the existence of a definite value x even if the measurement M is not performed, whereas UR asserts the existence of a single result x only if the corresponding measurement is performed.

In the present paper, I use counterfactual definiteness always in this strong sense, which is to be distinguished from, and placed in opposition to, UR. No satisfactory derivation of nonlocality, or the existence of faster-than-light influences, can be based upon such a CFD assumption: a failure of this assumption is (at least in my opinion) far more likely than the existence of a faster-than-light influence.

In the literature, the term counterfactual definiteness is applied sometimes to the strong and objectionable CFD concept described above, which contradicts orthodox quantum thinking, and sometimes to the property UR, which does not . It might be reasonable to use the terms strong CFD and weak CFD for these two different properties, but in the present work I use CFD to denote the "strong" property that is definitely contrary to orthodox quantum thinking.

The EPR argument can be formulated in terms of CFD, instead of "physical reality." The argument has three steps. In the first step one proves, for any specified pair of incompatible measurements M_1^B and M_2^B that might be performed in a space-time region R_B, the following result:

$$QT + LOC_1 \Rightarrow CFD(M_1^B) + CFD(M_2^B). \tag{1}$$

This says that the conjunction of QT and an assumption called LOC_1, which will be defined presently, entails the conjunction of the two CFD properties. The second step of the argument is the observation that

$$CFD(M_1^B) + CFD(M_2^B) \Rightarrow QTInc. \tag{2}$$

This says that the conjunction of the two CFD conditions entails that the quantum theoretical description is not a complete description of physical reality. The third step is to combine (1) and (2). This gives the EPR result

$$QT + LOC_1 \Rightarrow QTInc. \tag{3}$$

To prove (1), a pair of measurements M_1^A and M_2^A one of which might be performed in R_A is considered. Their alignments are chosen so that if M_1^A and M_1^B are both performed, then QT predicts that some common value x_1 will appear for the result of each, and similarly for M_2^A and M_2^B. Consequently, if M_1^A is performed, then the appearing result x_1 is the value x required to make $CDF(M_1^B)$ true. On the other hand, if M_2^A is performed, then the appearing result x_2 is the result x required to make $CFD(M_2^B)$ true.

The condition LOC_1 is defined as follows: "For either value of i, the truth of $CFD(M_i^B)$ cannot depend upon which of the two alternative possible measurements is performed in R_A. The analog with A and B interchanged also holds.'' Since R_B is spacelike separated from R_A, this property LOC_1 appears to be an immediate consequence of the idea that there is no faster-than-light influence of any kind. We thus have the following situation: QT implies that if M_1^A is performed then $CFD(M_1^B)$ is true, and that if M_2^A is performed then CFD (M_2^B) is true. But LOC_1 says that the truth of these CFD properties cannot depend upon which measurement is performed in R_A. Thus result (1) follows.

Result (2) is an immediate consequence of the fact that quantum theory cannot represent conjunctively well-defined values for two incompatible measurements. Combining (1) and (2) one obtains the EPR result (3).

Bohr (1935) defended the completeness of quantum theory by rejecting the EPR locality assumption. Being unwilling to countenance the possibility of a mechanical influence of the required kind, Bohr introduced the curious idea of an epistemological disturbance. The opportunity for this was provided by the occurrence in the original EPR argument of the concept of ''physical reality'': the original EPR locality condition was formulated as the requirement that the ''physical reality'' in R_B cannot depend upon what is done in R_A. But according to Bohr the whole idea of ''physical reality'' must be radically revised. In the context of quantum thinking, the ''physical reality'' associated with a physical system must be tied to what can be *predicted* about that system. But what can be predicted about the system in R_B depends on what we do in R_A. Thus the ''physical reality'' associated with R_B depends on what we do in R_A. Consequently, the EPR locality condition must, according to Bohr, be rejected.

This strained interpretation of physical reality raises a legitimate concern that Bohr's answer may be a palliative that cloaks an unexpected feature of nature, rather than exposing it. Bohr's rebuttal is less easily applied to the CFD form of the EPR argument.

Bell examined the consequences of accepting the EPR assumption, and, in particular, their first conclusion:

$$QT + LOC_1 \Rightarrow CFD(M_1^B) + CFD(M_2^B),$$

and the similar:

$$QT + LOC_1 \Rightarrow CFD(M_1^A) + CFD(M_2^A).$$

These two result combine to give:

$$QT + LOC_1 \Rightarrow CFD(M_1^A) + CFD(M_2^A) \\ + CFD(M_1^B) + CFD(M_2^B). \tag{4}$$

According to the right-hand side of (4), the results that each of the four alternative possible combinations of measurements would give, if they were to be performed, are well-defined numbers. Under this condition, we can formulate a further (but very closely related) consequence of the idea that there is no faster-than-light influence of any kind:

LOC_2: For each of the two values of i, the result that measurement M_i^A would have, if it were to be performed, is independent of which of the two possible measurements M_j^B is performed in R_B, and vice versa.

Applying this condition, we can write the result that would appear in R_A if M_i^A were to be performed as x_i, independent of j, and the result that would appear in R_B if M_j^B were to be performed as y_j, independent of i. The central arithmetical point of Bell's work is that if in the spin-correlation experiment (with initial spin-singlet state) we select an appropriate quartet of possible measurements M_1^A, M_2^A, M_1^B, and M_2^B, then the assumption that the results x_i are independent of j, and that the results y_j are independent of i, is incompatible with the predictions of quantum theory. This result is expressed by

$$CFD(M_1^A) + CFD(M_2^A) + CFD(M_1^B)$$
$$+ CFD(M_2^B) + LOC_2 \Rightarrow \sim QT. \tag{5}$$

If the conjunction of the two no-faster-than-light-influence conditions is denoted as LOC_3,

$$LOC_1 + LOC_2 \equiv LOC_3, \tag{6}$$

then (4) and (5) combine to give

$$QT + LOC_3 \Rightarrow \sim QT, \tag{7}$$

or equivalently,

$$QT \Rightarrow \sim LOC_3. \tag{8}$$

This is the *CFD* version of the EPR-Bell result.

3. Eliminating CFD

The occurrence of *CFD* in the above proof is a carry-over from the original program of EPR, which was to prove *CFD* as a step in the proof of the incompleteness of quantum theory. Our present aim is totally different from that of EPR: we are trying to examine the validity of the no-faster-than-light-influence premise of EPR, not trying to prove the incompleteness of quantum theory. Since the idea of no-faster-than-light-influence is not intrinsically tied to the concept that nature assigns values to the results of unperformed quantum measurements, it would seem more satisfactory to eliminate completely this probably false concept.

In philosophy and theoretical physics we deal with theories about the world, not with the world itself. Theories should be tested, and quantum theory has been tested extensively. Thus we take the predictions of quantum theory as our data, and ask how the requirement of compatibility with these data affects ideas about nature. Our data are, more specifically, the predictions about the correlation parameter. There is, first of all, a prediction for the average measured value of this parameter. This predicted average value depends upon the alignments of the two devices. Moreover, as n, the number of pairs of correlated signals in the two devices, increases to infinity, the predicted dispersion of the measured values about the predicted average value of the correlation parameter shrinks to zero. Thus we may define an "allowed" domain consisting of values of the correlation parameter that are within, say, 0.01 of the predicted average value. Then for any positive number ε, no matter how small, one can take n large enough so that the total predicted probability that the measured result will lie outside the allowed region is smaller than ε.

Quantum theory predicts only probabilities, and hence, in some strict sense, makes no physical predictions at all: every prediction refers in some strict sense only to an infinite-n limit that can never be physically realized. This problem lies at the root of the controversies that rage in the foundations of probability theory. To cope with this basic problem with probabilities, I adopt the common strategy of setting to zero some "sufficiently small" probabilities. In particular, I shall consider the domain of the correlation parameter that lies outside the "allowed" domain described above to be *rigorously excluded*. The "sufficiently small" probability that is thereby replaced by zero can be made arbitrarily small by taking n sufficiently large.

Another way to describe this tactic is to say that we are trying to find what properties can be deduced about nature if it is required to be compatible with this slightly distorted version of quantum theory. This slightly distorted version is for all practical purposes equivalent to quantum theory, and can be made arbitrarily close to quantum theory (in terms of these predicted probabilities) by taking n sufficiently large. By this tactic, I have introduced a certain weak element of determinism into the analysis: the values of the individual results are not fixed deterministically, but the correlation parameters are "determined" to lie inside the allowed domains. This restriction is the basis for the introduction in what follows of the concept *could* in place of *would*.

In the situation under consideration, M_1^A and M_2^A are two alternative possible measurements, one or the other of which will be performed in R_A, and M_1^B and M_2^B are two alternative possible measurements, one or the other of which will be performed in R_B. The general theoretical framework is based on the assumption:

Unique Results: For each value of i, if M_i^A were to be performed *then* some single result x would appear in R_A; *and*, for each value of j, if M_j^B were to be performed, *then* some single result y would appear in R_B.

We are considering a general theory, or idea about nature, that is required to be at least approximately compatible with the predictions of quantum theory for the four spin-correlation experiments under consideration. This is all we know about the theory. We have no information about its form or inner structure, but want to determine whether it is compatible with the idea that there are no-faster-than-light influences of any kind. The *QT* requirement on our general theory, or idea of nature, is that the result appearing under any condition we might set up must conform to the (slightly modified) prediction of quantum theory, least approximately (say to within 0.01 in the correlation parameter). Thus the *QT* requirement is:

QT: For any pair (i,j), if (M_i^A, M_j^B) were to be performed, then the appearing pair (x,y) must be an "allowed" pair: i.e., the appearing value $c(x,y)$ of the correlation parameter can differ from the mean value $\bar{c}_{ij}$ predicted by quantum theory by no more than 0.02.

Our objective is now to show that these two assumptions cannot be reconciled with the idea that no influence of any kind can act outside the forward light cone (i.e., propagate faster than light).

Before proceeding to the demonstration it may be helpful to describe the key problems that have to be dealt with. The first is to specify the meaning of 'no-faster-than-light influence'. In the argument of EPR, the concept *physical reality* was introduced, and it was the "physical reality" that was not influenced. But that approach uses the concept *physical reality*, which we have not introduced. Moreover, even if we can specify in a satisfactory way exactly what it is that is not "influenced" we must still specify in a nonprejudicial way (and without using the ideas of determinism or *CFD*) what it means for that thing to be "uninfluenced" by something else.

The second problem is to cope with the fact that the proof requires one to deal "simultaneously" with all four *alternative* possible measurements, only one of which can be performed. But our assumption *UR* asserts that a single result is specified only if the associated measurement is performed. Since only one of the four alternative possible measurements can be performed, it might seem that we lack the logical ingredients needed to bring all four measurements simultaneously into consideration. Stated differently, if we could use the concept of "what *would* happen" in all four cases, then we could make direct use of the earlier arguments. But we have only the weaker requirements on "what could happen," and it is perhaps not an altogether trivial matter to show that these weaker requirements are sufficient.

A third problem concerns the sensitivity of the argument to a "small" logical detail, namely the location or position of the conditional clause "if measurement M is performed." In particular, there is a crucial logical difference between the following two statements:

(1) If M is performed, then there is an x such that the result of M is x, and

(2) There is an x such that if M is performed, then the result of M is x.

A replacement of (1) by (2) would be a logical error tantamount to the introduction of CFD. Given the fact that the introduction of CFD would allow us to tie into the earlier proof, there is an *a priori* danger of committing this logical error. Indeed, Paul Teller has suggested that any proposed proof of the kind I am setting forth must contain such a logical error, since only in this way can one tie the argument into the earlier proof, which required CFD. The presentation to be given here displays the fine logical details in order to show explicitly that no logical error of this kind has been committed.

The ensuing argument is based on logic, not physics, and the laws of logic must be respected. Thus the conjunction: (If X then Y) and (If not-X, then not-Y) which means that X is equivalent to Y, does not imply: (Y and not-Y), which is a contradiction. Similarly, the conjunction: (If $i = 1$, then there exists an x_1 with property P_1) and (If $i = 2$ then there exists an x_2 with property P_2) does not allow one to conclude that (There exists an x_1 with property P_1) and (There exists an x_2 with property P_2).

On the other hand, it is correct to infer (inference 1) from (There is an x_1 such that if $i = 1$ then x_1 *could* appear) and (There is an x_2 such that if $i = 2$ then x_2 *could* appear) that (There is a pair (x_1,x_2) such that if $i = 1$ then x_1 *could* appear and if $i = 2$ then x_2 *could* appear). Also, one can infer (inference 2) from (If $i = a$, then it is false that there exists no x_1 such that if $i = a$ then property P holds) that (There exists an x_1 such that if $i = a$ then property P holds). This may be proved by considering separately the two possibilities $i = a$ and $i \neq a$. Here, and in what follows, the condition (If $i = 1$) is equivalent to (If M_1^A were to be performed), etc. The two inferences described above are used in the derivation of property A described below.

Quantum theory makes predictions for each of the four cases, and our general theory is required to be compatible with all four conditions. This demand is incorporated into the logical framework by considering the variables i and j to be independent free variables: each is allowed to range freely over its two allowed values, and all conditions are required to hold for each of the four allowed combinations of the pair (i,j). Thus we obtain conditions for $i = 1$ *and* conditions for $i = 2$. Of course, one cannot simultaneously have $i = 1$ and $i = 2$. That is a contradiction. Rigorous attention to this fact is required to ensure that no passage is made from conjunctions of properties or requirements that hold only under mutually exclusive conditions to a simple conjunction of these properties. That would be the logical error mentioned above, and would introduce, effectively, an assumption similar to *CFD*, which is exactly what must be avoided.

The condition that there be no faster-than-light influence of any kind is formulated as follows:

> LOC: For all that pertains to the generation of the result of a measurement performed in R_A, the choice made by the experimenter in R_B can be considered not to exist, *and* for all that pertains to the generation of the result of a measurement performed in R_B, the choice made by the experimenter in R_A can be considered not to exist. This condition LOC is a conjunction of two conditions, which are called the first and second parts of LOC. They will be symbolized as LOC_A and LOC_B.

Condition LOC is always applied in conjunction with the following rule of inference:

> RI: If a condition X can be satisfied *only* by having something depend nontrivially upon something that does not exist, then that condition X cannot be satisfied: nothing can be *forced* to depend upon something that does not exist. That would be a contradiction in terms.

4.   *Proof that LOC is incompatible with QT*

LOC entails the following two properties:

Property A 1: There must exist a value x_1 such that: (1), if (M_1^A, M_1^B) were to be performed, then the result appearing in R_A *could* be x_1, and (2), if (M_1^A, M_2^B) were to be performed, then the result appearing in R_A *could* be x_1.

Property A 2: There must exist a value x_2 such that: (1) if (M_2^A, M_1^B) were to be performed, then the result appearing in R_A *could* be x_2, and (2) if (M_2^A, M_2^B) were to be performed, then the result appearing in R_A *could* be x_2.

Property A 1 is proved in the following way: Suppose there were no x_1 satisfying the two conditions imposed in property A 1. Then if M_1^A were to be performed, the result appearing in R_A would be *forced* to depend upon a choice that by virtue of *LOC* can be considered not to exist. But then *RI* implies that the supposition made at the beginning of this paragraph cannot be satisfied. Then inference 2 of Section 3 yields property A 1. Property A 2 is proved in a similar way.

Note: It is essential to use 'could', not 'would', in properties A 1 and A 2. Using the latter would involve determinism or CFD.

The properties A 1 and A 2 combine, with the aid of an inference similar to inference 1 of section 3, into:

Property A: There must exist a pair (x_1, x_2) such that, for each value of i, if (M_i^A, M_1^B) were to be performed, then the result appearing in R_A *could* be x_i, and if (M_i^A, M_2^B) were to be performed, then the result appearing in R_A *could* be x_i.

There is the similar:

Property B: There must exist a pair (y_1, y_2) such that, for each value of j, if (M_1^A, M_j^B) were to be performed then the result appearing in R_B *could* be y_j, and if (M_2^A, M_j^B) were to be performed then the result appearing in R_B *could* be y_j.

Property A can be stated in the following form:

There is a pair (x_1, x_2) such that: (1) if $i = 1$, then nature *could* produce in R_A the value x_1, independently of the value of j; and (2) if $i = 2$, then nature *could* produce in R_B the value x_2, independently of j.

This property can be restated in the following form:

Property A': There exists a pair (x_1, x_2) such that nature *could*, if $i = 1$, produce in R_A the result x_1, independently of j, and, if $i = 2$, produce in R_A the result x_2, independently of j.

There is an analogous property B'. Since both of these properties follow from *LOC* we may conclude that *LOC* implies:

Property A' + B': There exists a quartet (x_1, x_2, y_1, y_2) such that:
(1) Nature *could*, if $i = 1$, produce in R_A the result x_1, independently of j, and, if $i = 2$, produce in R_A the result x_2, independently of j; and
(2) Nature *could*, if $j = 1$, produce in R_B the result y_1, independently of i, and, if $j = 2$, produce in R_B the result y_2, independently of i.

Part (1) of this property $A' + B'$ comes from the first part of *LOC*, taken by itself. Part (2) of property $A' + B'$ comes from the second part of *LOC*, taken by itself. That is, part (1) comes from the condition of no influence from R_B to R_A, with influence from R_A to R_B allowed, and part (2) comes from the condition of no influence from R_A to R_B, with influence from R_B to R_A allowed.

It is obvious that no conflict with quantum theory can arise from property $A' + B'$ since quantum theory is certainly compatible with each of the two parts of *LOC* taken by itself. This can be seen by noting that the predictions of quantum theory can be explained either by a model with collapses in R_A and R_B induced by the measurement in R_A or by a model with collapses in R_A and R_B induced by the measurement in R_B. Each model satisfies one part of *LOC* but not the other. *LOC* imposes, however, the more incisive *conjunctive* requirement of no influence in either direction. From this conjunctive requirement we would like to derive the *conjunctive* form of property $A' + B'$, namely:

> **Property C:** There exists a quartet (x_1, x_2, y_1, y_2) such that nature *could*:
> (1) if $i = 1$ produce in R_A the result x_1, independently of j, and if $i = 2$ produce in R_A the result x_2, independently of j; and
> (2) if $j = 1$ produce in R_B the result y_1, independently of i, and if $j = 2$ produce in R_B the result y_2, independently of i.

The first part of *LOC* implies that condition (1) of property C holds: if this property were to fail then, for one or the other of the conditions that might be set up in R_A, the result appearing under those conditions would be *forced* to depend upon a choice that according to the first part of *LOC* can, in this context, be considered not to exist. Hence the first part of *LOC*, taken by itself, entails that nature must be *allowed*, under either condition that might be set up in R_A, to produce a result in R_A that is independent of j.

Spelled out in detail this condition (1) of property C demands the existence of a set $(x_1, x_2, y_{11}, y_{12}, y_{21}, y_{22})$ such that the following four conditions are all satisfied:

1. If $(i,j) = (1,1)$, then the appearing pair (x,y) *could* be (x_1, y_{11}).
2. If $(i,j) = (2,1)$, then the appearing pair (x,y) *could* be (x_2, y_{21}).
3. If $(i,j) = (1,2)$, then the appearing pair (x,y) *could* be (x_1, y_{12}).
4. If $(i,j) = (2,2)$, then the appearing pair (x,y) *could* be (x_2, y_{22}).

The second part of *LOC* entails condition (2) of property C, which demands the existence of a set $(x_{11}, x_{12}, x_{21}, x_{22}, y_1, y_2)$ such that the following four conditions are all satisfied:

1. If $(i,j) = (1,1)$, then the appearing pair (x,y) *could* be (x_{11}, y_1).
2. If $(i,j) = (2,1)$, then the appearing pair (x,y) *could* be (x_{21}, y_1).
3. If $(i,j) = (1,2)$, then the appearing pair (x,y) *could* be (x_{12}, y_2).
4. If $(i,j) = (2,2)$, then the appearing pair (x,y) *could* be (x_{22}, y_2).

These two separate parts of property C arise from the two parts of *LOC* taken separately. However, *LOC* imposes its two subconditions conjunctively: it demands that both conditions be satisfied together, or simultaneously.

Now our assumption *UR* requires, for each of the four possible values of (i,j), that if (M_i^A, M_j^B) is performed then some single pair (x,y) must appear. If several conditions are imposed on this pair (x,y), then they are imposed upon a single pair (x,y): this single pair cannot be two different pairs. Thus, if the two parts of LOC are imposed simultaneously, or together, then there must exist a pair of sextets, $(x_1,x_2,y_{11},y_{12},y_{21},y_{22})$ and $(x_{11},x_{12},x_{21},x_{22},y_1,y_2)$, such that the following four conditions are all satisfied:

1. If $(i,j) = (1,1)$, then the appearing pair (x,y) *could* be (x_{11},y_1) and (x_1,y_{11}).
2. If $(i,j) = (2,1)$, then the appearing pair (x,y) *could* be (x_{21},y_1) and (x_2,y_{21}).
3. If $(i,j) = (1,2)$, then the appearing pair (x,y) *could* be (x_{12},y_2) and (x_1,y_{12}).
4. If $(i,j) = (2,2)$, then the appearing pair (x,y) *could* be (x_{22},y_2) and (x_2,y_{22}).

This condition is equivalent to property C.

To eliminate any possible ambiguity arising from a misinterpretation of the word 'could' that appears in the foregoing proofs I give here also an equivalent set-theoretic formulation of the proof that LOC implies property C. Let A_{ij} be the set of possible results (x_{ij},y_{ij}) allowed (by a theory) in case the measurement (M_i^A,M_j^B) is performed. Let $A \equiv A_{11} \otimes A_{21} \otimes A_{12} \otimes A_{22}$. Then condition (1) of property C, which arises from LOC_A alone, asserts that $A \cap L_1$ is nonempty, where L_1 is the set defined by the conditions $x_{11} = x_{12}$ and $x_{21} = x_{22}$. Similarly, condition (2) of property C, which arises from LOC_B alone, asserts that $A \cap L_2$ is nonempty, where L_2 is defined by the conditions $y_{11} = y_{21}$ and $y_{12} = y_{22}$. The conjunction of conditions LOC_A and LOC_B is represented by the intersection of the corresponding set-theoretic conditions L_1 and L_2, and demands that $A \cap (L_1 \cap L_2)$ be nonempty. This is property C.

To complete the proof of the incompatibility of LOC with QT, it need only be shown that property C implies the failure of QT. This is easily done (Stapp 1968, 1971, 1977, 1988b). The ·quantity x_1 is an ordered set of n integers $(x_{11},x_{12}, \ldots x_{1n})$, $x_{1i} = \pm 1$, and similarly for x_2, y_1, and y_2. If the two possible settings of the angles in R_A are $\theta_1^A = 0°$ and $\theta_2^A = 90°$, and the two possible settings of the angles in R_B are $\theta_1^B = 0°$ and $\theta_2^B = 45°$, then the predicted mean value

$$\frac{1}{n} \sum_{k=1}^{n} x_{ik}y_{jk} = \bar{c}_{ij} = -\cos(\theta_i^A - \theta_j^B),$$

gives for the angles chosen,

$$\frac{1}{n} \sum_{k=1}^{n} x_{1k}y_{1k} = -1, \tag{1}$$

$$\frac{1}{n} \sum_{k=1}^{n} x_{1k}y_{2k} = -1 / \sqrt{2}, \tag{2}$$

$$\frac{1}{n} \sum_{k=1}^{n} x_{2k}y_{1k} = 0, \tag{3}$$

and

$$\frac{1}{n} \sum_{k=1}^{n} x_{2k}y_{2k} = -1 / \sqrt{2}. \tag{4}$$

Neglecting, at first, the small departures from the mean values one obtains from (1) and the properties $|x_{1k}| = |y_{1k}| = 1$ the equation

$$x_{1k} = -y_{1k}. \tag{5}$$

Consider then the quantity:

$$\frac{1}{n} \sum_{k=1}^{n} (x_{1k} + x_{2k} + \sqrt{2}y_{2k})^2$$

$$= \frac{1}{n} \sum_{k=1}^{n} (x_{1k})^2 + \frac{1}{n} \sum_{k=1}^{n} (x_{2k})^2 + \frac{2}{n} \sum (y_{2k})^2$$

$$+ \frac{2}{n} \sum_{k=1}^{n} x_{1k}x_{2k} + \frac{2\sqrt{2}}{n} \sum_{k=1}^{n} x_{1k}y_{2k}$$

$$+ \frac{2\sqrt{2}}{n} \sum_{k=1}^{n} x_{2k}y_{2k}. \tag{6}$$

Using the above equations we obtain the value:

$$1 + 1 + 2 + 0 - 2 - 2 = 0. \tag{7}$$

On the other hand, the conditions $|x_{ir}| = 1$ entail that every term in:

$$\frac{1}{n} \sum_{k=1}^{n} (x_{1k} + x_{2k} + \sqrt{2}y_{2k})^2$$

is strictly positive, and that the sum is at least $(2 - \sqrt{2})^2$. This contradicts the result (7), which claims zero. This large discrepancy cannot be undone by considering the small allowed deviations from the equations (1), (2), (3), and (4). Thus property C is incompatible with QT, and we have therefore proved that $LOC \Rightarrow \sim QT$.

K. Kraus (1985) has challenged this conclusion. He showed quantum theory to be compatible with a certain condition, and claimed thereby to demonstrate the complete compatibility of quantum theory with the idea that influences propagate no faster than light. Kraus' condition did not involve the crucial quartet structure obtained above. He supplied no proof that his condition entails compatibility with no faster-than-light influences. His implicit claim that it does has been refuted by exhibiting a model that manifestly entails faster-than-light influences and satisfies his condition (Stapp 1988c).

M. Redhead (1987b) has objected to an earlier version of this argument on the grounds that the crucial Matching Condition, or quartet condition, comes from the objectionable *CFD* condition, which, he argues, is not derivable from a locality condition that he proposes. The present work is an answer to Redhead's objection. It shows that the crucial Matching Condition arises from a satisfactory locality assumption without ever assuming, deriving, or using in any way, the objectionable *CFD* assumption.

The argument given above does not use the concept *would*, which would bring in *CFD*: it uses, instead, *could*. The formulation of locality in terms of *could* has been meticulously described above. Briefly stated the essential point is that the (EPR) idea of locality demands that any local theory *must allow* the results that, according to *UR*, will, under appropriate conditions, appear in a region to be independent of which experiment is performed in the other region. This demand can be expressed still more briefly as the requirement that the results pertaining to either region *could* be independent of which experiment is performed in the other region. But the word 'could' must then be construed in a way that entails that the theory *must allow* the required independence. That is, 'could be independent' must be construed to mean 'is allowed by the theory to be independent'. This specification entails that 'could be' is not necessarily equivalent to 'might be'. In particular, the locality principle does not imply only that the results in one region *might* be independent of the experiments performed in the other region. This condition, as normally interpreted, does not demand that the theory *must allow* the required independence. Thus the passage from the 'could be' form to the 'might be' form of the locality principle is improper: the original meaning is lost. To protect against such improper reasoning, which might appear to destroy the entire line of argument, it is best to use, instead of the brief formulations of locality given in this remark, the meticulous formulation set forth earlier.

5. Connection with Bell's theorem

The quartets defined in property *C* play a central role in the proof given in section 4. Similar-looking quartets occur in Bell's deterministic hidden-variable theorem. However, the logical status of my quartets is completely different from those of Bell. His arise from holding a hidden variable λ fixed at some specified value, and varying the choices of the two experimenters. This generates a quartet of values that may be interpreted as the set of results that *would* be obtained under the various alternative possible experimental conditions if the hidden variable had the specified value. The existence of these Bell quartets thus depends on the assumed existence of some new entities, the hidden variables, and the occurrence of well-defined quartets of values expresses the idea that the results of unperformed experiments are specified within nature.

In view of these strong ontological and counterfactual presumptions, it is evident that Bell's original theorem, by itself, places in no jeopardy whatever the idea that influences propagate no faster than light. The natural interpretation of Bell's result is simply that these ontological and counterfactual presumptions are incorrect. This conclusion conforms completely with the views of orthodox quantum physicists, who had already rejected those ideas on other grounds.

The prior rejection of those ideas by quantum physicists was based, at least in part, on a remarkable feature of economy displayed by quantum theory. This feature of economy pertains to a close connection between our capacities at the physical level to perform alternative possible measurements conjunctively, and the capacity of the theory to represent the results of certain measurements conjunctively: the theory apparently is incapable of representing conjunctively the results of precisely those combinations of measurements that cannot be performed conjunctively. The theory has, therefore, a natural alignment with the physical situation, and economically provides no values simultaneously for quantities that are in principle not observable simultaneously.

This feature of economy lies at the heart of quantum theory, and the success of the theory is construed by quantum theorists as evidence that nature herself probably leaves unspecified the results of unperformed experiments to the extent that their representation is outside the capacity of quantum theory. Since the data incorporated into Bell's quartets is precisely the kind of information that exceeds in principle the capacity of quantum theory, it comes as no surprise to orthodox quantum physicists that hidden-variable models of this kind cannot conform to QT. The locality requirement on the models, though needed for the proofs, seems to be incidental.

When in the past I have referred to the profound implications of Bell's theorem, it was always in a context in which the theorem in question was essentially the one described in the preceding section. That theorem contains no reference either to hidden variables, or to the idea that nature might specify the results of unperformed quantum measurements. So no escape from the nonlocality conclusion is provided by a rejection of those ideas. Nor is there any demand made in the proof for an *explanation* of quantum correlations. And no assumptions are made about the nature of physical reality, beyond the demands contained in the assumptions that the choices of the experimenters can be regarded, in this context, as free variables, and, for each of the four possible measurements under consideration, that *if* that measurement were to be performed, *then* nature would specify a single result.

If these two assumptions are accepted, together with *QT*, then the no-faster-than-light-influence idea expressed by *LOC* must fail. These three assumptions are completely in line with orthodox quantum thinking. The only thing outside normal quantum theory is the concept *LOC*, which is, however, formulated without introducing any new ideas, and which expresses in a pure form the requirement that there be no faster-than-light influence of any kind.

DO CORRELATIONS NEED TO BE EXPLAINED?

ARTHUR FINE

1. The experiment

Consider a generic correlation experiment of the sort proposed by EPR (Einstein, Podolsky, and Rosen 1935), later refined by Bohm and Aharonov (1957), and that is the subject of the Bell theorem (1964). In the experiment, pairs of objects are produced from an on-line source in the same, briefly interactive (joint) state. (Hopefully without prejudice, we refer to the objects as "particles.") After emission the pairs separate, each particle in a pair moving off to a different, spatially separated wing of the experimental apparatus, the A wing and the B wing. In these wings, for the simple ("2x2") experiment that we discuss, each particle is measured in one of two ways: A_1 or A_2 in the A wing, and B_1 or B_2 in the B wing. Using the language of quantum theory we sometimes say that what is being measured are "observables", i.e., physical parameters, such as components of position or momentum in the same direction (the original experiment of EPR), or spin components in different directions in the plane transverse to the "path" of the particle (the Bohm-Aharonov experiment).

Suppose that the observables are sufficiently coarse-grained to yield just two possible outcomes for each measurement, which we code as "positive" ($+$) or "negative" ($-$). Suppose also that the different observables measured in a wing are incompatible (as they are in the instances above), so that no two measurements in either wing can be performed at the same time. A run of the experiment consists of selecting one observable to be measured in each wing, and then measuring the particles in a pair accordingly, as successive pairs come off the production line and enter the wings. So far as possible, we try to make the measurements on the same pair at the same time. For simplicity assume that each run consists of the same number of successive measurements. In a *correlation experiment,* all four possible runs are made; that is, A_1 is measured with B_1, A_1 is measured with B_2, A_2 is measured with B_1, and A_2

175

is measured with B_2. Of course the runs need not be completed as a whole unit, back to back. We could shift between measurements, moving to and fro on line, and then separate out the distinct runs after the fact (Aspect et al. 1982).

In each wing the data from successive measurements of the same observable consist of a sequence of pluses and minuses that, in a typical case, might look like this:

$$+ - + + - - + - + + - + + + - + - - + - + - + - -$$

In any run there will be roughly the same number of positive as of negative outcomes. Moreover, the interspersing of the positive with the negative outcomes occurs at random. Thus for any one of the measurements, the probability for a particular outcome (whether positive or negative) is ½. A typical run for a sequence of 25 pairs might show outcomes like this:

$$+ - + + - - + - + + + - - - + - + + + - - - + - +$$
$$+ - + - - - + + + + + - - + - - + + + - - - + - -$$

In this run, roughly half the outcomes of each measurement are positive and half negative. But in 40% of the pairs (i.e., in 10 out of the 25) a positive result of one measurement is accompanied by a positive result in the other. This is certainly a significant correlation between the positive outcomes, since if each particular outcome occurs at random (and so has probability ½), then one might expect that the odds for a positive outcome in both wings would be 1 in 4; that is, one might expect it to be just the product of the separate probabilities. This expectation, which amounts to looking for no correlation between the positive outcomes, would yield roughly 6 out of 25 positive pairs, not the 10 out of 25 displayed above. Typically, the runs in a correlation experiment yield significant correlations between positive outcomes. The correlations, just like the 50% chance of a positive for the individual outcomes, are stable and regular. Experiments show the same correlations in nearly all largish runs, when the same measurements are made. These stable correlations are grounded in the laws of the quantum theory, which predicts them accurately as functions of the common state in which the pairs are produced, and of the particular measurements in a run.

Below I will take up the question of whether one ought to be puzzled by these correlations in and of themselves, that is, whether the mere fact of correlations between sequences of randomly occurring events inherently calls for explanation. Not surprisingly, I will suggest that the answer depends on what attitude one takes toward explanation, and toward the reliable predictions of the quantum theory. Even those who may not find such correlations inherently puzzling, however, might still encounter some difficulty with the relationships that can obtain between the correlations arising in different runs of the experiment. This is the subject of the Bell inequalities, to which we now turn.

2. The Bell inequalities

Consider the correlations that arise in certain quantum correlation experiments. (The numbers below are the joint probabilities, rounded to one decimal place, for a spin experiment where A_1 corresponds to measuring the vertical spin component and A_2 the horizontal; where B_1 splits the 90° angle between the As, and B_2 is the component 135° away from A_2, and perpendicular to B_1.) Suppose that the run in which A_1 is measured with B_1 yields a correlation of $4/10$ between positive outcomes, as in the example in section 1 above. I will write it this way:

$$P(++/11) = 4/10.$$

Using the same conventions, suppose that the other runs yield:

$$P(||/12) = 4/10,$$
$$P(++/21) = 4/10,$$
$$P(++/22) = 1/10.$$

Thus all the runs, except for the one where A_2 is measured with B_2, regularly yield the same correlation of $4/10$ between the plus outcomes in the two wings, and this last run reliably turns up a correlation of $1/10$. Of course, the individual outcomes of any single measurement are still roughly half plus and half minus, as usual, which we could write (without having to pay attention to the particular run) as:

$$P(+) = 1/2.$$

As emphasized, the correlations that obtain between the positive outcomes for fixed measurements in the different wings are stable and reliable, depending only on the particular measurements performed. Although the two measurements in a single wing cannot both be done at the same time, we might still wonder whether that kind of stability of correlations between outcomes would continue to hold if we could conceive somehow of combining the outcomes of measurements in a single wing, at least hypothetically. We might, for example, pursue the following train of thought.

Suppose we were to perform the B_1 measurement in the B wing, say with A_1 being done in the A wing, then we know we would get some definite sequence of positive and negative B_1 outcomes (indeed, ones that overlap with the A_1 positives in 4 instances out of 10, on the average). But now suppose that, rather than B_1, we imagine measuring B_2 instead. Again, we would obtain some definite sequence of outcomes (whose positives would also overlap with those of A_1 an average of 4 instances out of 10). Surely, then, for the same, fixed population of pairs, we can at least imagine a string of hypothetical B_1 outcomes and a string of hypothetical B_2 outcomes. Of course, since the population of pairs is finite, whatever particular sequences of outcomes turns up and regardless of whether they are repeatable or not, there would be some definite overlap (or other) between the positive outcomes, pair by pair, had we measured B_1, and the positives had we measured B_2. Or so we must imagine. Indeed, we can say something quite definite about the possible overlap between the B positives.

We know that out of ten emitted pairs, on the average, we get five positives and five negatives in every measurement. Given the correlations assumed above we can see that there are constraints on the outcomes, as indicated below, where next to each measurement we list what an average run of ten would be like in a typical case.

$$A_1: \quad - \ + \ - \ + \ + \ + \ - \ - \ + \ -$$
$$B_1: \quad . \ X \ . \ X \ X \ X \ . \ . \ X \ .$$
$$B_2: \quad . \ X \ . \ X \ X \ X \ . \ . \ X \ .$$

Next to B_1, four of the Xs must be plus, and this is also true for four of the Xs next to B_2, in order that the correlations:

$$P(++/11) = P\ (++/12) = \tfrac{4}{10}$$

be satisfied. *This means that in an average run of ten, there must be at least three overlaps among positive outcomes between* B_1 *and* B_2, since there is no way to distribute four objects among five places twice over without at least three duplications occurring.

Another constraint on the possible overlap between the B positives arises if we go through a similar line of reasoning for an average run of ten, where it is A_2 that is measured in the A wing. In this situation we get, for example:

$$A_2: \quad + \ + \ - \ - \ - \ + \ - \ + \ + \ -$$
$$B_1: \quad X \ X \ . \ . \ . \ X \ . \ X \ X \ .$$
$$B_2: \quad X \ X \ . \ . \ . \ X \ . \ X \ X \ .$$

Next to B_1, four of the Xs must be plus, in order to satisfy the correlation

$$P(++/21) = \tfrac{4}{10}.$$

Next to B_2, exactly one of the Xs must be plus, in order that

$$P\ (++/22) = \tfrac{1}{10}.$$

So, among the Xs at most one plus occurs in common between B_1 and B_2. Outside the Xs there remains only one other plus next to B_1. Even if it overlaps with a plus next to B_2, that would contribute at most one additional plus in common. *Thus in an average run of ten there cannot be more than two overlaps between the positive B_1 and B_2 outcomes.* We saw above, however, that there must be at least *three* overlaps in order to satisfy the other two correlations!

This contradiction shows that the quantum correlations do not fit together in a way that would allow a stable distribution of outcomes between hypothetical measurements in the same wing. This, finally, is where the Bell inequalities come in. For satisfaction of those inequalities by the correlations generated in the four runs of a 2x2 experiment are the necessary and sufficient conditions for the correlations to be consistent with such a stable, hypothetical distribution of outcomes of incompatible measurements (Fine 1982b). The Bell inequalities simply require that if we take the correlations for any three of the runs, add them up, and then subtract the correlation generated in the remaining fourth run, we get a nonnegative number that is no bigger than 1. If we apply this rule to the correlations above, then one possibility is

$$4/10 + 4/10 + 4/10 - 1/10 - 11/10 > 1.$$

Thus, as we have seen, these correlations do not fit together according to the rule. I might just add that to derive the Bell inequalities one simply carries out the counting argument that produced the contradiction above, under the stipulation that the overlap between B_1 and B_2 when measured with A_1 be consistent with that overlap when measured with A_2. This is possible just in case the Bell inequalities hold.

The reasoning that produced the contradiction depends on treating incompatible measurements differently from how they are treated by the quantum theory. It depends on thinking that even where one cannot carry out two measurement-operations at the same time, one can still imagine one having been done rather than the other, comparing the hypothetical results, and projecting that comparison in a lawlike and stable way into a variety of different circumstances. The trouble, I would suggest, comes not from the exercise of our imagination in a counterfactual way, but from the lawlike projection of the hypothetical results of the comparison. Contrary to this particular exercise in fancy, the quantum theory itself does not contain the resources for describing the overlapping results of such hypothetical measurements in a lawlike way. That is, it does not allow the joint distribution of observables to be defined as a function of the state of a system (and so projected lawfully), unless that state is one for which the observables are actually compatible (Fine 1982a). Thus the imaginative exercise can be viewed as providing a good and appropriate test for whether the boundaries of the projectable, according to quantum theory, are properly drawn. Assuming that the quantum correlations, for instance those above, are correct for the experiments to which they relate, it seems that the contradiction in projecting the hypothetical overlap from one measurement context to another shows that the limits drawn by quantum theory as to what correlations are stable is quite appropriate.

To be sure, sometimes one can do better, for instance where the experimental correlations do satisfy the Bell inequalities; but not, it would seem, systematically. This is a nice and modest judgment. A theory that has proved itself extremely reliable experimentally turns out to set the boundaries for what joint probabilities can be projected from different finite runs just right. Thus those who may have been concerned with the connections among the correlations in the different runs of an experiment, and in particular with why they fail the Bell inequalities, may feel relieved. What these inequalities require is a lawlike projectibility of distributions for the outcomes of hypothetical and incompatible measurements, a projectability that well exceeds what the theory provides. Indeed the failure of the Bell inequalities is a nice demonstration that, with respect to joint distributions, in general the theory *cannot* provide, even in principle, what it *does not* provide. Moreover if the theoretical predictions are correct, nature does not provide for such projectability either.

Not everyone, however, is satisfied with these modest judgments. Instead many seem disposed to view the quantum correlations as inherently puzzling, and to see in their failure to satisfy the Bell inequalities a sign either of some new and quasi-mysterious physical process (maybe funny "influences" or odd "passions"), or an experimental demonstration of the end of the era of realist metaphysics. To be sure the idea of an experimental refutation of metaphysics is a charming (if oxymoronic) concept, and I for one would be happy to see realist metaphysics fade away. But I am afraid that the philosophical disposition in this case is grounded in the metaphysics of determinism, which seems to me no improvement, and that the puzzle over the correlations seems to require a similarly inflated, realist-style, essentialism in regard to explanations. We begin there.

3. *Explanations*

What surprises or puzzles is relative to context, which includes at least psychological set and background beliefs. Explanation is similar. Accordingly, what counts as an explanation can be expected to depend on particular features of the context of inquiry. Similar context dependency is to be expected in what calls for ("requires" or "needs") explanation. Moreover, there is no reason to presume any general or uniform concept of explanation (or of what "needs" an explanation) that necessarily cuts across and unifies the various contexts. If one holds to such a nonessentialist attitude, then the general question of whether correlations require explanation is not a particularly useful question to pursue, for one anticipates that the answer will only be: "Sometimes yes, and sometimes no."

Instead it would be good to keep an open mind on the issue of what needs explanation as it arises in the case of particular correlations set in the context of particular beliefs and frames of reference. Thus, the "natural" or zero state of mind with regard to the correlations between outcomes in different wings of a correlation experiment should be no different from what it is for the randomness of the outcomes in each wing separately; namely, unless features are present that point to the correlations (but not the randomness) as in need of explanation, they are not.

I think it is fair to say that this is not the usual attitude toward correlations. General treatments set them in an essentialist framework, one that takes the mere existence of correlations as calling, all by itself, for an explanation. The explanatory resources, moreover, are usually pretty sparse. Either correlations are taken to be coincidental (or "spurious," i.e., not genuine in the sense of not indicative of a "real" connection between the correlates) or they are taken as signs of an underlying causal relation. In the latter case, the relation can either be direct, with one of the correlates causally connected to the other; or indirect, where the correlation is mediated by a network of common causal factors; or some combination of the two.

The framework of common causal factors involves two distinct requirements. One is that the initial correlation be derived by averaging over the contribution to the joint outcome made by each of the causal factors separately. The other is the requirement that no residual correlations remain when each factor is held fixed; that is, that relative to each causal factor, the outcomes are stochastically independent, and so their joint probability is just the product of their separate probabilities relative to the factor.[1] This last assumption embodies the essentialist conception that correlations inherently need to be explained, so that it would not be proper to use some correlations to explain others. An essentialist would hold that such an "explanation" could not be complete.[2]

When essentialists exhaust their explanatory resources (that is, treat

---

[1] Following Reichenbach (1956), this requirement of conditional stochastic independence was called "screening-off." See Salmon (1984) and Suppes (1984). In the Bell literature, the interpretation of this condition was prejudiced by calling it "locality." Fine (1981, 1982b) suggested the term 'factorizability' in order to free up the discussion.

[2] I believe this is one of the motivations for Jarrett (1984) calling this requirement "completeness." One should not confuse this terminology with Einstein's, where he charges the quantum theory with descriptive incompleteness. See Fine (1986) for the several senses in which Einstein used this term.

correlations that seem to be neither coincidental nor causal), they find themselves in the position of holding that the correlations are mysterious, for they
need to be explained *and* they cannot be explained. Clearly something has to
give. They might retreat from essentialism by expanding the conception of a
common cause, admitting some basic residual correlations to be used in
explaining others. Or they might withdraw the initial presumption that the
correlations in question do require explanation. Thus without essentialism
there would be several degrees of freedom.

The only essentialist alternatives, however, are to reexamine the possibility of coincidence or to propose new causal connections to do the explanatory job. For the quantum-correlations experiments, these options are rather
constrained. The correlations arise between outcomes of measurements performed in spatially separated wings of the apparatus. We may suppose that the
measurement-events are spacelike, that is, they are not separated by time
enough to signal the results in one wing of the experiment to the other. Thus
there can be no direct influence between the outcomes, unless that influence is
conveyed between the wings with a speed faster than that of light. Neither is
there a network of common background causal factors to produce the correlations. For the existence of such common causes would imply that the correlational data satisfy the system of Bell inequalities, and we may suppose that
experiments have been chosen for which this is not the case.[3] The correlations
obtained in these experiments are grounded in the quantum theory. They are
stable, predictable and, by arranging for the appropriate measurements to be
made, controllable in advance. This sort of correlation could hardly be called
coincidental. Indeed, then, for an essentialist faced with the quantum correlations, something has to give.

One possibility would be to break with ordinary physics and to introduce the concept of superluminal influences. This would amount to inventing
new physics, hopefully integrating it with the old. Such a move certainly
cannot be ruled out *a priori,* but neither should it be accepted on that basis.
That is, one should be suspicious of the argument that since the quantum
correlations stand in need of explanation and since superluminal influences
provide the best (and perhaps even the only) explanation for them, therefore
one ought to accept the introduction of such influences. One does not need to
have reservations about the strategy of inference to the best explanation in
general, as I do, to be reserved about the soundness of this particular application. For we can break the argument at the very initial stage simply by
stepping way from the essentialist conception of explanation and the general

---

[3]A common-cause explanation, as outlined above, is what the foundational literature calls a
factorizable stochastic hidden-variables model. Fine (1982b) shows that there exists such a
stochastic model for the data of a correlation experiment if and only if there is a deterministic
model for the same data. For a 2x2 experiment, this is equivalent to the satisfaction of the Bell
inequalities, as discussed in section 2.

suspiciousness of correlations on which it rests. We might also wonder, even were physical speculation allowed to be driven simply by one's essentialist cravings, whether the speculation has to be quite so conservative. In particular, why stick to "influences" propagating in space-time? The algebraic and topological structures of recent string and supersymmetry theories surely provide rich resources for reconceptualizing the experiments, resources that do not involve the prerelativistic worldview of things simply moving faster than light. Who knows, these other ways might even connect with progressive physics?

4. Locality

Since the explanationist case for superluminal influences seems entirely *ad hoc* and *a priori* (Latin sins than which not much could be worse), is there perhaps some other and better way to make a case? There is this: The Bell inequalities follow from several different sets of assumptions. If we could derive them from the *denial* of faster-than-light influences, then the violation of the inequalities in the correlation experiments would entail the existence of the influences that had been denied. Let us call the desired premise, whose denial entails the existence of superluminal influences, LOC (for "locality"). Whatever the precise formulation of LOC might be, it is clearly a principle that denies certain kinds of influence (or dependence) between the outcomes of measurements in one wing of the experiment and what happens in the other wing. It follows that a principle denying *any* influence between happenings in different wings would imply LOC. Hence the Bell inequalities could be derived from such a strong principle, if they could be derived from LOC itself. Call such a strong principle SLOC. Were SLOC consistent with the denial of the Bell inequalities, then the inference from LOC to those inequalities would fail, and so would this case for superluminal influences. But at least the *relative* consistency of SLOC with the failure of the Bell inequalities is well known. It is consistency relative to the quantum theory. For SLOC is actually built into the quantum theory, according to which there *is* no influence between the two wings of the experiment, that is, no physical interaction of any sort that is represented by terms in the Hamiltonian of the composite system at the time one or the other component is measured. As Bohr (1935) emphasized in his response to EPR, "There is in a case like [EPR] no question of a . . . disturbance of the system under investigation during the last critical stage of the measurement procedure."[4] Of course,

---

[4] It is true that Bohr (1935) goes on to talk about "an influence on the very conditions which define the possible types of predictions regarding the future behavior of the system." The word 'influence' here could be misleading. For Bohr is not referring to what is at issue above; namely,

correlations violating the Bell inequalities are also built into the quantum theory, as is well known; hence the case for superluminal influences fails here just as it did previously.

Despite Bohr's authority, not everybody will be persuaded by this argument; although to me Bohr's strategy here of calling on our best theories and respecting what they say seems pretty sensible. Let me, however, try again, concentrating this time on how one is to understand the idea of no-influence, or independence, in the case of the correlations. Suppose we do an experimental run where we measure A_1 in the A wing and either B_1 or B_2 in the B wing. The no-influence idea is that whatever one does in the B wing makes no difference to what happens in the A wing. Now what does happen in the A wing? I want to conceive of it this way: the A_1 measurement is carried out and *in a perfectly random way* an outcome (either plus or minus) occurs. I emphasize the randomness, for if we conceive of the outcomes as predetermined (in the sense that for every measurement there is an outcome such that if we were to perform the measurement that outcome would result), then the Bell inequalities automatically govern the statistics of the experiment (well, given some other reasonable assumptions; see Halpin 1986).

For definiteness, suppose that the measurement in the A wing turns up plus, and that in fact B_1 is measured in the B wing. The no-influence idea of SLOC is that what happened in the B wing (i.e., the performance there of the B_1 measurement) did not influence what happened in the A wing (i.e., the performance there of the A_1 measurement, yielding the plus result). Let us just assume that there is no wing-A-measurement to wing-B-measurement influence. (We might think to insure this by making the choice of what measurement is being performed depend on random selections made separately in each wing. But one can readily see that, in the context of this discussion, relying on this to rule out influences just begs the question!) The possible dependency left over would be between the B_1 measurement itself and the outcome of the A_1 measurement. However, we have supposed that the A_1 outcome is random. Nothing determines that particular outcome, and there are no factors on which it depends. So randomness implies that the outcome does not depend on which measurement is performed in the B wing, not even on *whether* a measurement is performed there.

Thus if we adhere to the idea that the measurement results are truly random, and we rule out influences affecting which particular measurements are made in the wings, we automatically have a framework in which the

an ''influence'' from measurements performed in one wing that affects the particular outcomes in the other wing. Rather, Bohr is pointing to his positivist conception of predication (what it means to attribute a property to an object), and its material presuppositions involving measurement preparations. That is a wholly different topic.

strong locality assumption, SLOC, is satisfied. It remains to show that in such a framework the Bell inequalities can fail. But that is easy. All we require are four pairs of random sequences of pluses and minuses corresponding to the four measurement runs in a correlation experiment, that carry the four quantum joint probabilities for the outcomes. Such random sequences are obtainable from the data of any actual correlation experiment by discarding all the pairs of results where one or the other of the detectors failed to register a result. For experiments in which the Bell inequalities fail (like the one discussed in the first section), the sequences show the consistency of SLOC with that failure. Hence the attempt to derive the Bell inequalities from even weaker locality assumptions, like LOC, breaks down.

Several objections can be raised to this line of argument:

Objection (1). The argument contains an undischarged assumption: namely, that performing a measurement in one wing does not influence what measurement is performed in the other wing.

To take this objection seriously would be to open up the possibility of what John Bell (Davies and Brown 1986, p. 47) calls "superdeterminism." It is a skeptical hypothesis that, once opened, could not easily be laid to rest. But like other skeptical hypotheses, it would require a lot of work, I believe, to get it going. Briefly, in the absence of a theory of "influences" between measurements performed, which shows how to integrate them with ordinary experimental practice, and in the absence of specific reasons to entertain suspicions about the presence of such influences, I think we can pass them by. Merely skeptical doubts ought not to stand in the way of judgments based on otherwise sensible arguments.

Objection (2). The argument seems to equivocate on the term "random." Sequences of numbers (or pluses and minuses) can be said to be random in the technical sense of a table of random numbers. This is the sense used at the end of the argument where the data-set from correlation experiments is cited as being random. But earlier that term is used to designate measurement-outcomes that are not determined at all, and hence independent and uninfluenced. This is a different sense of the word.

True. The idea precisely was to treat sequences of experimental data, which are random in the sense of random numbers, as representing outcomes of measurements where nothing at all determines or influences any particular outcome. The idea is to impose an indeterminist framework on the quantum experiments in order to demonstrate that, within such a framework, the strongest locality assumptions are perfectly compatible with experimental results that violate the Bell inequalities. If that demonstration holds up, then it will follow that those who see locality at issue in the violation of those inequalities do so only on the basis of additional determinist assumptions or presuppositions.

Objection (3). Strong locality requires that the particular outcome in one wing not depend on what is being measured in the other wing. This implies that the very same outcome would have occurred in a given wing regardless of whether one or another measurement were carried out in the opposite wing. So the sequence of A_1 outcomes would be the same whether B_1 or B_2 were measured with it, according to strong locality. Similarly, the outcomes of A_2 measurements would not vary between different B wing measurements. Suppose that we measure A_1 with B_1 and then A_2 with B_2. The sequence of B_2 outcomes in the second measurement might very well have been obtained had B_2 been measured instead with A_1, and in that case the A_1 outcomes, according to strong locality, would have been just whatever they originally were. Thus the A_1 outcomes fit together with those of the B_1 and the B_2 measurements to form a trio of sequences from which correlations for A_1 with B_1 outcomes as well as for A_1 with B_2 outcomes can be calculated. Similarly, the A_2 results fit together with the above outcomes from the two B wing measurements to form another trio of sequences from which correlations both for A_2 with B_2 and also for A_2 with B_1 outcomes can be calculated. But one of your very own theorems (says my knowledgeable interlocutor) shows that there are compatible trios from which correlations can be calculated just in case the correlations satisfy the Bell inequalities (Fine 1982a, 1982b). Hence the argument from strong locality to those inequalities goes through. No sort of determinism is involved, and randomness is of no avail in blocking it.

The preceding argument uses much more than strong locality. It infers from the requirement that the outcome in one wing not depend on what is being measured in the other wing (which strong locality does maintain) that the very same outcome would have occurred in a given wing regardless of whether one or another measurement were carried out in the other wing. Strong locality denies that there are any influences from the circumstances regarding measurements carried out in one wing to the actual outcome obtained in a measurement performed in the other wing. It says that the measurements carried out in the one wing make no difference whatsoever to the outcome obtained in the other. The outcome does not depend on them in any way at all.

What is the logic of this independence assertion? Does it imply that what did happen would have happened anyway (because nothing would have changed)? Surely it does, if the outcome in a wing is the result of stable local circumstances there. For then, according to strong locality, switching the measurement elsewhere would not affect those stable circumstances, which would therefore issue in the same result. But what if there are no such stable, local determinates? Then it would seem that although indeed nothing relevant would change had we switched measurements, that fact alone is entirely compatible with the occurrence of a different (undetermined) outcome.

To take a somewhat remote example. It is probably true that the color of my car does not influence my luck at poker. But it scarcely follows that had I a car of a different color, my luck would not turn out to be somewhat different from what it in fact is. Indeed there are many things not dependent on the color of my car, any number of which might in fact have been different had I a car of a different color. The inference from "no influence" to "things would have been just the same" requires supplementary assumptions in order to go through. To insist on it, as the above objection does, is to introduce the principle that where nothing relevant to an outcome changes, the outcome itself could not change. This is just a version of the idea that change requires a cause. Thus the objection relies on determinism, or something in that neighborhood. Indeterminism blocks it, hence it blocks this route from strong locality to the Bell inequalities.

Objection (4). One does not need the strong inference from "no cross-wing influence" to "the same outcome would have occurred." We can run the argument from the weaker principle that if there is no influence from the measurement performed in one wing to the outcome in the other, then had we switched measurements the same outcome *might* have (or *could* have) occurred. To take up the automobile example, if the color of my car were different, then although I am correct in noting that indeed my luck at poker might have been different, it might also have stayed just the same. So the principle proposed here is weaker than the one above. Moreover, this principle does follow from strong locality; for surely if stability of outcomes is not even possible, were the measurement to have been different, then the measurement does influence the outcome.

There is no doubt that the "might/could" principle is weaker than the "would" one. What is certainly doubtful is whether the restriction of a logical possibility should count as an "influence" of the sort intended by a physical locality principle. Fortunately, we need not get bogged down in a metaphysical squabble over what does or does not count as a real influence since, contrary to the objection, when the argument is weakened as suggested it no longer goes through.

To see this let us call different runs in the experiment *adjacent* (recall that a run consists of a sequence of simultaneous measurements of one observable in one wing and another in the other wing) if they have the measurement of some one observable in common, e.g., the A_1B_1 and the A_2B_1 runs are adjacent (having B_1 in common), as are the A_2B_1 and A_2B_2 runs, which have A_2 in common. We can adapt some terminology first introduced by Schrödinger (1935b) and say that a correlation experiment is *entangled* just in case there are adjacent runs in the experiment whose sequences of outcomes for the shared observable are different. (For present purposes, we still suppose that the total number of outcomes is the same for each run.) The point of the third and fourth objections was to try to use strong locality to get unentangled experiments, for the correlations in such experiments satisfy the Bell inequalities (Fine 1982c). The determinist principle used in the third objection as a supplement to strong locality does entail that some experiments would be unentangled. But the weak "might/could" version of that principle in the fourth objection only implies that some experiments *might* be unentangled. Thus some experiments might yield statistics that do satisfy the Bell inequalities.

If we pick experimental arrangements whose statistics, according to quantum theory, do not satisfy those inequalities, then the argument in question only shows that one could in principle fail to verify the quantum statistics in such experiments, provided strong locality holds. I believe this to be a perfectly valid argument, but it does not show that strong locality implies the Bell inequalities, as it claimed to do, nor that strong locality conflicts in any way with the quantum theory. After all, the conclusion of an argument can be no stronger than the premises; so "might/could" *in* means "might/could" *out*. Nor can one see anything especially anomalous in the weak conclusion that the data might not be quantum mechanical. After all any experiment could fail to verify any set of predictions. We hardly need principles like strong locality to learn that. Moreover the jeopardy in quantum mechanics is double to begin with, since the predictions it yields are probabilities, and it is well understood that probabilistic predictions always have some actual likelihood of failure in any finite experiment. Nothing in this argument, however, suggests that the quantum experiments *do* fail or *would* fail; just that they might.

Objection (5). You have just identified the puzzle here that the previous objections were trying to get at. It is to understand why the quantum statistics, unlike others with which we are familiar, cannot be exhibited in unentangled experiments. For although the number of unentangled experiments is extremely small relative to the entangled ones (for a fixed largish number of outcomes in each run), we might suppose that among all the possibilities realized by various experiments sometimes we happen to find that as the data accumulate all the adjacent runs do show the very same outcomes, term by term, for the measurement they have in common, i.e., that the experiment is not entangled. But if the data are quantum-mechanical, this admittedly small possibility is entirely ruled out. How can that be? How can it happen that if we were to move among adjacent runs, by changing a measurement in one wing, the new sequence of outcomes in the opposite wing must differ somewhere along the line from the earlier sequence? Surely this is nonlocality, an influence from events in one wing generating changes in the other.

Is it? The argument shows this: Where the experimental outcomes are governed by probabilities that fail to satisfy the Bell inequalities, the way the data accumulates involves a shift in some outcomes between at least one pair of adjacent runs. If the individual outcomes in a measurement sequence occur at random and independently of one another, then the odds of repeating the same sequence of n outcomes twice are 2^{-2n}. (For a pretty short run of length 50 (i.e., for $n = 50$), we already have that 2^{-100} is smaller than 10^{-31}, an impressively small number!) Thus the odds for there *not* being a shift in some outcomes between the several adjacent runs in a correlation experiment are negligible, regardless of the experimental data. If the correlational data do violate the Bell inequalities, these negligible odds vanish.

It is somewhat tedious to show that even this tiny difference is compatible with strong locality, but maybe it is worth the effort if it will help remove even tiny residual doubts. So consider two runs in such an experiment. First we measure A_1 with B_1 and then we measure A_2 with B_2. In accord with the drift of this objection we now imagine some alternative possibilities, which are sketched out below:

	First Run	Second Run
Actual	A_1/B_1	A_2/B_2
Hypothetical	$[A_2] / [B_1]$	$[A_1] / [B_2]$

Possibilities
- [1] Actual A_2, 2nd run = Hypothetical $[A_2]$, 1st run.
- [2] Actual A_1, 1st run = Hypothetical $[A_1]$, 2nd run.
- [3] Actual B_1, 1st run = Hypothetical $[B_1]$, 1st run.
- [4] Actual B_2, 2nd run = Hypothetical $[B_2]$, 2nd run.

Suppose in the first run we had measured A_2 instead of A_1, then it is entirely possible (although *extremely* unlikely) that the sequence of outcomes in the hypothetical A_2 measurement in this run would have matched exactly the actual sequence of outcomes of the A_2 measurement in the second run (cf. possibility [1] above), and that the sequence of B_1 outcomes would just have been a repeat of the original (cf. [3]). Similarly, if in the second run we had chosen to measure A_1 instead of A_2, then it is entirely possible (although again *extremely* unlikely) that the sequence of outcomes in the hypothetical A_1 measurement would have matched exactly the actual sequence of outcomes of the A_1 measurement in the first run (cf. [2]), and that here again the sequence of outcomes in the B wing, this time of B_2, would simply have been a repeat of the original (cf. [4]).

If, however, we had chosen *both* to measure A_2 instead of A_1 in the first run *and* to measure A_1 instead of A_2 in the second run, then, assuming that the correlational data do not satisfy the Bell inequalities, at least one of these four possibilities would have to go. Either the hypothetical sequence of outcomes of the A_2 measurement in the first run would differ from the actual A_2 sequence in the second run [1], or the hypothetical sequence of outcomes of the A_1 measurement in the second run would differ from the actual A_1 sequence in the first run [2], or the actual sequences of outcomes of the B_1 measurement in the first run (where A_1 was measured with B_1) would differ from the hypothetical sequence of outcomes (where A_2 would have been measured with B_1) [3], or the actual sequence of outcomes of the B_2 measurement in the second run (where A_2 was measured with B_2) would differ from the hypothetical sequence of outcomes (where A_1 would have been measured with B_2) [4]. Only the latter two alternatives, corresponding to the failure of [3] and [4] above, could possibly suggest a challenge to strong locality, for only they involve any possible dependence between measurements made in one wing and outcomes in the other.[5]

Consequently, we can derive the failure of strong locality from the failure of the Bell inequalities only if we make some additional assumptions. We have to assume something that necessitates that were we to shift the measurement in the A wing in both runs, the new sequences of outcomes there would match, across the runs, both of the old sequences of outcomes, i.e., that both possibilities [1] and [2] are realized. Clearly, nothing can guarantee the occurrence of both of these highly unlikely possibilities short of determinism; that is, the requirement that if the local conditions do not change, then a repeat of the same measurement would produce exactly the same outcome each and every time. Hence here once again we see that it is only the combination of strong locality with determinism from which the satisfaction of the Bell inequalities follows. Locality alone is not enough. To put it differently, in answer to the objection, it is true that correlation experiments whose statistics fail to satisfy the Bell inequalities are entangled. It is not true that such entanglement implies the existence of any nonlocal influences or dependencies.

When all these objections are considered together, they seem to add up to no more than the original cry, which was that if nonlocal influences (or the like) are not invoked to explain the tangled statistics of the correlation experiments (i.e., statistical correlations that can only arise in tangled experiments), then how are we to explain them at all? I have tried to suggest the way out above. Let me try to reinforce the suggestion below.

---

[5]It is important to stress the *possibility* here, for it is by no means clear that the association between changing measurements in one wing and changing some outcome or other in the second wing represents a real influence or dependency from one wing to the other. Whether and when such associations count as real influences is an issue that we have only just begun to investigate. The answer in this case is not yet in.

5. *Indeterminism*

If we adopt an indeterminist attitude to the outcomes of a single, re-peated measurement, we see each outcome as undetermined by any factors whatsoever. Nevertheless we are comfortable with the idea that, as the mea-surements go on, the outcomes will satisfy a strict probabilistic law. For instance, they may be half positive and half negative. How does this happen? What makes a long run of positives, for example, get balanced off by the accumulation of nearly the *very same number* of negatives? If each outcome is really undetermined, how can we get *any* strict probabilistic order? Such questions can seem acute, deriving their urgency from the apparent necessity to provide an explanation for the strict order of the pattern, and the back-ground indeterminist premise according to which there seems to be nothing available on which to base an explanation. If one accepts the explanationist challenge, then one might be inclined to talk of a "hidden hand" that guides the outcome pattern, or its modern reincarnation as objective, probability-fixing "propensities."

This talk lets us off the hook, and it is instructive to note just how easily this is accomplished. For if propensities were regular explanatory entities, we would be inclined not just to investigate their formal features and conceptual links, but we would make them the object of physical theorizing and experi-mental investigation as well. However, even among the devoteés of propen-sities, few have been willing to go that far. The reason, I would suggest, is this. Once we accept the premise of indeterminism, we open up the idea that sequences of individually undetermined events *can* nevertheless display strict probabilistic patterns. When we go on to wed indeterminism to a rich proba-bilistic theory, like the quantum theory, we expect the theory to fill in the details of under what circumstances particular probabilistic patterns will arise. The state/observable formalism of the quantum theory, as is well known, discharges this expectation admirably. Thus indeterminism opens up a space of possibilities. It makes room for the quantum theory to work. The theory specifies the circumstances under which patterns of outcomes will arise and which particular ones to expect. It simply bypasses the question of how any patterns *could* arise out of undetermined events, in effect presupposing that this possibility just is among the natural order of things. In this regard, the quantum theory functions exactly like any other, embodying and taking for granted what Stephen Toulmin (1961) has nicely called "ideals of natural order." What then of correlations?

Correlations are just probabilistic patterns between two sequences of events. If we treat the individual events as undetermined and withdraw the burden of explaining why a pattern arises for each of the two sequences, why not adopt the same attitude toward the emerging pattern between the pairs of outcomes, the pattern that constitutes the correlation? Why, from an indeterminist perspective, should the fact that there is a pattern *between* random sequences require any more explaining than the fact that there is a pattern internal to the sequences themselves?

We have learned that it is not necessary to see a connection linking the random events in a sequence, some influence from one event to another that sustains the overall pattern. Why require a connection linking the pairs of events between the sequences, perhaps some influence that travels from one event in a pair to another (maybe even faster than the speed of light) and sustains the correlation? We have explored part of the answer above. Our experience with correlations that arise in a context in which there generally are outcome-fixing circumstances has led us to expect that where correlations are not coincidental, we will be able to understand how they were generated either via causal influences from one variable to another or by means of a network of common background causal factors. The tangled correlations of the quantum theory, however, cannot be so explained.

The search for ''influences'' or for common causes is an enterprise external to the quantum theory. It is a project that stands on the outside and asks whether we can supplement the theory in such a way as to satisfy certain *a priori* demands on explanatory adequacy. Among these demands is that stable correlations require explaining, that there must be some detailed account for how they are built up, or sustained, over time and space. In the face of this demand, the tangled correlations of the quantum theory can seem anomalous, even mysterious. But this demand represents an explanatory ideal rooted outside the quantum theory, one learned and taught in the context of a different kind of physical thinking. It is like the ideal that was passed on in the dynamical tradition from Aristotle to Newton, that motion *as such* requires explanation. As in the passing of that ideal, we can learn from successful practice that progress in physical thinking may occur precisely when we give up the demand for explanation, and shift to a new conception of the natural order. This is never an easy operation, and it is always accompanied by resistance and some sense of a lost paradise of reason. If we are to be serious about the science that we now have, however, we should step inside and see what ideals *it* embodies.

The quantum theory takes for granted not only that sequences of individually undetermined events may show strict overall patterns, it also takes for granted that such patterns may arise between the matched events in two such sequences. From the perspective of the quantum theory, this is neither surprising nor puzzling. It is the normal and ordinary state of affairs. This ideal is integral to the indeterminism that one accepts, if one accepts the theory. There was a time when we did not know this, when the question of whether the theory was truly indeterminist at all was alive and subject to real conjecture. Foundational work over the past fifty years, however, has pretty much settled that issue (although, of course, never beyond *any* doubt). The more recent work related to EPR and the Bell theorem has shown, specifically (although again, not beyond *all* doubt), that the correlations too are fundamental and irreducible, so that the indeterminist ideal extends to them as well. It is time, I think, to accept the ideals of order required by the theory. It is time to see patterns *between* sequences as part of the same natural order as patterns *internal* to the sequences themselves.

A nonessentialist attitude toward explanation can help us make this transition, for it leads us to accept that what requires explanation is a function of the context of inquiry. So when we take quantum theory and its practice as our context, then we expect to look to *it* to see what must be explained. This leads us to the indeterminist ideal discussed above, and to the "naturalness" of (even distant) correlations. There is a small bonus to reap if we shift our thinking in this direction. For the shift amounts to taking the correlations of the theory as givens not in further need of explanation and using them as the background resources for doing other scientific work. One thing they can do is to help us understand why the theory has correlational gaps. From the very beginning, one wondered about the incompatible observables and why one could not even in principle imagine joint measurements for them. After all, as Schrödinger (I believe) first pointed out, in the EPR situation, one could measure position in one wing and momentum in the other and, via the conservation laws, attribute simultaneous position and momentum in both wings.

The conventional response here has been to point out that only the direct measurements yield values that are predictively useful. (See note 4) Not everyone has been happy with the positivism that seems built into this response. But if we recall the discussion in section 1, then we see that (at least in part) there is a better response at hand. For we have seen how the correlations that the theory does provide actually exclude the possibility that there could be any stable joint distributions for incompatible observables in those states where the correlations are tangled. This shows us that there is no way of augmenting the theory with values for incompatible observables, and distributions for those, that would follow the same lawlike patterns as do the distributions of the theory itself. To put it dramatically, the shadow of the given correlations for compatible observables makes it impossible to grow stable correlations for the incompatibles. There is a sense, then, in which there would be no point in trying to introduce more for incompatible observables than what the theory already provides.

This way of thinking turns the Bell theorem around. Instead of aiming to demonstrate some limitation or anomaly about the theory, this way proceeds in the other direction and helps us understand why the probability structure of the theory is what it is. That understanding comes about when we take a nonessentialist attitude toward explanation, letting the indeterminist ideals of the theory set the explanatory agenda. Such an attitude means taking the theoretical givens seriously, and trusting that they will do good explanatory work. Thus, in the Bell situation, we shift our perspective and use the given quantum correlations (and the simple sort of counting argument rehearsed in section 1) to explain why, even in principle, correlations forbidden by the theory cannot arise. Nonessentialism leads us to engage with our theories seriously, and in detail. In the end, that is how better understanding comes about.

What then of nonlocality, influences, dependencies, passions, and the like, all diagnosed from correlational data? As one good statistician remarked about the similar move from linear regression to causal connection, and as we have seen demonstrated above, *"Much less is true."*[6]

---

[6]"It is easy to think that when we . . . find a linear regression of y on x (a statistically significant regression), we have evidence that increasing x causes y to increase. *Much less is true*" (Moses 1986, 294).

BELL'S THEOREM, IDEOLOGY, AND STRUCTURAL EXPLANATION

R. I. G. Hughes

1. Ideology and explanation

Duhem regarded with skepticism the suggestion that the aim of scientific theory was to furnish explanations.[1] Explanations, on his view, are attempts to account for phenomenal laws in terms of prior metaphysical assumptions; it follows that,

> If an appeal is made, in the course of the explanation of a physical phenomenon to some law which that metaphysics is powerless to justify, then no explanation will be forthcoming, and physical theory will have failed in its aim.[2]

Further,

> no metaphysics gives instruction exact enough, or detailed enough to make it possible to derive all the elements of a physical theory from it.[3]

The conclusion he draws is not that physical theory is doomed to fail in its aim, but rather that to view this aim as the production of explanations is a mistake.

At first sight, the problems arising from Bell's theorem seem a striking corroboration of Duhem's views. On any sane account, quantum mechanics is a remarkably successful theory. Yet some of the events it successfully predicts not only resist explanation, they actually undermine a cluster of metaphysical beliefs. Specifically, putative explanations of EPR-type correlations

---

[1]P. Duhem, *The Aim and Structure of Physical Theory*, trans. P. P. Wiener (New York: Athenaeum, 1962), chap. 1.

[2]Ibid., p. 16.

[3]Ibid.

entail other results at odds with experience; quantum theory, on the other hand, yields not only the correlations but the other observed probabilities as well. We are led to inspect these "explanations" to see what excess metaphysical baggage is being carried aboard with them. Thus scientific theory, far from being derived from metaphysical beliefs, may require us to revise them.

Paul Teller, in his talk at the Notre Dame Conference, put this in the language of psychotherapy; we have to come to terms with the abandonment of our particularist presuppositions. I prefer to talk in terms of *ideology*, not because I want to propose a Marxist or a radical feminist analysis of the significance of Bell's theorem, but because I think that the notion of ideology has a very general epistemic application. We can use the concept to say useful things, even if we do not think that every ideology can be traced back to the social formation within which it was engendered (no pun intended).[4]

On the account I want to use, an *ideology* is a set of attitudes and assumptions which is shared by and promulgated within a social group. It is more general, and operates at a deeper epistemic level, so to speak, than a Kuhnian paradigm. It determines what the members of that group regard as *natural*, i.e., as requiring no explanation. (Thus, on the Marxist account the assumptions are often assumptions about *human nature*.).[5] An ideology is not made explicit, and its promulgation is usually by indirect means, appearing, for example, in the presuppositions of the problems that the group chooses to address. Elements of an ideology may well have been explicitly formulated as metaphysical theses at some time, usually a period of revolutionary change, and only later have become taken for granted. Thus a certain quota of Descartes' *Principles* may seem to us to be truisms, until we realize that precisely those statements which strike us as the most obvious are those which ran counter to the Aristotelian science Descartes aimed to overthrow. I use the term 'ideology', in fact, to emphasize that the assumptions it makes are always tacit.

Because it is not explicitly articulated, an ideology is not systematic, and may even be inconsistent. Indeed the Marxist analysis of historical change traces it to internal contradictions within the dominant ideology. Now a classical scientific ideology is not in any obvious way, internally inconsistent.[6] Nor is Bell's theorem an anomaly within our current scientific paradigm, namely quantum mechanics. Rather, the theorem points to a dislocation

---

[4]This account owes something to Althusser; see especially pp. 158–186 of Louis Althusser, "Ideology and ideological state apparatuses," in *Lenin and Philosophy, and Other Essays*, trans. B. Brewster (New York: Monthly Review Press, 1971). However, by failing to consider the socio-economic origins of the ideology I am looking at, I betray my *petit-bourgeois* interests.

[5]Compare the "self-evident" proofs of Bell's inequality in Michael Redhead (1987a).

[6]However, *ad hoc* assumptions may be needed to reconcile the thesis of determinism with other elements of this ideology: see Earman (1986a), chaps. 2, 3.

between theory and ideology. By examining it we become more self-conscious about this ideology, and in particular about the "natural" way to talk, which is where our ideology manifests itself. It may be that physicists who have grown up since the development of quantum field theory do not share, for example, our particularist ideology, and that their "natural" way of talking is not longer particularist. If so, philosophers are lagging behind them, but, by achieving in a laboriously self-conscious way what these physicists have acquired by becoming, as it were, native speakers of nonparticularism, they make clear exactly what conceptual shifts quantum theory does, and does not, force upon us (see Wessels, in this volume).

To return to Duhem, it seems that he gives us a diagnosis of why the EPR correlations should be regarded as inexplicable (as Arthur Fine's paper in this volume suggests). Briefly, the quantum behavior of pairs of particles violates the assumptions of our ideology; this ideology is implicit in our metaphysics, and, since all explanations are formulated in terms of this metaphysics, it follows that no explanation can be forthcoming.

But must every mode of explanation be metaphysical? Carl Hempel, for example, gave a purely formal account of explanation;[7] his Deductive-Nomological model represents an explanation as a deductive argument whose premises are laws of nature and particular facts, and whose conclusion is the explanandum. But when we confront the EPR correlations, the D-N model of explanation gives us little help, since there are no obvious candidates for the "laws of nature" from which statements of these correlations can be derived. In fact we seem to have two choices: either we regard these statements as themselves laws of nature, in which case they furnish their own "explanations," or we derive them from more general "theoretical principles" which together function as the explanans.[8] But in the latter case the proffered "explanation" will merely rehearse the quantum-theoretic derivation of the statements of correlation. Neither alternative does much to increase our understanding, and this, surely, is the aim of any explanatory act.[9]

For someone who seeks to explain the EPR correlations, the dilemma is this. Either the putative explanation will go beyond quantum theory into ideology, or it will merely present the derivation of the correlations that the theory provides. And if quantum theory is indeed "the blackest of black-box theories" (Fine 1982c, 740), then this presentation will not yield understanding.

---

[7]Carl G. Hempel, "Aspects of scientific explanation," in *Aspects of Scientific Explanation and Other Essays in the Philosophy of Science* (New York: Free Press, 1965).

[8]Ibid., p. 343.

[9]See Peter Achinstein (1983), p. 16. On the lack of "instructions" for the formulation of explanations, see pp. 53–56.

2. Structural explanations

I want, however to suggest another possibility: that we can open the black box and inspect its inner workings. Rather than regarding the theory simply as a means of generating experimental predictions, we should look carefully at the models which the theory employs.

By talking of "models" I already betray a preference for the so-called "semantic view" of theories; indeed my suggestion can be put as follows, that we see what implications the semantic view of theories has for theoretical explanation. On the semantic view, theories present a class of mathematical models, within which the behavior of ideal systems can be represented.[10] It stands in contrast with the axiomatic view which regards an ideal theory as a (partially interpreted) formal system laid out in the Euclidean manner. The D-N model of explanation fits naturally with an axiomatic account of theories; in contrast, the semantic view presents the possibility of *structural explanations*.[11]

A structural explanation displays the elements of the models the theory uses and shows how they fit together. More picturesquely, it disassembles the black box, shows the working parts, and puts it together again. "Brute facts" about the theory are explained by showing their connections with other facts, possibly less brutish. This may strike some as an excessively vague prescription for generating explanations. However, to me this vagueness seems entirely appropriate; I think that there can be no formally specifiable set of "instructions" (Achinstein's term) for providing explanations, partly because understanding is a matter of degree.

An example of a structural explanation may be helpful. Suppose we were asked to explain why, according to the Special Theory of Relativity (STR), there is one velocity which is invariant across all inertial frames. (This velocity is actually the speed of light, and we often express the invariance by saying that the speed of light is the same for all observers; however, since I am here regarding STR just as a theory of space-time, that fact is irrelevant.[12]) A structural explanation of the invariance would display the models of space-time that STR uses, and the admissible coordinate systems for space-time that STR allows; it would then show that there were pairs of events, e_1, e_2, such that, under all admissible transformations of coordinates, their spatial

---

[10]For a general account and notes on the origin of this view, see van Fraassen (1980), chap. 3; for its application to quantum mechanics, see R. I. G. Hughes (1989), chap. 2.8.

[11]I find that this term has already been used by McMullin (see E. McMullin, "Structural Explanation," *American Philosophical Quarterly* 15 (1978): 139–147). However, we mean different things by it: McMullin uses it to describe an explanation in which the behavior of a complex *entity* is explained by reference to the structure of the *entity*.

[12]I owe this point to David Malament.

separation X bore a constant ratio to their temporal separation T, and hence that the velocity X/T of anything moving from e_1 to e_2 would be the same in all coordinate systems. It would also show that only when this ratio had a particular value (call it "c") was it invariant under these transformations. This kind of explanation would not fit the D-N model in any obvious way; nor is it a causal explanation: no "shrinking of rods" or "slowing down of clocks" is appealed to. It simply shows how the explanandum gets built into the models of space-time that STR postulates; it is a purely structural explanation.

My suggestion is that something similar is possible in the case of the EPR correlations, and that they can be made intelligible by a close inspection of the mathematical models that quantum theory employs. To this end, I will display these models in as nontechnical way as possible, assisted by a little hand-waving here and there.

According to quantum mechanics, the states of a system, and the observable physical quantities for that system, are to be represented within Hilbert spaces. To illustrate this representation, I will take a sphere of unit radius in three-dimensional real space (R^3); quantum mechanics uses complex spaces, some of finite, some of infinite dimensionality, but for present purposes the simplification will not be misleading. Quantum theory uses a three-dimensional space to represent an observable quantity (call it A) that has just three possible values, a_1, a_2, a_3. (Think of the components of spin of a spin-1 particle, which can take the values $+1$, 0, or -1.) When A is measured on a system, these values will each have a particular probability of occurrence, $p(a_1), p(a_2), p(a_3)$. What these probabilities are will depend on the system's *state*. It could be the case, for example, that $p(a_2) = 1$ and $p(a_1) = 0 = p(a_3)$; this would be an *eigenstate* of A. In general, however, the only constraint is that $p(a_1) + p(a_2) + p(a_3) = 1$, reflecting the fact that any measurement will yield one of exactly three values.

We represent these values of A by a set of orthogonal (mutually perpendicular) axes through the center of the sphere; call them x_1, x_2, x_3. Then any so-called *pure* state of the system can be represented within R^3 by a diameter d of the sphere which makes angles θ_1, θ_2, θ_3 with these axes, such that $p(a_1) = \cos^2\theta_1$, $p(a_2) = \cos^2\theta_2$, $p(a_3) = \cos^2\theta_3$. Because x_1, x_2, x_3 are mutually orthogonal, this fulfills the additivity condition. Note that, in this representation, the plane through the center of the sphere which contains, say, x_1 and x_2 represents the disjunction of values $a_1 \vee a_2$, which we can think of as the result of a nonmaximal measurement of A.

So far the representation is almost trivial. However, a key feature of quantum systems is that there are associated with them families $\{A_i\}$ of observables systematically related to each other. (As an obvious example, take the family $\{S_i\}$ of the components of spin of a particle.) Clearly each member of the family could be represented within a different sphere in the way I have described; the truly remarkable fact is that the whole family can be represented within the same sphere, with each set $\{a_{i1}, a_{i2}, a_{i3}\}$ of measurement-outcomes represented by a different triple $\{x_{i1}, x_{i2}, x_{i3}\}$ of orthogonal axes in such a way that, for each pure state of the system, there is a single diameter d of the sphere which will represent it and give us the probabilities associated with all the values of all the A_i.

I will refer to this fact, that a whole family of observables can be represented in this way, as *Representability*. To bring out quite how remarkable it is, Mielnik has shown that we can, in imagination, devise simple "schquantum systems," for which a family of two-valued observables exists, together with a simple recipe for giving probabilities to their values, but which cannot be represented in a two-dimensional Hilbert space.[13]

Representability tells us that we cannot think of the values of these observables just as properties which the system possesses, and which measurement will disclose. This follows from a well-known theorem due to Kochen and Specker (1967). Thus even a preliminary look at the Hilbert space models of quantum theory tells us that some of our classical presuppositions must be abandoned.

One well-known response to this has been to say that these properties just "hang together" in a funny way, and that we can retain talk of "properties" by moving to a quantum logic (Putnam 1969). At one time I could be heard saying such things myself (1981). However, I now think that the important insight of the "quantum logic program" has nothing to do with properties, talk of which is best rejected; rather it is that there is a real sense in which the probability-assignments of quantum theory do not conform to the standard mathematical account of probability (see Suppes 1966). For the orthodox account of probability, due to Kolmogorov, defines it as an additive measure on a field of subsets of a given set.[14] This field of sets has a particular algebraic structure, that of a Boolean algebra. But the set of the axes, planes, and so on within the sphere which represent possible measurement-outcomes is not a Boolean structure of this kind. It is instead an ortho-algebra consisting of a family of Boolean algebras "pasted together" in a systematic way. Thus quantum probability-functions, though perfectly well defined, are not functions of the kind described by Kolmogorov. I will refer to this fact as *Non-Orthodoxy*.

One result of *Non-Orthodoxy* appears when we look for the quantum

---

[13]See E. G. Beltrametti and G. Cassinelli (1981), pp. 204–207, or Hughes (1989), chap. 4.4.

[14]A. N. Kolmogorov (1933), *Foundations of the Theory of Probability*, trans. N. Morrison (New York: Chelsea, 1950).

analogue of conditionalization, that is, when we ask, "What is the probability of Q, given P?" In the orthodox case we have, provided that $p(P)$ is non-zero,

$$p(Q \mid P) = \frac{p(Q \ \& \ P)}{p(P)}$$

The quantum analogue of this is the "Lüders Rule."[15] I will use the sphere in R^3 to illustrate it. Assume that we have a state which is represented by a diameter d which is skew to the plane P representing the disjunction $a_1 \ v \ a_2$ of values of an observable A; if we conditionalize on this disjunction, then the Lüders Rule tells us that the effect is as though the state were now represented by the diameter d' of the sphere which is the projection of d on P, i.e., the diameter in P which lies in the plane containing d and x_3. The probabilities assigned by d' to each measurement-outcome representable on the sphere give the conditional probabilities we are looking for. Conditionalization is sometimes said to "project" d to d', and I will call the thesis that it does so *Projectability*.

Another set of results of *Non-Orthodoxy* has no analogue in classical probability theory. It concerns the representation of coupled quantum systems, and it is this set of results which, for our purposes, is so significant. Assume that we are dealing with two systems, a and b, each of which can be modeled within a sphere of R^3. Now when a and b form a coupled system, $C = a + b$, the behavior of C is modeled, not, as it were, by the dumbbell formed from the spheres used for a and b, but by a 9-dimensional hypersphere whose "diameters" represent the pure states of C. This hypersphere is called the *tensor product* of the spheres associated with a and b. The tensor product space, and the states in it, have sundry curious properties, as follows. (These the reader will have to take on trust.[16])

(i) Each pure state W of C can be uniquely "decomposed" into "reduced" states, w_a, w_b of a and b.

(ii) Even when W is a pure state, w_a, w_b may be "mixed states", that is, states which behave like weighted sums of pure states.

(iii) The converse of (i) does not hold; when w_a, w_b are mixed states, there can be more than one state W which reduces to them.

(iv) There are states W such that a and b are not statistically independent of one another.

This is in marked contrast with what happens in classical probability theory. We can form the product of two classical probability spaces, but, if

[15]This was first recognized by J. Bub (1977).

[16]Alternatively, consult J. M. Jauch (1968), chap. 11.

we do, then (i) will hold, but (ii), (iii), and (iv) will not (see Stairs 1983). In particular, (iii) tells us that we may have more information available from a specification of the state W of C than from the specification of the two reduced states, w_a, w_b of a and b, while (iv) tells us why we might expect correlations when we deal with coupled systems.

The four features of tensor product spaces, (i), (ii), (iii), and (iv), I will refer to collectively as *Entanglement*.

3. *A structural explanation for the EPR correlations*

We can now describe what happens in a (Bohm-style) EPR experiment in terms of the four theses, *Representability*, *Non-Orthodoxy*, *Projectability* and *Entanglement*. Recall that in this case we are dealing with spin ½ particles, so that there are only two possible results, plus or minus, of a measurement of any component S_i of spin.

When the pair is produced, the state of the coupled system can be represented in a Hilbert space (*Representability*). Its state W (the *singlet spin state*) is a pure state, but the reduced states of the two components are mixed states (*Entanglement* [i] and [ii]); the particles are, as we say, completely *unpolarized*, and, for each particle, the probability of obtaining the plus value in a measurement of any component, S_i, of spin is ½.

Now assume that a measurement of S_x (say) on particle a yields the result plus. Call this the event $(S_x^{(a)}, +)$. We know from *Entanglement* (iv) that this may affect the probabilities associated with particle b since the two systems are not statistically independent. In fact the b-probabilities are now as though system b were in a pure state, the eigenstate $x\_^{(b)}$ of spin which gives probability 1 to the event $(S_x^{(b)}, -)$. How is this to be understood?

The answer is that these probabilities are just what *Projectability* predicts. I will spell this out in more detail. Prior to the event $(S_x^{(a)}, +)$ the coupled system is in state W, which reduces to the unpolarized (mixed) states w_a, w_b. Conditionalization on $(S_x^{(a)}, +)$ yields new probabilities, which are just those given by the state W', where W' is the state obtained from W by the Lüders Rule, when that rule is applied to the tensor-product space. Surprisingly, applying the rule has an effect, not only on the probabilities associated with a but also on those associated with b: it turns out W' reduces to two pure states, namely the eigenstates $x_+^{(a)}$ and $x_-^{(b)}$. (We write: $W' = x_+^{(a)} \otimes x_-^{(b)}$.)[17]

This analysis not only tells us how *Outcome Independence (Jarrett-Completeness)* comes to be violated, and how we get *Perfect Correlation* (van

---

[17]The proof appears in Hughes (1989), chap. 8.8.

Fraassen's term); it also yields *Parameter Independence* (van Fraassen's *Surface Locality*) (this volume).

What I have said would need amplification before it was a full (structural) explanation of the EPR correlations; I have here just sketched out the general form such an explanation would take. Now not all the elements of this sketch are uncontentious. Indeed, I can imagine the objection being made that as long as the alleged explanation stays within quantum theory it is uninformative, and that where it goes beyond the theory, it is suspect. To counter this possible objection, I will briefly discuss the justification of each of the four principles I have put forward, and show how it would contribute to a fully realized explanation.

The least problematic of the four is *Representability,* certainly as far as the treatment of spin is concerned.[18] Not only are all components of spin representable in one Hilbert space, this is the way that quantum theory actually represents them. The same is true for all four clauses of *Entanglement*; I have simply highlighted four features of the mathematical theory of tensor-product spaces, the theory which quantum mechanics uses to represent composite systems. Thus, in enunciating *Representability* and *Entanglement,* I am doing no more than drawing attention to the models that quantum theory employs. Even someone for whom the theory was no more than a formal expression of the analogies which diverse phenomena had in common could subscribe to them. The provision of a structural explanation consists in making explicit what is involved in doing so.

The first move in this direction is made with the thesis of *Non-Orthodoxy*. Acceptance of *Non-Orthodoxy* makes us look, for example, at the results of using tensor-product spaces from a new perspective. In particular, it encourages us to compare the tensor products of two Hilbert spaces with the products of two classical probability spaces, as Stairs does in the paper I have already cited.[19] By making clear exactly what acceptance of tensor-product models commits us to, this helps to loosen further the hold of classical accounts on our minds.

As for the justification of the principle, to me it seems undeniable that, by using Hilbert space theory, quantum mechanics requires probability functions that are defined over non-Boolean structures. One could deny that this was *significant* by pointing out that (1) the ortho-algebras on which these functions are defined are themselves families of Boolean algebras, and that

---

[18]The case of the Hilbert space in which position Q and momentum P are represented is a bit less conclusive, since there are Hermitian operators on this space, like $PQ + QP$, which do not seem to represent genuine observable quantities.

[19]Note in particular, his remark that, "logic plays an *explanatory* role, . . . the non-standard aspect of quantum systems [helps] to account for their puzzling behavior" (Stairs 1983, 48) original emphasis. See also A. Stairs (1984).

(2) the restriction of a quantum probability function q to any one of these Boolean algebras is a perfectly orthodox (Kolmogorov) probability measure, and claiming that (3) each of these Boolean algebras should be considered independently. But the price to be paid for this move is that we would thereby lose any account of the relations between the probabilities defined on the individual Boolean algebras; that is to say, we would lose the precise account of the relationships between different observables that the Hilbert-space representation affords.

On more general, methodological grounds, we can ask whether an insistence on reinterpreting quantum theory in terms of Kolmogorov probability functions does not betray a failure to take seriously the models that the theory employs. Such an insistence would not only be a rejection of the very notion of structural explanation, but would make us unlikely to ask one of the most interesting foundational questions of all: Why do Hilbert-space representations work? As the example of the "schquantum system" shows, this is not a trivial question.

Lastly I come to *Projectability*. Of the four principles I have put forward, this provokes by far the most disagreement. The Lüders Rule (1951) was first suggested as an improvement of a rule proposed by von Neumann (1932, Chap. 5) to describe the transitions he took to be characteristic of the measurement process. According to von Neumann, an ideal measurement on a system would change the system's state in such a way that an immediate repetition of the measurement would be certain to yield the same result as before. This postulate is generally referred to as the "Projection Postulate." When a measurement is less than maximally specific, however, there are a number of transitions which would conform to this postulate; for measurements of this kind, the Lüders Rule and the rule which von Neumann originally suggested ascribe different final states to the system. To use the example given earlier, if a measurement of A were to yield the result $a_1 \, v \, a_2$, then the Lüders Rule tells us that the system's state after the measurement would be given by the diameter d', but von Neumann's rule gives a state which would ascribe equal probabilities to all the measurement outcomes represented by axes with the plane P. Both rules would conform to the postulate, since in either case a repetition of the measurement of A would yield the result $a_1 \, v \, a_2$ with certainty.

Those who accept the projection postulate now tend to prefer the Lüders Rule to von Neumann's. Various authors, however, have either (1) denied that the projection postulate holds for measurements (Margenau 1936; van Fraassen 1974), or (2) regarded it as no more than a "fortuitous approximation" (Teller 1983). My view is that it only rarely applies to measurements, many of which, after all, result not in a mere state-transition, but in the absorption and effective annihilation of the measured system. Instead, I follow Bub in regarding the Lüders Rule as the quantum analogue of conditionalization, and hence as a natural corollary of *Non-Orthodoxy*.

I acknowledge that this does not resolve all the difficulties surrounding *Projectability*. Its assessment remains one of the central problems of the interpretation of quantum theory.[20] In particular, consider the problem raised by Shimony (1986). Assume that, in an EPR-type experiment, the measurements carried out on the components *a* and *b* of the coupled system are simultaneous in the laboratory frame of reference. Then, according to STR, there is a frame of reference, F_a, in which the measurement-event associated with system *a* precedes the event associated with *b*, and another frame F_b in which the *b*-event precedes the *a*-event. *Projectability* then tells us that, within F_a, an *a*-event has occasioned a change of state of *b* prior to the occurrence of the *b*-event, but that, within F_b, if this change of state of *b* has occurred, it has occurred as a result of the *b*-event, and this has also occasioned a change of state of *a*. Thus, while no outright inconsistency results from giving *Projectability* an ontological significance, doing so makes the occurrence or nonoccurrence of a change of state frame-relative.

I grant this, and also grant that, for a realist, it poses a deep problem. But I would point out that, viewed as the rule of quantum conditionalization, the Luders Rule provides a beautifully elegant treatment of the notoriously problematic two-slit experiment (Bub 1979). It also adds to our understanding of the EPR correlations in the following way. As we have seen, *Representability* and *Entanglement* account for these correlations by giving the quantum-theoretic analysis of them. *Non-Orthodoxy* is an invitation to look at a particular aspect of this analysis, and to see how the results derive from the nonclassical nature of the probability functions involved. One could reasonably claim that this alone would constitute a structural explanation of them. However, we can reinforce this invitation by pointing out that the probabilities involved in the EPR experiments—including those that yield the correlations—are often conditional probabilities: we ask, what is the probability of a particular outcome for system *b*, given the occurrence of a specific outcome with system *a*? If, as *Non-Orthodoxy* tells us, we are dealing with nonclassical probability spaces, then we should surely see whether a revision of our rule of conditionalization is called for. Given that the obvious candidate for this revised rule is the Lüders Rule, how should we regard its success in giving the correct conditional probabilities except as a confirmation that our emphasis on nonclassical probability spaces was well founded?

I have presented quantum theory as a theory which postulates novel, nonclassical "possibility structures" (Bub's phrase), and I have also suggested that we can explain a set of phenomena by locating the events in

---

[20]I discuss the relation between structural explanation and interpretation below.

question within these structures. In the final part of this paper I will consider a fundamental objection to the second of these proposals.

4. An objection

Recall that, before discussing the EPR correlations, I gave as an example of a structural explanation one which involved the models of space-time appearing in STR. The view I want to consider starts by acknowledging that the notion of structural explanation was, in that instance, entirely appropriate, since STR deals solely with a particular structure, that of space-time. But quantum mechanics, the objection continues, is a very different kind of theory. Even if, for argument's sake, we allow that it involves reference to nonclassical possibility structures, it is not *about* those structures. It is about physical systems, and the physical parameters which characterize them.

Effectively, the objection claims that there can be no explanation of, for example, the EPR correlations without an interpretation of quantum mechanics. An *interpretation* would present a categorial framework—an ontology and an etiology—within which the behavior of quantum systems could be described.[21] That is, it would present an answer to the question, what kind of world would obey the laws of quantum mechanics? But, in response, we may ask how the demand for a categorial framework is to be met. Kochen and Specker's theorem shows us that the classical category of substance-attribute (or system-property) cannot be part of the categorial framework of the quantum world. So do other results, including Bell's theorem (see Mermin, this volume). In addition, Bell's theorem shows that causality is ruled out; we cannot even content ourselves with a causation that is merely statistical (see van Fraassen, this volume). If the major elements of a classical categorial framework are to be discarded, where are we to look for another?

Where else but within the models which the theory supplies? My own view is this will lead us to an interpretation in terms of *latencies* (*potentiae*, or *be-ables* if you prefer), but this is beside the point. Whatever the elements of a new categorial framework are, they will be identified within the mathematical structures used by the theory. A latency, for example (as whatever grounds a disposition to certain kinds of probabilistically governed behavior) would be precisely specifiable within a Hilbert space as a system's state. A new categorial framework would thus serve to emphasize the role of various features and relations shared by all the models the theory employs.

Such a framework would be a fully explicit, hence nonideological, metaphysics, and I take the search for one to be a valuable philosophical

---

[21]The term 'categorial framework' I owe to Stephan Körner.

project. (It would suggest, for instance, what ontological significance we should attach to conditionalization.) Nor do I underestimate the suggestive power of a successful interpretation for the practicing physicist. Further, since the metaphysical postulates within this interpretation would not be presupposed, but would have emerged from an examination of the theory itself, Duhem's indictment of "metaphysical" explanations would lose its force. However, it is not clear what useful explanatory work this interpretation would perform over and above that provided by a full articulation of the models the theory presents. For, since all the elements of the framework would have been discovered within these models, explanations in terms of this framework would essentially be structural explanations in another guise. A new categorial framework would not *account for* the four principles I have set out; rather, it would *restate* them in another vocabulary. It would be none the worse for that. It might even make them more acceptable. And in whatever way they were expressed, to show how all the troublesome phenomena of the quantum world follow from these manifest principles would be a very great step in philosophy, though the causes of these principles were not yet discovered.

RELATIVITY, RELATIONAL HOLISM, AND THE BELL INEQUALITIES

PAUL TELLER

Most of us feel baffled by the failure of the Bell inequalities. I suggest that this feeling arises from an unnoticed and very deeply seated presupposition we make about ontology. Once we make the presupposition explicit, we have an easier time reconciling the facts with our intuitions, and we see more clearly one of the ways in which quantum mechanics constitutes a "non-classical" theory. The discussion will also help us understand more clearly why there is, after all, no conflict between failure of the Bell inequalities and "relativity theory."

1. The determinate-value argument for the Bell inequalities

In its determinate-value form, the argument for the Bell inequalities is absurdly simple. Particles have a property called spin. We can measure for spin in any direction we like, but for a given particle we have to make a choice, for we cannot measure in two or more directions on any one given particle. Any measurement will yield exactly one of two values: "up" (+) or "down" (−). We now prepare two particles, the A particle and the B particle, in a special correlated state (the "singlet state"), and let them fly off in opposite directions. In any given direction an individual particle has a 50–50 chance of yielding a plus or a minus when measured. In addition, the spins are correlated. If we measure the A and B particles in the same direction, they always yield opposite results. If, in a sequence of such experiments, we measure the two particles in different directions the results are still correlated, though in more complicated ways.

This work was supported by NSF grant #SES 8519928 and comments from many colleagues.

Why, when we measure for spin in a direction, do we always get a definite result? The simplest assumption is that just prior to the measurement the particle has, for each direction in which we might measure, a definite property, its spin, which determines the outcome of the experiment should we choose to measure in that direction. This assumption is called *Determinateness,* the assumption that, whether or not we measure in a given direction, the particle has a determinate spin-value in that direction. We can also explain determinateness as *Counterfactual Definiteness,* the assumption that, for each direction there is a definite value, plus or minus, which is the value which we would find should we measure in that direction.

Locality provides a second natural assumption. If the particles are widely separated at the time of measurement, the value of the *A* particle's spin in a given direction should not depend on what the *B*-experimenter decides to do with the *B* particle. In particular, *A* values should be independent of the direction the *B* experimenter chooses to measure on the *B* particle.

These two assumptions suffice for an extraordinarily simple argument for the Bell inequalities. Given a run of experiments, and given the assumptions that the values are there and not dependent on measurement-settings on opposing wings of the experiment, simple application of set-inclusion demonstrates that the relative frequencies of these values in various directions must satisfy a Bell inequality. The probabilities provided by quantum mechanics do not satisfy this inequality. Many experiments have shown that the relative frequencies are as predicted by the quantum probabilities, in violation of the Bell inequalities.

The situation gets worse. The argument did use two assumptions, determinateness and locality. Apparently we should reject one of them. But both appear to be implied by relativity theory. Thus relativity theory, *all by itself* appears to imply the Bell inequalities, the violation of which then appears to show that relativity theory must be false!

The current Bell literature now widely accepts the contrary conclusion that there is no technical conflict between relativity and the violation of the Bell inequalities. But the reasons given proceed by showing that the failure of the inequalities cannot be exploited to send an information-bearing signal faster than light. These demonstrations do not make clear what is wrong with the pattern of argument I have been considering, so that doubt lingers. We need to examine the argument from relativity to determinateness and locality, to review the nature of relativity theory, and to apply these results to critically evaluate the overall argument.

2. The role of "relativity theory"

People believe that relativity applies because they believe that it requires all signals and causal influences to travel no faster than the speed of light—the *No Superluminal Propagation* principle. This principle is supposed to apply in the following argument.[1]

Step (1) states that:

(1) RELATIVITY (+ CORRELATION) → DETERMINATENESS OF VALUES

Recall that such values as there might be are correlated. If we perform a measurement on wing *B* we know that, should we also measure in the same direction on wing *A*, we will certainly get the value opposed to the one obtained at *B*. Thus, in the special circumstance that a measurement at *B* has taken place in direction *j*, we know that there is a determinate value—the value we would get if we were to measure—for the spin of *A* in the same direction. Now let us suppose (as I will henceforth always assume) that the measurements of *A* and *B* take place so far apart in space and so close together in time that light could not make it from one measurement to the other. (We say that the events are "spacelike related.") Then relativity, through the No Superluminal Propagation principle, tells us that what happens at *B* cannot influence what happens at *A*. In particular, whether or nor we decide to measure *B* in direction *j* cannot affect whether or not the value of *A* is determinate in that direction. Thus *A* must have a determinate value in that direction whether or not we measure *B* in that direction. The same holds for all directions and for the roles of *A* and *B* reversed. Thus (1).[2]

The second step states that

(2) RELATIVITY → LOCALITY OF DETERMINATE VALUES

"Locality" here means, as before, that the value on one side in a given direction does not depend on the measurement setting—on the direction in

---

[1]Many have taken this line of argument to be suggested by the original papers of EPR (1935) and Bell (1964). Fine also details a version of Einstein's in a letter to Schrödinger (1986, 36–37). Hellman has recently spelled the argument out in some detail (1987, 562–568).

[2]I should note that I have glossed over important assumptions in this and other aspects of the Bell arguments. Most importantly, I have assumed that when one measures for spin, one always gets a result and that, for a given direction, the results are perfectly correlated. Most of the Bell literature embraces these severe idealizations, apparently taking the overall experimental evidence as overwhelming support for saying that departures from the idealizations are entirely due to limitations in experimental technique. Fine's "prism models" (1981, 1982d) show how we must reevaluate these assumptions if such idealizations are rejected.

At the Notre Dame Conference, I made an issue of the fact that one gets determinateness trivially if one uses the fallacious form of argument I called the "Candy Bar Principle": from "If I were hungry, I would eat some candy bar" conclude "There is some candy bar which I would eat if I were hungry." I neglected to say that I learned this important fact from John Halpin, who has developed it admirably (1986).

which the experimenter chooses to measure—on the other side. This is again supposed to follow from the No Superluminal Propagation principle.

Step three just invokes the argument I mentioned earlier, that the consequents of (1) and (2) together imply the Bell inequalities. This step involves only unproblematic application of set inclusion. In summary, we have

(3) RELATIVITY $\rightarrow$ DETERMINATE VALUES + LOCALITY $\rightarrow$
BELL INEQUALITIES

Since the Bell inequalities fail, it looks like relativity must likewise fail!

I do not question (2). But (1) requires careful examination. To begin, we need to get clear on what this thing is which other writers on the Bell inequalities have unreflectively called "relativity theory."

3. The role of "relativity" examined

Universal contemporary practice presents relativity theory as what Michael Friedman and John Earman have called a *Space-Time Theory*. Such a theory describes the world in terms of a collection of space-time points each of which has its own value for each of several physical quantities, such as mass density and the electromagnetic field. What makes such a theory relativistic[3] is that laws take the same form for all inertially moving observers, while descriptions given by such observers must Lorentz-transform into each other. On this framework one may fill in a variety of laws, such as laws governing the motion of massive particles or Maxwell's laws for the electromagnetic field.

Saying this much already suggests a way to challenge (1). One could suggest that relativity only constrains actual values of quantities at space-time points. It has no application in the circumstance that a value is absent or indeterminate. But, as Allen Stairs pointed out to me, one may circumvent this argument by noting that if some quantity fails to take on a definite value at a point, that is itself a definite fact about the point. If relativity constrains all the facts about space-time points, it should also constrain this absence-of-value kind of fact about a space-time point. One might protest this claim. However, for the moment I will simply give it to the defender of (1) because we will discover much more important difficulties with applying the No Superluminal Propagation requirement. In the end, these difficulties will return to show that there is, after all, a way in which quantities fail to take on

---

[3] I will throughout be concerned only with special relativity. For reasons described in Don Howard's contribution to this volume, the situation gets worse for general relativity.

definite values at points, a way which quite clearly makes the No Superluminal Propagation requirement inapplicable.

This sketch of "relativity theory" reminds us of something which most
writers on Bell have ignored: There is no one theory, called "relativity theory." As Einstein put it from the beginning (e.g., 1919, 228; 1949, 51–53),
"relativity" should be taken to be a "principle-theory," that is a set of
general constraints which any more detailed theory must satisfy. The more
detailed theories do not even have to be framed as space-time theories, a point
to which I will return. In addition, there are relativistic space-time theories
which involve processes, such as "tachyons," that certainly appear to constitute some kind of faster-than-light causal propagation.[4] Thus (1) again
comes into question.

Trouble with tachyons and the like does not provide as good reasons for
calling (1) into question as I would like. Redhead (1983, 166 ff.) has argued
that help from tachyons cannot be ruled out, but we are still lacking any
promising specific proposals for help from this quarter. More generally, tachyons and other proposals involving relativistic superluminal propagation are
highly speculative, not otherwise subject to experimental tests, and in any
case outside the mainstream of accepted relativistic theories. There are more
important problems with (1).

All familiar cases of broadly accepted relativistic theories share two
features. They embody some form of *contact* or *local action*, that is the idea
that the values at one space-time point have direct causal connection only with
values at infinitesimally close points. And these theories require that causally
connected chains of events never proceed faster than light. I do not think that
anyone has ever made it clear when a sequence of events constitutes a causally
connected chain, providing yet another ground for quibbling with (1), which I
yet again will not pursue. In familiar cases, we know well enough which are
the genuine causal chains. I will call any theory which embodies these local
actions and no superluminal propagation requirements *Relativistic Causal
Theories*.

The force of (1) comes from the conviction that any respectable relativistic theory will be a relativistic causal theory. But we do not need to attack
with tachyons or the like to avoid (1). Our unhesitating acceptance of relativistic causal theories, which supports (1), involves an assumption so basic to
the thinking of most of us that we are not even aware we are making it.

---

[4] I will make no attempt to summarize the voluminous literature on tachyons. Let it just be
said that the implications are in dispute. For discussion of additional forms of relativistic superluminal action, see Earman (1986a, 66–77).

4. Particularism vs. relational holism

Forget, for the moment, the conceptual heartache of Bell and quantum mechanics. Return with me to a simpler time, to the mechanistic worldview as it was supposed to be delivered by classical physics. The world was thought to be a deterministic machine, like a great clock. If Laplace's superintelligence knew the exact state of the world-machine at one moment, he or she would be able to calculate its state at any other time.

I maintain that there is a further, generally unnoticed feature of this picture. Mechanism tacitly includes the idea that the world is composed of individuals—atoms, space-time points, or what have you. Each individual has its own roster of nonrelational properties—such as its mass or mass density and its charge. There are, to be sure, relations which hold between individuals, such as the fact that one object has a greater mass than a second object. But these relations do not, on this view, constitute further, irreducible facts. Laplace's superintelligence would not have to take into account all the nonrelational properties of all the individuals and then *in addition* take into account further relations holding between individuals. The fact that object 1 has a greater mass than object 2 is already subsumed by having taken into account the mass of object 1 and the mass of object 2.

Let me refine and name the view I am describing. *Particularism* states that the world is composed of individuals, that the individuals have nonrelational properties, and that all relations between individuals supervene on the nonrelational properties of the relata. This means that if two objects, 1 and 2, bear a relation R to each other, then, necessarily, if two further objects, 1′ and 2′, have respectively the same nonrelational properties which 1 and 2 have, then 1′ and 2′ will also bear the relation R to each other. I call such a relation a *Supervening Relation.*[5]

I claim that we tend unreflectively to presuppose particularism as a facet of our conception of the world, a facet which never gets explicitly stated and yet conditions all our thinking. In application to relativistic theories, particularism takes the form of supposing the theory to apply exclusively to space-time points and their nonrelational properties. Once we make this assumption, local action and no superluminal propagation follow right along. They do not have to, but before dodging them with maneuvers such as tachyons we should

---

[5]This characterization is incomplete because some relations, such as *x is a brother of y*, get expressed as two-place but are really three-place; much as *x is married*, a syntactic one-place relation, is really a two-place relation, *there is a y such that x is married to y*. I have introduced and discussed at a little greater length (1986) the ideas of particularism, nonsupervening relations (there called "inherent relations"), and relational holism, to be discussed below.

question the unspoken assumption, particularism, which sets the precondition for getting any relativistic causal theory off the ground. Indeed, we should question particularism because quantum mechanics itself gives us very strong reasons for rejecting it.

Quantum mechanics employs something called the *Superposition Principle*. A quantum mechanical state can attribute a property, p, to a particle. A second state can attribute p'. One can then always consider a third state, formed, literally, by adding the first two states. This superimposed state attributes neither p nor p' but some new property, q, which can be considered to be a property somehow resulting by superimposing the properties p and p'.

This strange situation gets much stranger when we consider pairs of objects. A first state may attribute the property x to particle A and the property y to particle B. A second state similarly attributes x' to A and y' to B. But what properties are attributed to A and to B by the state which we form when we add the first two states to form their superposition? The superposition attributes to A and B together a collective property—that is a relation which holds between them. But there are arguably no nonrelational properties of A and B on which the superposition-relation supervenes.[6]

Unless one takes a starkly instrumental attitude toward quantum theory, quantum theory tells us that particularism is wrong. It tells us that we must endorse what I call *Relational Holism,* the view that there are nonsupervening relations—that is, relations which do not supervene on the nonrelational properties of the relata.

Relational holism is compatible with relativity, at least it is if by relativity we mean the framework principle that the descriptions of inertial observers must Lorentz-transform into each other. This comes out most clearly in relativistic quantum field theory in which nonsupervening quantum states are attributed to collections of space-time points in a Lorentz covariant way.

I want now to apply these ideas to refine some older conceptions. The Bell literature tends to talk about some one idea of "locality." Particularism being presupposed, this can only mean local action and, in view of relativistic causal theories, local action comes down to the requirement of no superluminal propagation. But particularism can itself be thought of as a kind of locality—an *ontological locality* of values, stating that all properties are

---

[6]The singlet state, providing the example for the Bell inequalities, exemplifies exactly the sort of superposition under discussion. To argue that there are no subvening properties on which the singlet state supervenes, suppose there were. If the hypothesized subvening nonrelational properties are the spins, we get the wrong quantum state, as well as the determinate values needed to argue the Bell inequalities, without the help of (1). If these properties are some "deeper-lying" properties we can see them as the parameter, λ, in the argument of section 8, giving us factorizability and, once more, the violated Bell inequalities. Parts of the foregoing argument are spelled out elsewhere (Teller 1986, section 3).

nonrelational properties of individuals or are relations supervening on these. If ontological locality—particularism—is presupposed, our experience with relativistic theories makes it very plausible to expect that a relativistic causal theory should apply, so that what I will call *causal locality* should apply.

But when (or insofar as) particularism is denied, the idea of causal locality has no application. Causal locality concerns the lawlike connection between nonrelational properties applying (in this discussion) to space-time points. To say that causal locality has been violated most plausibly should be taken to mean that there are nonrelational properties of space-time points which are related in some other way—by action (lawlike dependencies) at a distance or through superluminal causal chains. On the other hand, when we are concerned with nonsupervening relations, this circle of ideas has no grip. There is no question of superluminal or distant action between nonrelational, definite values.[7]

All this is very hard to see as long as particularism has us in its conceptual vise. As long as particularism goes unstated but tacitly assumed, we can recognize only one kind of locality: causal locality.[8] The unacknowledged particularism results in our thinking that causal locality must hold because our

---

[7]Heywood and Redhead (1983) introduced a distinction between "OLOC" ("Ontological Locality") and "ELOC" ("Environmental Locality"). As I understand these conceptions, as they are explained in Redhead (1987b, 139–151), they constitute more specific instances of what I here describe as "Ontological" and "Causal" Locality. Redhead also sees ELOC as presupposing OLOC (1987b, 151) and sees, much as I do in sections 8 and 9 below, violation of outcome-independence as akin to a violation of OLOC (1987b, 106–107, 151). I hope to have advanced the discussion by characterizing the distinction more generally and by further clarifying its connections with other facets of the larger set of problems.

[8]One person—Einstein—seemed to appreciate the presumption of particularism. He also illustrates how tightly particularism binds. Howard (1985) calls our attention to Einstein's 1948 characterization of his "separability principle" as the view that physical things

claim an existence independent of one another, insofar as these things "lie in different parts of space." Without such an assumption of the mutually independent existence (the "being-thus") of spatially distant things, an assumption that originates in everyday thought, physical thought in the sense familiar to us would not be possible. Nor does one see how physical laws could be formulated and tested without such a clean separation. (Howard's translation [1985, 197–188] of Einstein [1948, 321–322]).

I suggest that my conception of particularism refines (in ways which undoubtedly go beyond Einstein's texts) the idea of separability described in this passage.

According to Howard, the later Einstein distinguished his separability principle from the principle of local action which states that "an external influence on *A* has no immediate effect on [spatially distant] *B*" (Howard 1985, 187–188, again translating Einstein's 1948). Howard's discussion suggests to me that Einstein at this time may have had a good grip on the idea that his principle of local action presupposed his separability principle. Fine's discussion of Einstein's thought about these principles in 1935–36 (Fine 1986, 36–37, 47, 50) seems to indicate that at that time Einstein had not at all distinguished these principles from each other or charted the experience with relativity makes alternatives to relativistic causal theories most implausible in a particularist world. In this way we find ourselves embracing causal locality applied to nonrelational values, then step (1) of the argument, and finally the Bell inequalities. When the Bell inequalities fail we are thrown into confusion. We can resolve the confusion by acknowledging and rejecting the assumption of particularism, as quantum theory shows we must.

5. The stochastic argument

Our expanded circle of ideas will prove its worth in clarifying the other argument for the Bell inequalities. One does not have to assume or argue determinate values in order to derive the Bell inequalities. Let us acknowledge that the spin-outcomes of measurement on A and on B are ultimately a matter of chance. It is possible, of course, that quantum mechanics is an incomplete theory, and that there are some further facts, summarized by the parameter, 'λ', which condition the probabilities for plus and minus outcomes of the measurements. We can cover any such possibility by talking about the probability $p^{AB}(x,y|i,j,\lambda)$: the probability for getting result x when spin is measured in direction i on particle A and the result y when spin is measured in direction j on particle B, all on the assumption that all unknown but determinate quantities have the value summarized by λ.[9]

Before our consciousness-raising about particularism and relational holism it was easy to think along the following lines. We suppose, as always, that the measurements are made so that light cannot make it from one measurement to the other. Under this circumstance, it would seem that the outcome on one wing cannot be affected either by the measurement setting or by the outcome on the other wing. So it would seem that the outcomes, x and y,

---

relations between them. In this respect, we have, on my account, continued to think as did Einstein in the 1930s. Much in the present paper bears considerable resemblance to ideas in Howard (1985).

[9]In my notation, I have written lambda as an explicit argument on which such probabilities are conditional instead of as a subscript as it appears in the notation in other contributions to this volume. I have done this because in what follows it will be very important to think of lambda as a condition on which such probabilities are conditional, and the alternative subscript notation is less forceful in keeping this conditionalizing role clearly in mind. λ cannot be allowed to cover too much. For example, it would be contrary to the intended analysis to let it include the spin outcomes, x and/or y. Some authors clarify by taking λ to include a complete description of the source of the two particles. More generally, we may allow it to include all facts, known and unknown, determinate in the overlap of the past light cones of the two spin measurement-events. Butterfield's contribution to this volume substantiates this characterization in fine detail, as well as supplementing other important parts of the argument of the present paper.

should be probabilistically independent of each other and of the measurement settings on the far wing. This would seem to require that:

(4) $p^{AB}(x,y|i,j,\lambda) = p^A(x|i,\lambda)p^B(y|j,\lambda)$ (Factorizability)

that is, that the joint probability be equal to the probability for getting outcome x when A is measured in direction i, conditional on λ, times the probability for getting outcome y when B is measured in direction j, again conditional on λ. Indeed, for years the whole Bell literature unquestioningly took factorizability to constitute locality and to be supported by the No Superluminal Propagation requirement of "relativity." Since factorizability implies the Bell inequalities and the Bell inequalities fail, we again seem to have a conflict with relativity.

6. Common causes, factorizability, and relativity

Let us think through in a little more detail the apparent conflict with relativity. How could a correlation between x and y occur? There could be a causal connection between the outcome or the measurement direction setting on one wing and the outcome on the other wing. This connection could determine an outcome completely, or in our stochastic setting, it might merely make the outcome somewhat more or less likely than it otherwise would have been. But this is not the only possible source of correlation. There could be a common cause, that is, some factor in the past which acts on both outcomes, x and y, making them more (or less) likely to occur together than if left to chance.

This idea of a common cause will be important to what follows, so it is worth examining in more detail. Sweethearts Jack and Jill are sadly separated for the month, Jack in New York and Jill in Los Angeles. As a romantic gesture, they have made arrangements so that each day at precisely 6:00P.M. Eastern Standard Time they will both buy an ice-cream cone, some days vanilla and some days chocolate, with a 50–50 probability for either flavor. However, though the flavors are picked randomly, they know that on any given day they will both be eating the same flavor. How have they arranged this mysterious togetherness? Every day an accomplice in Chicago flips a coin. If he gets heads, he mails Jack and Jill each a letter saying "chocolate," and if tails the letters say "vanilla." The coin toss is, in the relevant sense, the *Common Cause* of the correlated choice of flavors. Any flavor-correlation can be traced back to this common cause.

By conditionalizing on the common cause, the correlation can be "factored out" or "screened-off". This is true in the foregoing example but becomes clearer where the probabilities conditional on the coin-toss outcome are not zero or one. Let us suppose that Jack and Jill always order the flavor directed by the messages, but one time in ten the ice cream parlor, being out of stock of the chosen flavor, substitutes the other flavor. Conditional on the coin coming up heads, Jack and Jill each have a .9 chance of getting chocolate and a $.9 \times .9 = .81$ chance of both getting chocolate. Though the unconditional joint probabilities are still highly correlated, we see that they factor when conditionalized on the coin-toss outcome.

We can now return to explain why relativity seems to imply factorizability. In $p^{AB}(x,y|i,j,\lambda)$ any possible correlation due to a common cause has been "factored out" by making everything conditional on λ. Any correlation would then seem to have to result from some kind of direct causal link between A and B. But when the measurement-events occur sufficiently far apart in space and close in time, the No Superluminal Propagation requirement imposed by relativity theory seems to forbid any such link.

7. Analyzing factorizability

Since we are now trained to watch for the prejudice of particularism, much in this argument will already seem suspect. However, it is worth going over the argument in detail to see where and how the prejudice has operated. Jarrett (1984) has noted that factorizability is equivalent to the conjunction of two simpler conditions.[10] The first we call (following Shimony's terminology) *Parameter Independence*, because it says that the outcome on one wing (x for A and y and B) is independent of the measurement direction (the measurement "parameter"—i for A and j for B) on the other wing:

Parameter Independence:
$p^A(x|i,\lambda) = p^A(x|i,j,\lambda)$ and
$p^B(y|j,\lambda) = p^B(y|i,j,\lambda)$

The second condition, called *Outcome Independence*, says that, for given measurement directions on both wings, the outcome on one wing is independent of the outcome on the other wing. We say this by saying that the outcome, x for measurement on A is the same whether or not we conditionalize on y, the outcome for measurement of B:

Outcome Independence:
$p^A(x|i,j,\lambda) = p^A(x|i,j,y,\lambda)$, and
$p^B(y|i,j,\lambda) = p^B(y|i,j,x,\lambda)$.

---

[10]Jarrett's formulation, technically superior to what I present here, does not require tacit assumption of a probability distribution over the parameter settings. But what follows works well to introduce the ideas.

Parameter and outcome independence together imply factorizability, which in turn gives us the Bell inequalities. Since the Bell inequalities fail, at least one of parameter-independence and outcome-independence must also fail. Which one should we suspect?

Jarrett argued that, if parameter independence failed, one could in principle use that fact to send a superluminal message, a circumstance which he took clearly to be ruled out by relativity. How could one exploit failure of parameter independence to send a message? The measurement parameters, i and j, are under the experimenters' control. The B experimenter could exploit parameter-independence failure by systematically choosing a parameter setting to make his collaborator's probabilities at A higher. One event may not give the A experimenter very much to go on. But with sufficient repetitions the A experimenter will, with great reliability, be able to detect a relative frequency different from the one which he would be seeing if the B experimenter had made another choice of parameter setting.

Does relativity conflict with the possibility of sending such explicit superluminal messages? Yet again the answer is not clear-cut. We do not have sufficiently precise analyses of change of information, of causal chains, and of the connection between these. However, the capacity to convey information is as good a practical test as we have for the operation of a causal connection. Thus, for the sake of argument, I will accept the claim that, where a message can be sent, we have a causal chain, so that causal relativistic theories should apply, excluding faster-than-light signaling. Parameter-independence seems secure so that outcome-independence must be at fault.[11]

Could failure of outcome-independence likewise be used to send a superluminal signal? No, or at least not by the method which used parameter-independence failure. That method relied crucially on the fact that the parameter settings, i and j, are under the experimenters' control. To exploit outcome-indpendence failure in a similar way, the outcomes would have to be under the experimenters' control. But they are completely random.

No one has been able to suggest any explicit conflict between outcome-independence failure and "relativity." Still, failure of outcome-independence "feels" like some kind of locality failure. When a common cause has been ruled out, a robust correlation would seem automatically to count as some sort of connection between the locations of the correlated events. When causal

---

[11]Stairs has suggested an example which seems to show that, if laws of physics were very different, there could be superluminal signaling without any apparent conflict with relativity. However, the example turns on precisely the sort of nonsupervening relation which failure of outcome-independence provides for us, as I will argue below. Thus, if parameter independence should fail in the way that Stairs suggests, we will have exactly the same final conclusion which I will draw from the failure of outcome-independence.

locality is the only locality we know, it seems that the failure of locality which is involved in failure of outcome-independence must be some kind of failure of causal locality, and we cannot get away from the suspicion that there is some hidden conflict, or at least strain, with "relativity."

Now that we know that there is a second kind of locality, ontological locality or particularism, we have a new place to look in explaining our intuition that outcome-independence failure constitutes some kind of locality failure. Indeed, I will now argue that outcome-independence failure constitutes an instance of failure of particularism, which failure we know to be consistent with relativity as a framework principle and consistent with the circumstance that causal relativistic theories do not apply.

8.   Failure of outcome-independence as an instance of failure of particularism

I will proceed by arguing that particularism implies outcome-independence. At least it does, if one grants two assumptions, assumptions which I have been making throughout and which I want now to record more explicitly. I do not think I am alone in finding these assumptions compelling. For the first, the statistician Lincoln Moses wrote:

> When x and y have a high correlation, we can conclude that (1) x causes y, or (2) y causes x, or (3) some joint cause(s) influenced them both, or (4) some combination of these applies. (1986, 317)

As the reader must know by now, I do not think this is true without qualification. But when we assume particularism, applicable conceptions of causal connection and of requirements on explanation make such a condition seem compelling to most of us. I suggest that Moses was unwittingly presupposing particularism, never having considered any alternative. Thus he made his statement without qualification. Putting in the qualification of particularism and expressing it in our terminology, we have

> First Assumption: When particularism holds, nonaccidental correlations arise only through:
>   a) The action of a common cause, or
>   b) The action of a direct causal chain.

The second assumption I have made throughout:

> Second Assumption: When particularism holds, relativistic causal theories exclude superluminal causal propagation.

The argument now is easy. Consider the probabilities pertaining to the Bell experiment and assume particularism. By assumption 1 any nonaccidental correlations arise only through (1a) the action of a common cause or through (1b) the action of a direct causal chain. The action of a common cause has been eliminated by making everything conditional on λ, the list of all possible known and unknown common causes. The action of a direct causal chain is ruled out by assumption 2. Thus, there are no nonaccidental correlations conditional on λ. That is, outcome-independence holds.

Turning the argument around we see that, if one grants the assumptions about a particularist world, failure of outcome-independence must involve some kind of failure of particularism.

Some have found this argument so simple that they think it uninformative, with the conclusion too obviously packed into the assumptions. These critics make an important point, but they also miss what this argument does for us. Many find the assumptions extremely plausible. I believe that Moses spoke for all of us in taking assumption 1 to be an obvious truth, and for years all writers on Bell similarly endorsed 2. The assumption's restriction to particularism plugs loopholes in the argument which we have discovered in recent investigations, culminating with this study. All the same, the assumptions are substantive and need to be justified. This calls for some very difficult, and sorely needed clarification of our conceptions of cause and causal chains. I hope this work will be done.

Even so, this simple argument improves our understanding of how various parts of our yet insufficiently analysed conceptions fit together. In particular, the argument identifies the role to be played by particularism, a crucial part of the conceptual structure, and a part which previously had gone quite unnoticed.

9. Relational holism exemplified

I have argued that failure of outcome independence must involve some kind of failure of particularism. Since by "relational holism" I just mean a failure of particularism, that is, exemplification of one or more instances of nonsupervening relations, failure of outcome-independence should provide an instance of relational holism. I want now to try to show more intuitively what kind of instance of relational holism might be in question.

Let us think of probability in terms of objective chance or propensity. That is to say, an object having an objective probability for displaying a property (the display property) is for the object to have another property, a kind of "partially effective disposition." But, for objective probability, there is no "deeper" property which determines with certainty whether or not the display property will appear.

Arthur Fine has argued in the following way (1981, 1982c): We have a hard time swallowing this idea of objective probability. Our old determinist

prejudices and standards of explanation make us want to insist that there must
be some underlying circumstances which determine the outcome. But living
with quantum mechanics has made us more willing to put those intuitions
aside and to accept undetermined chance occurrences, mysterious though
such occurrences may seem. Now, let us suppose that we have a pair of
separated objects. Each has such an objective probability for a certain display
property. But in addition, the pair of objects displays an objective probability
for correlation. Such "random devices in harmony", as Fine called them,
together have a disposition to manifest their display properties in a correlated
way, without benefit of any common cause or direct causal chain. Why, he
challenged, should this be any more mysterious than the objective, undetermined chance behavior of the individual objects?

My reaction, needless to say, was negative. The display properties
couldn't just happen always, or usually, to come up together. There had to be
some explanation, some mechanism which accounted for the coordination.
This was long before I developed the distinction between particularism and
relational holism.[12] I was thinking of the world exclusively in particularist
terms. A particularist world has room for objective probabilities as nonrelational properties of individuals. But "random devices in harmony" are another matter. The two objects have *together* a propensity to turn up the display
property in unison. In other words, the fact of a correlation-propensity as an
objective property of the pair of objects is a relation which holds between
them. A particularist believes that there must be nonrelational properties of
the relata which underlie such a relation. And if there are such nonrelational
properties from which the relation arises it ought to be possible to explain the
relation in terms of the nonrelational, subvenient properties.

Relational holism sees the situation otherwise. The correlation—as an
objective property of the pair of objects taken together—is simply a fact about
the pair. This fact will arise from and give rise to other facts. But it need not
itself be decomposable in terms of or supervenient upon some more basic,
nonrelational facts. There need be no mechanism into which the correlation
can be analysed.

10. The lessons for explanation and ontology

We see here an important interplay between ontology and our standards
of explanation. It is said that in switching from Aristotelian to Newtonian

---

[12]This distinction followed from ideas I learned from Allen Stairs. Stairs's paper (forthcoming) also contains more specific ideas, to which the present section gives an alternative expression, especially his description of the singlet-state statistics as "nonlocal propensities that

physics, scientists simply shifted the standard of what needs explaining. For the Aristotelian, rectilinear motion needed explaining. Not so for the Newtonian. Moreover, it is said, there is no accounting for this change of standard of what needs explaining. It is just part of what it is to do Aristotelian or to do Newtonian physics. Similarly, some say that accepting the quantum correlations must involve a similar ungrounded shift in our standards of explanation. Before, correlations always called for an explanation. With quantum mechanics, we must simply change this standard.

Our analysis shows that there is more to be said. A particularist will feel the need to explain correlations because, for a particularist, the elements for such an explanation will inevitably be available. But relational holism takes there to be things in the world—nonsupervening relations—for which the elements of a particularist's explanation are simply not to hand. To change from a particularist ontology to that of relational holism enforces a correlative change in what kind of facts can be explained.

Study of the Bell inequalities has sharpened our understanding of the metaphysical lessons of quantum mechanics. We knew that quantum mechanics is "nonclassical." In the contrast between particularism and relational holism we now have a clarification of at least an important part of the departure from classical thinking. We learn that quantum mechanics has a broader subject matter than classical physics. This enables us to see that relativity is not shown to be false. Instead its relativistic causal variants apply correctly to the particularist subject matter for which they were designed, and broader, Lorentz-covariant theories apply where we are concerned with nonsupervening relations.

If "solving" the puzzles of the Bell inequalities meant somehow reconciling them with our old, particularist worldview, we have failed completely. If instead we try to read from the puzzles some more adequate ways in which to think about the world, we may have made some progress.

---

rest on nonlocal facts." Stairs and I are also eager to acknowledge inspiration from Shimony and many others.

HOLISM, SEPARABILITY, AND THE METAPHYSICAL
IMPLICATIONS OF THE BELL EXPERIMENTS

Don Howard

> The real difficulty lies in the fact that physics is a
> kind of metaphysics; physics describes "reality."
> But we do not know what "reality" is; we know it
> only by means of the physical description!
> —Einstein to Schrödinger, 19 June 1935

Seventeen years ago, just before the first experimental test of Bell's theorem, Howard Stein gave a paper in which he argued that "quantum mechanics poses no special problem of an epistemological kind," but that there is "a cluster of problems" concerning the "meaning" of the theory, problems "of a metaphysical . . . character," which "consist in unanswered questions *about the world*—the physical world" (Stein 1970, 93; see also Stein 1972). Stein was right. The problems that most interested him in 1970, namely, the measurement problem and wave-packet reduction, are not now in the forefront of our interests. But the history of subsequent work inspired by Bell's theorem demonstrates the truth of Stein's main point about the gaps in our understanding of the quantum world. In brief: We know that the quantum-

Much of this paper was written while I was a guest of the Center for Einstein Studies and the Center for the History and Philosophy of Science at Boston University; for providing me a home in this stimulating environment, I thank John Stachel and Robert S. Cohen. I wish also to thank Abner Shimony, Jon Jarrett, Paul Teller, Andre Mirabelli, and Klaus Hentschel, all of whom were patient with me in discussions of the topics treated here. Part of my work was supported under National Science Foundation Grant No. SES 8421040.

mechanical predictions regarding correlations between previously interacting systems violate the Bell inequalities in certain special cases; we know that these predictions thus differ from those given by what Bell called "local" theories; and we know that the quantum-mechanical predictions are confirmed by the Bell experiments. But we do not understand why—why a theory's being "local," in Bell's sense of the word, leads it to give the wrong predictions, nor why "nonlocal" quantum mechanics gives the right ones. Of course, some technical questions must be answered before we can achieve the desired understanding, but more than that is needed, for the puzzles about "nonlocality" are as much "metaphysical" as they are technical, and this in just the sense intended by Stein, inasmuch as they "lie beyond the present reach of [physics]." Understanding will come, therefore, only if we allow ourselves to indulge in a little metaphysics, only if we ask ourselves what "nonlocal" theories tell us about the world.[1] This is more than a little frightening to those of us who are the metaphysically repressed children and grandchildren of the Viennese diaspora. But duty calls, so sin, if sin we must.

What follows, then, is an attempt to tease out the metaphysical implications of Bell's theorem, its experimental tests, and, most importantly, its recent and revealing rederivation by Jon Jarrett (1983, 1984), all with the aim of understanding what kind of world would evince Bell "nonlocality." I will argue that the source of this "nonlocality" is not necessarily a violation of special relativistic locality constraints (the first-signal principle), but instead, perhaps, a kind of ontological holism or nonseparability (already hinted at in the orthodox quantum-mechanical interaction formalism), in which spatio-temporally separated but previously interacting physical systems lack separate physical states and perhaps also separate physical identities.

More specifically, I will argue that we confront here a possible violation of what I term the *spatio-temporal separability principle,* or just the *separability principle* for short. This is a fundamental ontological principle governing the individuation of physical systems and their associated states, a principle implicit in many classical physical theories. It asserts that the con-

---

[1]No sharp distinction of method or content between physics and metaphysics is intended here. I regard physics as aiming, first, to establish general principles (such as the relativity principle, the light principle, the first and second laws of thermodynamics) that function as constraints upon constructive models of the world. Developing the latter is the more properly metaphysical task; it is by the construction of models that we learn in what kind of world the physical principles can be realized. The two types of investigations are complementary, with the elaboration of new principles further constraining and thus guiding the search for models, and the development of new models helping to probe the limits of validity of the principles. If one prefers to view the construction of models as a task for physics itself, so be it; this is, after all, a question of terminology and thus of taste. But then I would insist on describing the constructive enterprise as the metaphysical moment or aspect of physics.

tents of any two regions of space-time separated by a nonvanishing spatio-temporal interval constitute separable physical systems, in the sense that (1) each possesses its own, distinct physical state, and (2) the joint state of the two systems is wholly determined by these separate states.[2] In other words, the separability principle asserts that the presence of a nonvanishing spatio-temporal interval is a *sufficient condition* for the individuation of physical systems and their associated states,[3] and that the states thus individuated exhaust the reality that physics aims to describe, that physical wholes are no more than the "sums" of their parts. In classical, prerelativistic physics, the analogous principle referred to spatial intervals and spatial separation, as opposed to spatio-temporal intervals, and thus should be called the *spatial separability principle*. But I will speak of just the separability principle, with the context indicating which version is intended.

There are two ways to deny the separability principle. The more modest concerns the individuation of states; it is the claim that spatio-temporally separated *systems* do not always possess separable *states*, that under certain circumstances either there are no separate states or the joint state is not completely determined by the separate states. I call this way of denying the separability principle the *nonseparability of states*. The more radical denial may be called the *nonseparability of systems*; it is the claim that spatio-temporal separation is not a sufficient condition for individuating *systems* themselves, that under certain circumstances the contents of two spatio-temporally separated regions of space-time constitute just a single system.

The separability principle must be distinguished from the *locality principle*. In its most general form, the locality principle (which is not to be confused with the Bell "locality" condition)[4] asserts that the state of a system

---

[2]How the joint state is determined by the separate states depends upon the details of a theory's mathematical formulation. At a minimum, the idea is that no information is contained in the joint state that is not already contained in the separate states, or, alternatively, that no measurement result could be predicted on the basis of the joint state that could not already be predicted on the basis of the separate states. I prefer to think of a physical state not as a cluster of definite properties (like the states of classical mechanics, which are representable by points in a phase space, corresponding to definite values of position and momentum), but more generally as a set of dispositions for the system to manifest certain properties under certain circumstances, which includes, as a special case, states conceived as clusters of definite properties. Accordingly, I define a state, λ, formally, as a conditional probability measure, $p_\lambda(x|m)$, assigning probabilities to measurement results, x, conditional upon the presence of measurement contexts, m. With states thus defined, to say that the joint state is wholly determined by the separate states is to say that the joint probability measure is the product of two separate measures.

[3]The presence of such an interval is also, of course, a necessary condition.

[4]The locality principle and the Bell "locality" condition both aim to express the same intuition about local action, but as I will argue in section 1, the Bell "locality" condition fails to do this in an unambiguous fashion. The terms, 'locality' and 'separability', have each been used

is unaffected by events in regions of the universe so removed from the given system that no signal could connect them. In classical physics, with no theoretical limit on signal velocities, that means any event simultaneous with the momentary state of the given system and separated from it by any finite spatial interval. The relativistic version of the principle asserts that a system's state is unaffected by events in regions of space-time separated from it by a spacelike interval. In either case, the aim of the locality principle is to rule out objectionable kinds of action-at-a-distance. In what follows, I will speak simply of the locality principle, allowing the context to determine whether the classical or the relativistic version is intended.

Locality assumes for its formulation the existence of separate states, but they need not be of the kind assumed by the separability principle; that is to say, they need not be such as to determine completely the joint state of every composite system to which the systems they characterize may belong as parts. Thus, it is possible to have a *local,* but *nonseparable* theory, quantum mechanics being the most important example.[5] The quantum theory is something of an exception, however, for many of our most important physical theories—among them general relativity and classical field theories, such as classical electrodynamics—satisfy both the locality and separability principles.[6] And the fact of their satisfying both principles is significant, for I will argue that all *local, separable* theories, including general relativity, are empirically false when applied to the kinds of microphysical interactions examined in the Bell experiments; or rather, that they would have to be false if one elaborated them into theories capable of describing such microphysical interactions. If one is unwilling to sacrifice locality, the assumption of separability must be recognized as the source of the difficulty. I will also argue that local, separable theories are fundamentally incompatible with quantum mechanics because of

---

in a variety of different ways in the literature on Bell's theorem and on the interpretation of quantum mechanics, so it is important to attend carefully to the definitions being given to them here.

[5]According to the quantum-mechanical interaction formalism, two previously interacting systems possess a joint state not representable as the product of separate states, at least until such time as one of the two systems undergoes a subsequent interaction, such as a measurement. See below, section 1 and section 2, n. 16, for more on the sense in which quantum mechanics is a local theory.

[6]Special relativity can also be given a field-theoretic formulation of the kind we associate with Minkowski, in which case it too would count as a local, separable theory. But for reasons to be elaborated below, I think it a mistake to build separability into special relativity, and so I will not include it among the class of local, separable theories. There are, of course, also some *nonlocal, separable* theories to be found chiefly among the nonlocal hidden variable theories. But they will not be discussed here. And it should be mentioned for the sake of thoroughness that one can imagine theories that are both *nonlocal,* and *nonseparable,* though why one would go to the trouble of constructing such a theory is not clear.

their separable manner of individuating systems and states. This last fact, especially, should be appreciated. For years we have worried that Bell's theorem and the Bell experiments, by exhibiting a kind of "nonlocality" in quantum mechanics, point to a conflict between quantum mechanics and special relativity. Now, however, we find that the conflict lies not there, but between quantum mechanics and general relativity, and that it concerns the fundamental issue of the manner in which the two theories individuate systems and states. This result is pregnant with implications for a variety of problems, not least of which is the quest for a unified fundamental theory incorporating all of the basic forces, including the strong and weak nuclear forces, electromagnetic forces, and gravitation.

All of these results point to the importance of understanding *nonseparability*. We confront here a radical physical holism at odds with our classical intuitions about the individuation of systems and states, and it is precisely this feature of the quantum theory that enables it to provide the correct predictions in the Bell experiments. But the quantum formalism by itself offers neither a deeper explanation of nonseparability nor an account of its larger significance for our understanding of the physical world. This is where physics stops and where metaphysics must show the way, at least until the path is clear enough to allow physics to proceed again.

1. Locality, separability, and the Bell experiments: A nontechnical summary of the formal issues

Bell's theorem (Bell 1964) concerns a simple experiment in which one measures correlations between observables of two spatio-temporally separated, but previously interacting systems, here labeled A and B.[7] At the heart of the theorem is the Bell "locality" condition, which aims to capture the intuition that measurement results in each of the two "wings" of the Bell experiment depend only upon circumstances in the local environment of the measurement-event in that wing. This condition takes the form of a requirement that the joint probability for obtaining one result in the A-wing and another in the B-wing be the product of the separate probabilities for those results, the argument being that if the result in one wing is determined solely by local circumstances in that wing, then it is statistically independent of the result in the other wing, so that the joint probability is calculated according to the ordinary product rule for the compound probability of independent

---

[7]For a sketch of Bell's theorem and its experimental tests, see James T. Cushing, "A background essay," this volume, and Clauser and Shimony (1978). A thorough, recent discussion may be found in Redhead (1987b, 82–118).

events.[8] Bell's theorem asserts that the predictions of any theory whose description of the interaction satisfies this "locality" condition must necessarily satisfy, in turn, a certain inequality, the "Bell inequality," which is violated in special cases by the predictions of the quantum theory.[9]

In the experimental tests of Bell's theorem, culminating in the Aspect experiments (Aspect, Dalibard, and Roger 1982), the quantum-mechanical predictions have been consistently confirmed, sometimes with striking precision.[10] These empirical violations of the Bell inequality, taken together with Bell's theorem, thus entail a violation of the Bell "locality" condition by nature itself as well as by quantum mechanics. But here is where the puzzles begin, because as the example of the quantum theory shows, Bell "nonlocality" apparently need not involve a violation of special relativistic locality constraints.

Little progress was made in understanding this state of affairs until Jon Jarrett (1983, 1984) proved that the original Bell "locality" condition is really a conjunction of two logically independent conditions. The first of these requires the stochastic independence of a measurement result in one wing from the selection of an observable to be measured in the other wing. Jarrett calls it "locality," arguing that it is more deserving of the name than the Bell "locality" condition, since, as he claims, it is entailed by the first-signal principle of special relativity. Shimony (1986) recommends the more neutral term, "parameter independence." The other condition, which Jarrett calls "completeness"[11] and Shimony (1986) calls "outcome independence," as-

---

[8]In the version relevant to the present discussion, this condition is:

$$p_\lambda^{AB}(x, y|i, j) = p_\lambda^A(x|i) \cdot p_\lambda^B(y|j),$$

where x and y represent measurement outcomes, i and j the observables measured, in the A and B wings, respectively (for the notation, see Cushing, "A background essay,"). Correlations between the two measurement results are not excluded, indeed they are expected, given that the measured observables are assumed to satisfy a conservation principle; one merely assumes that the correlations are the result of prior programming, as it were, from the time of the interaction (an instance of a "common cause"), and not the result of any current distant conspiracy between the two wings.

[9]In this paper, the original Bell "locality" condition and its cousins are all called Bell "locality," deliberately ignoring the differences among them (the quotation marks being employed to distinguish Bell "locality" from the different notion of locality to be defined below). Similarly, the term "Bell inequality" refers, indifferently, to the various different versions of the inequality, and the term "Bell experiments" to all of the experimental tests of Bell's theorem. See Redhead (1987b, 82–118) for a discussion of some of the distinctions that are here suppressed.

[10]For a survey of the experimental results through 1978, see Clauser and Shimony (1978); an up-to-date survey is found in Redhead (1987b, 107–113).

[11]This is not the happiest choice of terminology. As is noted by Shimony (1984a, 226), a theory like quantum mechanics can fail to satisfy this condition and still be "complete" in the

serts the stochastic independence of the measurement result in one wing, not
from the observable chosen for measurement in the other wing, but from the
result obtained there.[12] On Jarrett's analysis, a violation of the Bell inequality
need *not* entail relativistic nonlocality, because it may result *either* from a
violation of the Jarrett locality condition, which would perhaps entail relativistic nonlocality, *or* from a violation of his completeness condition. Quantum mechanics, for example, violates completeness but satisfies Jarrett
locality.

But while significant progress has thus been achieved, some puzzles
remain. For one thing, the connection between Jarrett locality and special
relativity is not as clear as it might be. Jarrett's own argument is that violation
of his locality condition in the case of spacelike separated measurement events
makes possible superluminal signaling, so that special relativistic prohibitions
on the latter entail the locality condition. I find it more helpful to note that
satisfaction of what is here called the locality *principle* directly entails satisfaction of Jarrett's locality *condition* in such cases, if one assumes that measurement results are completely determined by the state of the measured
system and those circumstances in its immediate environment constituting the
measurement context.

More troublesome by far, however, is the fact that the *physical*
significance of Jarrett's completeness condition and the *physical* significance
of its violation in nature and in the quantum theory are not at all clear. What
are the physical conditions needed to secure the independence of a measurement outcome in one wing from the outcome in the other? And how would
one explain physically the opposite circumstance, the dependence of an outcome in one wing upon the outcome in the other wing?

This is where separability enters the picture, for Jarrett's completeness
condition turns out to be equivalent to what I call the *separability condition*,
which simply asserts that each of the two previously interacting systems in the
Bell experiments possesses its own physical state, the joint state being the
product of these separate states (Howard 1987).[13] It should not be surprising

---

sense that its description of the joint state of A and B may contain all possible information. It is
also not clear that Jarrett's ''completeness'' is the same as that intended by the Einstein,
Podolsky, and Rosen (1935, 777) ''completeness condition.''

[12]For an outline of the proof of Jarrett's theorem, see Cushing, ''A background essay.''

[13]The existence of the separate states follows straightforwardly from the identifications:

$$p_\alpha^A(x|i, j) = p_\lambda^A(x|i, j) \text{ and } p_\beta^B(y|i, j) = p_\lambda^B(y|i, j),$$

where α and β represent the separate states of the systems in the A and B wings, and λ represents
the joint state. (Recall that I define a state as a conditional probability measure assigning
probabilities to outcomes conditional upon the presence of global measurement contexts [see
above, n. 2]; here the relevant global contexts for the measurements in the A-wing and in the B-

that separability plays a part here, since the most novel, nonclassical feature of the quantum-mechanical interaction formalism is precisely its denial of the separability of the states of the two systems. Nevertheless, in the original proof of Bell's theorem, as in the proof of Jarrett's theorem, a single joint state for the two systems was assumed, in the belief that one thereby achieved a greater generality (see, in particular, Bell 1964, 196). But this generality turns out to be spurious: Any theory whose predictions satisfy the Bell inequality tacitly assigns separate physical states to the two systems, such that the joint state is the product of the separate states, whether or not that fact is explicitly recognized in the formalism of the theory (Howard 1987).[14]

The separability *principle* provides sufficient grounds for the satisfaction of the separability *condition*, just as the locality *principle* provides sufficient grounds for the satisfaction of the Jarrett locality *condition*. Since there is a nonvanishing spatio-temporal separation between the two measuring events in the Bell experiments, the spatio-temporal separability *principle* implies that the systems involved are indeed two, and that they possess separate physical states, the joint state being wholly determined by these separate states. The separability *condition* is a formal statement of the latter circumstance, the existence of separate physical states and a factorizable joint state. Its violation would entail that the two systems do not possess separate

---

wing are both determined by the choice of both parameters (i, j).) The essential step consists in noting that Jarrett's completeness condition:

$$p_\lambda^A (x|i, j, y) = p_\lambda^A(x|i, j) \text{ and } p_\lambda^B(y|i, j, x) = p_\lambda^B(y|i, j),$$

is equivalent to the factorizability condition (my separability condition):

$$p_\lambda^{AB} (x,y|i,j) = p_\alpha^A(x|i,j) \cdot p_\beta^B(y|i,j)$$

(only the definition of conditional probability is required). The separability condition plus the Jarrett locality condition (with α and β, respectively, in place of λ):

$$p_\alpha^A(x|i,j) = p_\alpha^A(x|i) \text{ and } p_\beta^B(y|i,j) = p_\beta^B(y|j),$$

together yield immediately the Bell-type "locality" condition (what Jarrett calls "strong locality") in the form:

$$p_\lambda^{AB}(x,y|i,j) = p_\alpha^A(x|i) \cdot p_\beta^B(y|j).$$

For a critical discussion of this analysis, see French (to appear).

[14]The Bell-type "locality" condition entails Jarrett locality if one assumes that the separate probabilities are defined as the marginals of the joint probability, for example:

$$p_\alpha^A(x|i,j) = \Sigma_y \, p_\lambda^{AB}(x,y|i,j)$$

(which definition also yields the existence of separate states, α and β); and then Bell-type "locality" together with Jarrett locality entail in an obvious way that the separate states (probability measures), α and β, satisfy the requisite factorizability condition.

physical states of such kind that the joint state is a product of the separate ones, and thus implies at least what I call the nonseparability of states.[15]

Let me summarize the formal situation. With the help of the Bell and Jarrett theorems, it can be shown that any theory whose account of interactions satisfies both the Jarrett locality condition and the separability condition yields predictions for certain correlation measurements that satisfy the Bell inequality. But in the Bell experiments the Bell inequality is violated in special cases, and in these cases quantum mechanics gives the right predictions. It follows that one (or both) of the locality and separability conditions is violated, which, in turn, implies that one or both of the locality and separability principles must be denied. Quantum mechanics denies the latter.

In the remainder of this paper, I will focus almost exclusively upon the violation of the separability principle. My reasons for leaving the locality principle untouched are partly theoretical, deriving from the special theory of relativity, and partly methodological, deriving, as we shall see, from considerations of the conditions necessary for theory testing. And it is hardly irrelevant that our one correct theory of microphysical interactions, the quantum theory, is a local, nonseparable theory. But most important among my reasons for focusing on nonseparability is simply the fact that I believe it to be the more interesting way out of the Bell experiments, the way more likely to yield new insights that will be useful in our search for a more comprehensive fundamental physical theory.

2.   Field theories and separability

As far as I can determine, Einstein was the first to point out the fundamental role of the separability principle in field theories. His reflections on the quantum theory led him to distinguish two principles that are essentially the same as the locality and separability principles, and to conclude that their conjunction entails the incompleteness of quantum mechanics. The argument is simple. Consider the kind of physical situation investigated in the Bell experiments, involving measurements upon two previously interacting systems, A and B. If A and B, having between them a spacelike interval, are separable, then each possesses its own physical state. If, furthermore, the locality principle is satisfied, if, that is, the state of B is unaffected by events in the vicinity of A, then the physical state of B remains the same regardless of what we choose to do with A. But quantum mechanics assigns different ψ-functions to B, depending upon the parameter measured on A and the result of

---

[15]Whether or not it also drives us to consider the more radical nonseparability of systems themselves is discussed below.

that measurement.[16] Therefore, if we agree that completeness requires the assignment of one and only one theoretical state (ψ-function) to a system in a given physical state, then quantum mechanics is incomplete.[17]

In the course of what may be his clearest statement of this argument, Einstein (1948) wrote:

> If one asks what is characteristic of the realm of physical ideas independently of the quantum theory, then above all the following attracts our attention: the concepts of physics refer to a real external world, i.e., ideas are posited of things that claim a "real existence" independent of the perceiving subject (bodies, fields, etc.). . . . Moreover, it is characteristic of these physical things that they are conceived of as being arranged in a space-time continuum. Further, it appears to be essential for this arrangement of the things introduced in physics that, at a specific time, these things claim an existence independent of one another, insofar as these things "lie in different parts of space." Without such an assumption of the mutually independent existence (the "being-thus") of spatially distant things, an assumption which originates in everyday thought, physical thought in the sense familiar to us would not be possible. Nor does one see how physical laws could be formulated and tested without such a clean separation. Field theory has carried out this principle to the extreme, in that it localizes within infinitely small (four-dimensional) space-elements the elementary things existing independently of one another that it takes as basic, as well as the elementary laws it postulates for them.

---

[16]This does not mean that quantum mechanics violates the locality condition. In the sense of "state" defined above (see n. 2), the quantum mechanical "state" of B—that is, the probabilities for the possible outcomes of measurements on B, given various measurement contexts—depends not upon the choice of the parameter to measure on A, which would entail violation of the locality condition (see above, n. 13), but only upon the outcome of the measurement on A. Thus, quantum mechanics violates not locality (parameter independence), but separability (outcome independence = Jarrett's "completeness" condition). It may appear, nevertheless, that the state of B is changed by "events" in a distant region of the universe, namely, by the outcome of a measurement performed there, so that while no violation of the locality *condition* obtains, the locality *principle,* is violated. But it should be noted, first, that the "separate" states that we assign to A and B according to the quantum interaction formalism are dependent upon the joint state, which furnishes, in principle, the only correct description of A and B, and that the "separate" state of B is not changed by any "events" in the vicinity of A that do not also change the joint state. Second, it should be noted that the outcome of a measurement on A is not a "distant event" in the same way that, say, setting the parameter to be measured at A is, since this outcome is a function not only of circumstances in the local environment of A, but also of the joint state of A and B, a state that bridges the gap, as it were, between the two systems.

[17]See Howard (1985), where I argue that, in his correspondence with Schrödinger, Einstein repudiated the EPR incompleteness argument in the summer of 1935, a few weeks after its publication, favoring from that time on the incompleteness argument sketched here, an argument differing significantly from the EPR argument.

For the relative independence of spatially distant things (*A* and *B*), this idea is characteristic: an external influence on *A* has no *immediate* effect on *B*; this is known as the "principle of local action," which is applied consistently only in field theory. The complete suspension of this basic principle would make impossible the idea of the existence of (quasi-) closed systems and, thereby, the establishment of empirically testable laws in the sense familiar to us. (Einstein 1948, 321–322; author's translation)

Einstein's "principle of local action" and his "assumption of the mutually independent existence of spatially distant things" correspond, respectively, to the locality and separability principles.

Below I will consider the connection that Einstein suggests between these two principles and the possibilities of formulating and testing physical theories. For now I want to consider Einstein's comments about the manner in which field theories express the "assumption of the mutually independent existence of spatially distant things," the separability principle.

A field theory typically assumes as its fundamental ontology a set of points, a manifold in the parlance of the mathematician, together with a topology and a metric defined upon the points of that manifold. Partly for reasons of mathematical convenience, the topology is taken to be identical to that of a corresponding mathematical continuum—three-dimensional (R^3) in the case of classical field theories, four-dimensional (R^4) in the case of general relativity.[18] One does one's physics by first defining upon each of the points of this physical manifold mathematical structures representing the physical structures fundamental to that particular field theory, and then postulating fundamental laws governing the time-evolution of these mathematical structures (at least in the classical case) and the functional dependence of their values at any one point upon the values at other points. Thus, classical electrodynamics postulates a continuous, three-dimensional spatial manifold (once taken to constitute the aether), and defines at each of its points vectors representing the electric and magnetic fields, vectors whose evolution and functional relationships are governed by Maxwell's equations. General relativity postulates a continuous, four-dimensional space-time manifold, and defines upon its points the metric tensor and the stress-energy tensor governed by Einstein's gravitational field equations.[19]

---

[18]Or at least the topology of any suitably small piece of the physical manifold is assumed to be identical with that of a piece of the appropriate mathematical continuum.

[19]How much of the structure defined upon the manifold is deemed to have physical content depends upon the particular field theory under consideration. For example, in classical electrodynamics the metrical structure, which determines the geometry of the manifold, is considered nonphysical, part of the *a priori* conceptual background of our physical theory, whereas general relativity invests this metrical structure with physical content. But for our immediate purposes, such differences are inessential.

But while different field theories may postulate different structures, what is essential to all field theories is that *some* structure is postulated and that this structure is assumed to be well defined at every point of the manifold. It is also an essential characteristic of field theories that the structure thus defined is taken to exhaust the physical reality that the theory aims to describe. To know the strength of the electric and magnetic fields (or the corresponding potentials) at every point of space in a given region is to know all there is to know about the electromagnetic field within that entire region. Similarly, to know the values of the ten components of the metric tensor at every point within a given region of the space-time manifold is to know all there is to know about the gravitational field in that region. In this sense, field theories are radically reductionistic: the whole reality of a field in a given region is contained in its parts, that is to say, its points.[20]

One consequence of this last characteristic of field theories will emerge with special significance. It is that when one sets about describing physical interactions within the framework of a field theory, the only way to do it is in terms of functional relationships among the structures separately well defined at each of the points involved in the interaction. Thus, the value of the electric field at point A may be changed by virtue of an interaction between the field at this point and the field at point B (typically a point immediately adjacent to A, the field at B itself interacting with one of its immediate neighbors, C, and so on); but the interaction can consist in nothing more than such a change in the value of the field *here* because of the value of the field *there*.

Einstein's point about separability and field theories is now twofold. First, in taking the field, understood as a continuous manifold of points, as the basic reality described by a theory, we tacitly assume that each point of the manifold constitutes a separate physical system. Thus my reading of Einstein's comment about "localiz[ing] within infinitely small . . . space-elements the elementary things existing independently of one another that it takes as basic." Second, by assuming that the fundamental structures defined on the manifold (like the vectors representing electric and magnetic fields, or the metrical and stress-energy tensors) are well defined at every point and that they exhaust the reality described by the theory, we tacitly assume that a separate physical state is assigned to each of the point-systems and that the joint state of any set of such point-systems is wholly determined by the states of its constituents. Thus, my reading of Einstein's remark about "localiz-[ing] . . . the elementary laws it postulates for them [the point-systems]," since the fundamental structures (states) are what the fundamental laws govern.

---

[20]This is not to deny that the value of a field at one point may be functionally dependent upon the values at other points. But even in such a case the value of the field is well defined at each point of the underlying manifold, and that is the property essential to a field theory.

Putting these two remarks together, we can now understand the larger point Einstein intended to make about field theories. It is that by modeling a physical ontology upon the ontology of the mathematical manifold, we take over as a criterion for the individuation of *physical* systems and states within field theories the mathematician's criterion for the individuation of *mathematical* points. This criterion is the existence between two points of a nonvanishing interval, which gets interpreted as a three-dimensional spatial interval in classical electrodynamics, and as a four-dimensional spatio-temporal or metrical interval in general relativity. In this way, field theories—as understood by Einstein—necessarily satisfy the separability principle.[21]

Einstein also remarked that field theories carry out the separability principle "to the extreme." What he means is simply that the field-theoretic criteria of individuation yield an ontology of infinitesimal point-systems. But this extreme is not required by the separability principle, which demands only that the presence of a nonvanishing spatio-temporal interval be a sufficient condition for the individuation of systems and states. One way to avoid the field-theoretic extreme is to admit physical systems only of finite magnitude in one's ontology. Thus, in classical mechanics, where the three-dimensional spatial manifold plays the role of a container, a background against which physical events are played out, and where the systems described may be of any finite size whatsoever, spatial separation is still, implicitly, a sufficient condition for individuation. Another way to avoid the field-theoretic extreme is to assume that there is a minimum finite spatial or spatio-temporal interval, as one does in "finite" or "discrete geometries." As long, however, as one takes the presence of intervals of this size or larger as a sufficient condition for individuation, the separability principle is still respected.[22]

A crucial assumption necessary to secure the possibility of the field-theoretic way of implementing the separability principle (or, for that matter,

---

[21]The conception of field theories outlined here is similar in most respects to what is called in recent literature a "space-time theory"; see, for example, Friedman (1983, 32–70). The one important difference is that I do not insist that the states assigned to each point of the field correspond to a set of wholly definite properties. The historically important field theories, like classical electrodynamics and general relativity, have that form, but the basic field-theoretic structure is more general, allowing for states incorporating intrinsically indefinite properties (such as propensities), as long as the states themselves are definite, in the sense of being mathematically well defined. What is important is not the definiteness or indefiniteness of the properties, but the criteria whereby the systems and states are individuated. I should also note that my conception of field theories has even more in common with, indeed it is very nearly identical with the point of view that Paul Teller calls "particularism" (see Teller, this volume).

[22]This last observation implies that one does not deny the separability principle merely by assuming a discrete as opposed to a continuous manifold, as in theories postulating the existence of smallest possible "atoms" or "quanta" of space and time. Questions about separability cut deeper than the questions raised in the old debate over continuous versus discrete space and time.

the possibility of *any* way of implementing it) is that the spatial or spatio-temporal intervals whose presence is taken to be a sufficient condition for individuation are in some sense *objective*. This is no problem in classical field theories, where the full Euclidean structure is taken for granted, and with it the objectivity of all spatial intervals. In general relativity, however, matters are more complicated, because here the metrical structure of space-time is incorporated into the physics of gravitation, with the consequence that spatial intervals lose the objectivity they possessed classically. But the spatio-temporal or metrical interval: $ds^2 = g_{ij}dx_i dx_j$, is objective in general relativity, since it is invariant under arbitrary continuous coordinate transformations; this is why it takes the place of the spatial interval in the general relativistic version of separability. Indeed, one who demands separability in a physical theory may see in this circumstance an argument for covariance with respect to the group of continuous transformations as at least a minimum necessary condition on a physical theory, because enlargement of the transformation group threatens to deprive ds^2 of its invariant status.

One can, of course, employ the field-theoretic apparatus for the sake of its mathematical covenience, without thereby assuming that physical reality is represented by an ontology of separable point-systems. For the field can be regarded as an approximation to the physical reality being described, as in hydrodynamics, where the discontinuous molecular microstructure of a fluid is ignored for reasons of mathematical convenience. But in order to adopt this attitude—regarding the continuous field as approximating a reality with a different microstructure—it is necessary that one have in reserve an alternative criterion for the individuation of microsystems. In hydrodynamics, this criterion is provided by the atomic-molecular theory of the constitution of matter. The problem takes on a different aspect, however, in the case of field theories regarded as fundamental theories, where there is, by hypothesis, no other level of structure that could provide alternative criteria of individuation. Lacking such, it is hard to imagine any criterion other than that implicit in the structure of the mathematical manifold. This is not to say that there can be no alternative criteria of individuation; the point is rather that the criteria offered by the mathematical manifold seem more natural for lack of an evident alternative.[23]

---

[23]Notice that the ontology of field theories does not exclude the possibility of composite systems made up of sets of point-systems. But it does imply, first, that the state of any such composite system is completely determined by the separate states of its constituent point-systems, and, second, that under all circumstances the composite system is decomposable, in theory, into spatially (or spatio-temporally) individuated parts that are separable in the sense of possessing their own separate states that determine collectively the state of the whole, there being no theoretical limit to this decomposition excepting the ideal limit represented by the fundamental point-systems themselves.

Against the background furnished by the field-theoretic embodiment of the separability principle, the locality principle—Einstein's "principle of local action"—takes its traditional place, asserting that the state assigned to any point-system will be unaffected by events in "distant" regions, meaning, in the principle's relativistic versions, any events separated from the given point-system by a spacelike interval. The locality principle is thus an essential supplement to the separability principle, necessary to secure the traditional aim of field theories: elimination of action-at-a-distance, and with it the kind of ghostly conspiracies between events in different regions of the universe that could give rise to causal anomalies.

We commonly regard locality constraints as deriving from the first-signal principle of special relativity. Should the latter therefore be included among the theories whose employment of the typical field theoretic ontology convicts them also of endorsement of the separability principle? Special relativity can be formulated as a field theory; Minkowski was the first to do it in a formally satisfactory way. In this version, special relativity is necessarily a separable theory, the basic difference between special and general relativity being then simply that the former assumes a flat, quasi-Euclidean metric, and the latter a non-flat, variable metric. But the Minkowski formulation is only one version of special relativity, and it is an historical accident that we associate this formulation with the theory itself.

For our purposes, it is better to think of special relativity as an instance of what Einstein (1919) called a "theory of principle," consisting not of a constructive model—the manifold and metric of the Minkowski formulation—but of a set of regulative principles providing constraints on any possible constructive model. In the case of special relativity, these regulative principles are (1) the principle of (special) relativity itself, which asserts, in one version, the kinematic equivalence of all inertial reference frames, or that physical laws take the same form in all inertial frames, and (2) the light principle, which asserts that in an inertial frame the velocity of light is a constant, independent of the velocity of the source relative to that inertial frame. The first-signal principle and, thus, the locality principle, are arguably implied by (1) and (2), whereas the separability principle is not.

Now that we have a better understanding of how general relativity and other field theories satisfy the separability principle, let us reconsider more carefully the strong claims made above to the effect that (1) the Bell experiments imply the falsity of any fundamental microtheory based upon general relativity, and that (2) any such microtheory would be incompatible with quantum mechanics. My point is really a very simple one. To take general relativity—in its field theoretic formulation—as a basis for a fundamental microtheory is to take the ontology of general relativity as the starting point for the ontology of one's microtheory. It is to assume that, at root, the only reality is the space-time manifold and the mathematical structures (metric tensor, stress-energy tensor, etc.) defined upon the points of that manifold. And that means, most importantly, respecting the criteria of individuation for systems and states implicit in general relativity. In short, it means that one's microtheory will satisfy the separability principle.

One may want to define additional structures upon the points of the space-time manifold, say in order to explain interactions other than those mediated by gravitational and electromagnetic forces, but as long as these structures are well defined for every point of the manifold, and are understood as determining completely the relevant properties of any composite system, the separability principle will be satisfied. If one then employs this microtheory to explain the interactions investigated in the Bell experiments, one will, of necessity, assign separate states to the two interacting systems, of such a kind that the joint state is wholly determined by those separate states, and thus, one's description of the interaction will satisfy the separability condition. This already implies a fundamental incompatibility between such a theory and the quantum mechanical explanation of the interaction in question.

If, in addition to the assumption of the criteria of individuation inherited from general relativity—which entails satisfaction of the separability principle—one assumes locality, then the microtheory's account of the Bell interaction will necessarily give the wrong predictions for the correlation measurements in the Bell experiments, since it necessarily satisfies both the separability and the locality conditions. Thus, any fundamental microtheory built in this fashion upon the foundation of the field-theoretic space-time structure embodied in general relativity will be both empirically false and incompatible with quantum mechanics.

It might be argued that quantum field theories represent a counterexample to the point just made, inasmuch as they seem to combine the basic ontology of the space-time manifold with a typically nonseparable structure of quantum mechanical states. But if what I have argued up until now is correct, this is an impossible combination. And, in fact, the ontological picture of quantum field theories is not at all that clear, precisely because of its attempt to marry the field and particle ontologies at a fundamental level. The enterprise seems to succeed, after a fashion, for the quantum theory of free fields, the various states of which (aside from the vacuum state) can be identified with systems of noninteracting particles. But as soon as one attempts to describe interactions in the context of quantum field theory, the many notorious difficulties that have beset the program from its inception in the 1930s (e.g., the infinite self-energy of the electron, arising from its interaction with its own electromagnetic field) begin to set in, difficulties that can be remedied only by *ad hoc* expedients like renormalization. That the difficulties start here should come as no surprise, however, given the foregoing analysis; for it is precisely in the context of interactions that nonseparability rears its head.

3. Arguments for separability (and for locality)

In the long quotation above, Einstein argued that the separability principle is necessary because "without such an assumption of the mutually independent existence (the "being-thus") of spatially distant things . . . physical thought in the sense familiar to us would not be possible. Nor does one see how physical laws could be formulated and tested without such a clean separation" (Einstein 1948, 321). He followed this with an argument for the necessity of the locality principle, an argument that tied locality to the possibility of "establish[ing] . . . empirically testable laws in the sense familiar to us" (322).

But before looking into these arguments more closely, I want to consider another comment of Einstein's. It dates from March 1948, around the time when Einstein wrote the article containing the previous quotation. The occasion was Max Born's having sent to Einstein the manuscript of his Waynflete lectures (Born 1949), seeking Einstein's reaction to his discussion of Einstein's attitude toward quantum mechanics. Einstein responded with a number of what he himself characterized as "caustic marginal comments" (quoted in Born 1969, 221), and at the end of the manuscript he wrote the following:

> I just want to explain what I mean when I say that we should try to hold on to physical reality. We are, to be sure, all of us aware of the situation regarding what will turn out to be the basic foundational concepts in physics: the point-mass or the particle is surely not among them; the field, in the Faraday-Maxwell sense, might be, but not with certainty. But that which we conceive as existing ("actual") should somehow be localized in time and space. That is, the real in one part of space, *A,* should (in theory) somehow "exist" independently of that which is thought of as real in another part of space, *B.* If a physical system stretches over the parts of space *A and B,* then what is present in *B* should somehow have an existence independent of what is present in *A.* What is actually present in *B* should thus not depend upon the type of measurement carried out in the part of space, *A;* it should also be independent of whether or not, after all, a measurement is made in *A.*
>
> If one adheres to this program, then one can hardly view the quantum-theoretical description as a *complete* representation of the physically real. If one attempts, nevertheless, so to view it, then one must assume that the physically

real in *B* undergoes a sudden change because of a measurement in *A*. My physical instincts bristle at that suggestion.

However, if one renounces the assumption that what is present in different parts of space has an independent, real existence, then I do not at all see what physics is supposed to describe. For what is thought to be a "system" is, after all, just conventional, and I do not see how one is supposed to divide up the world objectively so that one can make statements about the parts. (Einstein to Born, 24 March 1948, in Born 1969, 223–224; author's translation)

Part of this passage recapitulates in abbreviated form the argument that I earlier attributed to Einstein, in which the incompleteness of quantum mechanics is said to follow from the conjunction of the locality and separability principles. And the second paragraph evaluates the prospects for escaping this conclusion by denying locality. What most interests me, however, is the last paragraph, where Einstein considers the denial of separability.

Einstein's assertion that if separability is denied "then I do not at all see what physics is supposed to describe" echoes his previously quoted remark to the effect that the separability principle is a necessary condition for the possibility of formulating a physical theory. But now he adds a supporting argument. He says, first, that the concept of a "system" is conventional, by which I take him to mean that a criterion of individuation is, logically, a convention, dictated neither by empirical considerations, nor by *a priori* principles. Since we must therefore choose a criterion of individuation, so Einstein implies, we must at least choose an *objective* one. And, concluding, he suggests that the separability principle provides the only imaginable or conceivable objective criterion. Einstein is thus giving a *methodological* justification for the *physical* principle of separability—some scheme of individuation is needed if we are to formulate our theories—but the methodological argument rests upon a further *physical* assumption, namely, that spatio-temporal separation is the only conceivable objective criterion of individuation.

No one will deny the need for objective criteria of individuation. But there may be debate about Einstein's claim that the choice of a criterion is conventional, and there should be debate about the claim that separability is the only imaginable or conceivable objective criterion. What lies behind these two claims? Since Einstein himself offers no further explanation, let me offer a hypothetical reconstruction of his reasoning.

The thesis of the conventionality of criteria of individuation has both a global and a local context in Einstein's thinking. The global context is Einstein's articulation and defense, for at least the previous thirty years, of a conventionalist philosophy of science, conventionalist in roughly the holistic,

Duhemian sense, similar in its essentials to the view that we now associate with the Quine of "Two Dogmas of Empiricism" (Quine 1951).[24] From this point of view, any assertion in a larger body of theory may be adjusted so as to secure the accommodation of the whole theory to the available evidence, since it is only the whole theory that stands the test of experience. That is to say, no part of a theory is granted immunity from revision on such grounds as its alleged *a priori* necessity, nor is the choice of the features to be revised forced upon us by experience. In short, every individual proposition belonging to a theory has the status of a convention.

This version of conventionalism must be contrasted with the Schlick-Reichenbach version, which confines the conventions to the coordinating definitions or bridge-principles—deeming these devoid of physical or empirical content—and maintains that the remaining genuinely empirical assertions each meet the test of experience individually (see, e.g., Schlick 1936, and Reichenbach 1924, 1–9). For our purposes, the point of the contrast between the two kinds of conventionalism is that the Duhem-Einstein-Quine variety accords the status of a convention not only to definitions, but also to assertions possessing physical content.[25] Thus, it is possible for Einstein to regard the choice of a criterion of individuation, a choice with abundant implications for the way we do physics, as a matter of convention.

The local context for Einstein's ascription of conventional status to criteria of individuation is his commitment to field theories. Ignore for the moment the implicit criterion of individuation that field theories borrow from the continuous mathematical manifold, and think of the "field" as an undifferentiated "stuff" filling space (or space-time). To do physics at all, we must somehow divide this undifferentiated "stuff" into distinct physical systems that will serve as the subjects of predication for our physics. But if this "field of stuff" is the fundamental physical reality, if, that is, no extrinsic criteria for the individuation of systems and states are found in another layer of structure, then the "field" does not of itself fall apart, as it were, along any inherent lines of division. Neither logic, nor *a priori* principle, nor experience compel us to partition the field in a given way.

Our choice of a partition has, therefore, the logical status of a convention, determined only by considerations of mathematical and physical convenience. Mathematical convenience is achieved by a partitioning that permits the employment of familiar tools, like the differential calculus. Physical con-

---

[24]For further development and documentation of this theme, see Howard (1984, 1988).

[25]In casting doubt upon the analytic-synthetic distinction, this variety of conventionalism questions also the legitimacy of a principled distinction between definitions and empirical propositions. For Einstein's questioning of the latter distinction, see Einstein (1936, 316) and Howard (1988).

venience is achieved by one that conduces to the overall simplicity of our physical laws. The only conceivable *a priori* constraint upon the choice of a partitioning, or a criterion of individuation, is that the criterion be objective.

What then of Einstein's claim that the separability principle represents the only imaginable objective criterion of individuation? Since I will argue below that there are other objective criteria, it is important to understand how Einstein reached this conclusion. Notice, first, that he did not say that separability is the only *possible* objective criterion, but that it is the only *imaginable* or *conceivable* one, arguing that if separability is given up, then "I do not *see* how one is supposed to divide up the world objectively so that one can make statements about the parts." Imaginability or conceivability are subjective matters; we are each endowed with different powers. But *Einstein* did not suffer from a weak imagination. These capacities, though subjective, are conditioned by objective historical factors—to a large extent, what we can imagine or conceive depends upon how our imaginings and conceivings have been schooled, and most importantly upon the models with which we have been outfitted. So to understand what Einstein could or could not imagine or conceive, we must look to the relevant history.

And there is an interesting history, going back at least to the beginnings of atomism among the Greeks. There is an inherent logic of atomism that drives one, inevitably, regardless of where one begins, to three conclusions. The first of what I might call these "lines of force" in atomism leads through the distinction between primary and secondary qualities, to the Cartesian and Newtonian conclusion that only the "numerical" or "mathematical" properties of physical bodies count as objective, primary qualities. And even these are gradually pared away until one is left with a purely spatial property, such as *position,* as the sole objective criterion for distinguishing physical systems.

The second line of force leads through the doctrine of the divisibility of matter to the conclusion that no finite physical structures can be ultimate or fundamental, that any finite system must have concealed within it a deeper structure, more basic parts that can be taken apart, at least in theory, so that nothing short of the infinitesimal point-particle can be fundamental. And the third line of force leads through the impossibility of explaining interaction in terms of contact action between perfectly elastic ("hard") finite atoms, to the conclusion that the "spaces" between atoms must be filled continuously by something capable of mediating interactions.

All three conclusions met with criticism in their day. Leibniz was the most forceful critic of the first, arguing from the relational doctrine of space to the conclusion that position has no absolute significance and thus cannot serve as the ground for distinguishing physical systems. He stated this conclusion most clearly in a fragment from around 1696: "All things which are different must be distinguished in some way, and in the case of real things position alone is not a sufficient means of distinction. This overthrows the whole of purely corpuscularian philosophy."[26] But none of the criticisms prevailed at the time, and with the emergence of the field-theoretic point of view in the late eighteenth and nineteenth centuries, in the work of Boscovich, Faraday, and Maxwell, the three lines of force of classical atomism found their ultimate expression.[27]

Einstein inherited this tradition, and his remarks about the separability principle as the only objective criterion of individuation must be seen against that background. His one major departure from the tradition, of course, was his siding with Leibniz in favor of the relational theory of space (or space-time). But he did not follow Leibniz all the way to the conclusion that adoption of the relational point of view deprives us of our last possible objective criterion of individuation. For while position loses its absolute, objective status from the relational point of view, Einstein saw where Leibniz did not that a frankly *relational* property, namely, spatial or spatio-temporal separation, in the guise of the metrical interval, can take the place of position as an objective ground for individuation, since it is a relativistic invariant.

Nevertheless, Einstein's way of seeing the world was shaped by this tradition, so much so that the *only* alternative he could find to position as a ground of individuation was another spatial (or, again, spatio-temporal) property. This constraint was made even more severe by Einstein's having collapsed the distinction between matter and space-time, between matter and geometry. For Einstein, all physical properties are, from a fundamental point of view, geometrical properties. The metrical interval being the only invariant among the geometrical properties, and hence the only objective property, means that it is the only candidate as a ground for individuation.

---

[26]G. W. Leibniz, "[Sur le principe des indiscernables]" (1696), in *Opuscules et fragments inédits de Leibniz*, ed. L. Couturat (Paris: Alcan, 1903), p. 8; English translation "On the principle of indiscernibles," in *Leibniz: Philosophical Writings*, ed. G. H. R. Parkinson (London: Dent, 1973), p. 133. Berkeley was perhaps the most famous critic of the second line of thought, that leading to the existence of infinitesimals, though, of course, this criticism was directed primarily against infinitesimals in mathematics. He contended, among other things, that a composite whole could not contain an infinite number of parts (*The Analyst* [Dublin and London, 1734], reprinted in *The Works of George Berkeley*, vol. 3, ed. A. C. Fraser [Oxford: Clarendon Press, 1901], pp. 1–60; see especially Queries 5 and 19). The essential arguments in this controversy are preserved in Kant's second antinomy (*Kritik der reinen Vernunft* [Riga: Hartknoch, 1781], pp. 434–443), and they survive to this day in debates over continuous versus discrete geometries.

[27]I mean quite deliberately to deny the common representation of the difference between atomistic theories and field theories as a fundamental metaphysical difference. I see in field theories the inevitable culmination of the inherent logic of atomism; they represent atomism carried to its logical extreme—a sea of infinitesimal atoms, any two atoms having between them a continuum of other atoms.

Thus Einstein's argument for the claim that the separability principle provides the only conceivable objective criterion of individuation, and that it is therefore necessary for the possibility of *formulating* physical theories. What about his related argument concerning both the separability and locality principles, to the effect that each is necessary for the possibility of *testing* physical theories—again a *methodological* argument? Recall his exact words. First, regarding separability: "Nor does one see how physical laws could be formulated *and tested* without such a clean separation." And then, regarding locality: "The complete suspension of this basic principle would make impossible the idea of the existence of (quasi-) closed systems and, thereby, the establishment of empirically testable laws in the sense familiar to us." Einstein gives us here again no further explanation of his reasoning, so we have to do more reconstruction.

Think about testing from an abstract point of view. We do physics by first dividing the world into parts that we call systems, then ascribing states to these systems, and then, finally, postulating laws governing the evolution of these states and their functional relationships. When we test, we look for some property thought to belong to a system in a given state because our laws imply the presence of that property—say a value of spin, position, or linear momentum—as the result of the system's evolution from an earlier state to the present one. We seek such properties through measurement, or more generally, observation. And if we are realistic in our attitude toward measurement, we assume that the result of a measurement is determined by the state of the measured system, at least to an extent sufficient to license inferences from measurement results to the presence or absence of the property sought.

Why would the separability principle be necessary for testing, understood thus? Einstein's answer is, I think, a simple one. Some method for individuating systems is necessary, for if one did not individuate, if one did not divide the world into parts, then the reference of any claims we might make about the world would be indefinite, or rather, the reference would comprehend the whole universe. One might make measurements, but *to what* would one ascribe the properties thought to be revealed by those measurements? Thus, some scheme of individuation is necessary. Believing that the presence of a spatio-temporal interval is the *only* objective basis for individuation, Einstein concludes that the separability principle is necessary to secure the possibility of testing as well as formulating theories.

What then about Einstein's claim that the locality principle is also necessary? The locality principle says that a system's state is not influenced by events in "distant" regions of space (or space-time). Suppose that we have suspended the locality principle, allowing such influences; suppose furthermore that we perform a measurement to test a claim about a system's being in a specific state, its possessing a specific property; and suppose finally that we get a result other than the predicted one. Does this mean that our claim about that system's being in that state was wrong? Not at all. It could always be the case that the measurement result was affected in some unforeseen fashion by one of these "distant" influences. Unless they are screened off, we cannot trust the measurement results to give us reliable information about the state of the observed system. Thus, Einstein says that the locality principle is necessary in order to secure the existence of closed systems, and *therefore* also to secure the possibility of testing theories.

Such an argument for the necessity of the locality principle does not aim to establish any particular upper bound on signal velocities as a necessary condition for theory testing. Instead, what it requires is a kind of theoretical closure. The project of physical science could withstand the discovery of super-luminal signals, for example, as long as a theory were developed to account for them. What Einstein's argument requires is that current physical theory establish *some* upper bound on signal velocities in order to secure the possibility of its own testing, recognizing full well that one possible—and not at all unreasonable—response to results inconsistent with the theory is to raise this upper bound. Testing becomes impossible only if the theory in question establishes *no* upper bound, for then one cannot define the concept of a "closed system."[28]

With Einstein's claim that the locality principle is a necessary condition for theory testing, I am in complete agreement, for exactly the reasons that Einstein gives. I agree as well with his claim that some scheme for individuating systems is necessary in order to formulate and test scientific theories. But I disagree with the more specific claim that the spatio-temporal separability principle is necessary, because I doubt Einstein's claim that it provides the only imaginable objective criterion of individuation. Agreeing with him on this last point would entail one's declaring the quantum theory, which violates the separability principle, to be, in effect, a fundamentally incoherent theory (and I suspect that such a worry lay behind Einstein's reservations regarding the quantum theory's candidacy as an acceptable fundamental theory). But this is a step that I do not feel compelled to take.

---

[28]Notice the kind of argument that Einstein does not use to justify the locality principle. Historically, the concept of local action, a stepchild of the concept of contact action in the mechanistic worldview, was preferred because the alternative was thought to be inconceivable. Thus, both Hume and Berkeley argued that there is no clear idea corresponding to the concept of a force of gravity. But we now understand that the argument was really circular. Nonlocal action is inconceivable because we cannot conceive—what?—a mechanism that would explain it by the mediation of local effects. But this circularity was not so obvious to earlier generations of thinkers. And thus the felt need for aethers and other metaphysical anesthetics to dull the pain of a broken worldview. No—Einstein's argument is not that nonlocal action is inconceivable. It is easy to imagine such effects. The problem is that their existence would make hash of physical science.

4. The possibilities for a nonseparable ontology

Where do we stand now? To begin with, the empirical evidence of the Bell experiments forces us to give up either the separability principle or the locality principle. But there are good theoretical and methodological reasons for retaining the latter, since the locality principle is arguably entailed by special relativistic constraints on action-at-a-distance, and since it is also arguably a necessary condition for securing the testability of our theories. Let me now add two more reasons. First, Aspect's sophisticated version of the Bell experiment (Aspect, Dalibard, and Roger 1982), in which the orientations of the analyzers and thus the observables to be measured are switched while the particles are in flight, provides us with a lower bound on the speed of the superluminal signals whose existence would be entailed by denial of the locality principle. In theory, one could raise that lower bound arbitrarily, either by increasing the distance between the analyzers in the two wings of the experiment, or by increasing the frequency with which the analyzer orientations are switched. There is, however, no a priori reason to expect that the troublesome quantum correlations leading to violations of the Bell inequalities will disappear at any specific value of this lower bound, and so the strategy of denying the locality principle threatens to turn into what Lakatos called a degenerating research program.

Second, the locality principle partakes more of the character of those high-level principles, like the conservation of energy, the second law of thermodynamics, and the light principle—I like to call them regulative principles—that Einstein (1919) said should serve as constraints in the search for constructive theories, whereas the separability principle partakes more of the character of a constructive hypothesis. Like Einstein, I believe that ultimate understanding is provided only by a constructive theory; but, also like Einstein, I believe that any particular constructive hypothesis should bow to the authority of regulative principles that, like the locality principle, enjoy considerable empirical substantiation. And so I would argue that the locality principle ought to be given the benefit of the doubt. The burden of proof should fall upon those who prefer nonlocality to nonseparability.

That leaves repudiation of the spatio-temporal separability principle as the only alternative. But some criterion for individuating systems and states seems to be necessary if we are to formulate and test physical theories. So the question becomes, what are the alternatives to the separability principle, which means, more specifically, what kinds of comprehensive, fundamental nonseparable theories can we imagine?

Some constraints control our imaginings. Most importantly, the pattern of correlations revealed by the Bell experiments and predicted by the quantum theory must be reproduced by any acceptable fundamental theory, these correlations themselves playing the role of a kind of regulative principle guiding the search for constructive fundamental theories. And some features of general relativity should probably be preserved, if only in the macroscopic limit, such as the principle of general covariance (which may prove to be an extremely weak constraint) and the insight into the connection between geometry and gravitation. On the other hand, the field-theoretic criterion of individuation cannot play a fundamental role—at least not for the individuation of states. But within these constraints, we should give imagination free rein.

One might first ask whether or not quantum mechanics itself could serve as the starting point for a fundamental theory, since it is a local, nonseparable theory. What is more, it seems to supply an objective criterion of individuation alternative to the one provided by the spatio-temporal separability principle. For it appears that at least an operational criterion of individuation is available precisely in the *nonexistence* of the troubling kinds of quantum correlations that lead to violations of the Bell inequalities. Thus, any two regions of space-time between which such correlations did not exist would be accounted separate systems possessing their own separate states that wholly determine the joint state.

But there is a problem, because in theory any two regions of space-time between which there is a timelike separation have to be regarded as being in interaction with one another, owing to the pervasiveness of gravitational and electromagnetic forces. So there ought to be quantum correlations between *any* two such regions of space-time, however weak those correlations might be and however difficult they might be to detect experimentally. In many cases the correlations will be so weak as to be practically negligible, and thus we may have here a practical criterion of individuation. But at the level of fundamental theory, the proposed criterion of individuation—the nonexistence of quantum correlations—is almost no criterion at all, since the correlations are so widespread.

In fact, this problem is not merely an objection to regarding quantum mechanics as a satisfactory fundamental theory, it raises questions about the very coherence of quantum mechanics as it is ordinarily employed. Strictly speaking, the quantum theory of interactions implies that we should write down one grand nonfactorizable state function for the whole of the forward light cone of any event. Of course we do not do this, but we have *no* fundamental principle that justifies our ignoring this radical nonseparability of quantum states. As a practical matter, we say that the correlations are negligible in most cases, the fundamental theoretical problem being swept under the rug.

Another reason why quantum mechanics itself cannot serve as a model for a satisfactory fundamental theory is that while it individuates *states* in a nonseparable fashion, it nevertheless implicitly individuates *systems* according to the criterion assumed in the spatio-temporal separability principle. Will this work? My answer is, first, that I do not see the point of individuating systems and states differently. What is a system if it has *no* set of properties that it can call its own? In what sense can we even talk of a system if we cannot predicate anything of it alone? One might reply that systems individuated after the fashion of the separability principle still do have their own properties, such as mass or charge. But these properties are not sufficient to distinguish two otherwise identical systems in the same way that spatio-temporal separation is thought, classically, to be adequate for individuation. Moreover, it would be unwise to distinguish systems on the basis of properties like charge and mass when we lack a fundamental theory of those very properties, at least a fundamental microtheory for them.

There is, however, a still more serious objection to the reactionary strategy of clinging to spatio-temporal separation as a criterion of individuation for systems when it has been abandoned as a criterion of individuation for states. It is that quantum nonseparability infects even the spatio-temporal location of interacting systems, so that there is no objective basis for asserting, in the case of two interacting systems, A and B, that system A is at position x^A and system B at position x^B. At best, one can assign definite probabilities to the possible values of the *relative* separation between the two, but even then one cannot say which particle is which.

From one vantage point, this should not be surprising. In a fundamental theory of the kind sought by Einstein, at least, the aim—not yet achieved in general relativity—is to absorb *all* of the physical properties of systems into the geometry and topology of the universe; or, from another point of view, to absorb the geometry and (at least some of) the topology of the universe into the physics of the systems inhabiting that universe. We should expect that in such a fundamental theory *all* of the defining features of the space-time manifold would be candidates for inclusion among the properties determined by the physical states of our systems, just as the metrical structure of space-time is determined by the distribution of mass-energy in general relativity. But then, if the nonseparable manner of individuating states is to be the norm for all physical properties, it should affect also the spatio-temporal location of systems. The mistake is thinking that the structure of the space-time manifold can be insulated from the nonseparability that affects the rest of our physics, so that this manifold stands alone as a ground of individuation. And this is an objection not only to regarding quantum mechanics as a fundamental theory; it is objection to any attempt to retain the spatio-temporal separability of systems after having abandoned it for states.[29]

---

[29]There are additional unsolved problems of interpretation in quantum mechanics to which one might also point as objections to according fundamental status to the quantum theory, such as the measurement problem and wave-packet reduction. But I do not cite them here, because I do

If quantum mechanics itself is not a candidate for the kind of fundamental nonseparable theory that we seek, what other possibilities exist? In particular, what are the possibilities for a theory that would be nonseparable both in the way it individuates states and in the way it individuates systems? We might, of course, be driven back to the tacit quantum-mechanical strategy of a nonseparable scheme for individuating states combined with a more classical field-theoretic scheme for individuating systems. But let us try the more radical program, if only to see how far it can be pressed before it fails.

Notice that we pass here from the realm of analytical metaphysics—where we try to uncover and assess the metaphysical implications of existing physical theory—into the realm of speculative metaphysics—where we try to extend existing physical theory, guided by the insights gained through the earlier analysis. Such speculations make no pretense to being themselves adequate physical theories. They are, instead, just the philosopher's hints and suggestions, there for the physicist to do with as he or she sees fit.

One possibility that comes quickly to mind involves a reconsideration of the nature and role of the metrical interval in field theories modeled upon general relativity. What enabled Einstein to regard the existence of a nonvanishing metrical or spatio-temporal interval as a criterion of individuation for systems and states is the fact that, within general relativity, the metrical interval is invariant under arbitrary continuous coordinate transformations. But what if we could construct a theory within which what appears from one point of view to be a non-null metrical interval separating two regions of space-time appears to be a null interval from another point of view? By "point of view" I mean here neither a reference frame nor a coordinatization, but something more like the kind of interaction between the two regions.

The result would be a *contextual* criterion of individuation—two systems distinguished from the point of view of one kind of interaction, say a gravitational one, may count as but one system from the point of view of another kind of interaction, say one governed by the strong nuclear force. Only if the two regions were separable from all points of view would they be considered totally separable, constituting two genuinely distinct physical systems.

---

not believe that they are the deep problems they are frequently taken to be. It is not often enough stressed that both problems concern interactions, this being, again, the context in which quantum nonseparability is evinced. The problem of measurement arises only because the nonfactorizability of the post-measurement joint object-apparatus state leads to the apparatus's being afflicted by the same indefiniteness that originally afflicted the object. For its part, wave-packet reduction is held to occur only when we perform an observation, which is to say only when the system in question interacts with another system. Neither of these problems will find a satisfactory solution until we have a better understanding of nonseparability.

How might this idea be realized mathematically? Perhaps we need a structure like the ordinary four-dimensional space-time manifold upon which we construct several overlapping geometries, a gravitational geometry, an electromagnetic geometry, and so forth, each with its own metric. I am tempted to suggest that gravity would be responsible for the macrogeometry of the universe, the other forces giving rise to microgeometries on ever smaller scales, geometries responsible for tiny "wrinkles" in the macrogeometry. But, of course, there cannot be such a simple hierarchy of scales, for quantum mechanics implies that the nonseparability arising from electromagnetic or weak nuclear interactions can affect regions of space-time "separated" in the sense of the gravitational geometry by intervals of arbitrary magnitude.

Another way to realize the idea of individuations relative to different kinds of interactions involves a higher-dimensional theory, say one in which gravity is responsible for the geometry of the ordinary four dimensions of space and time, with other forces being responsible for the geometries of the higher dimensions. Thus two systems might be separable within the first four dimensions, that is to say, separated by a nonvanishing metrical interval that is invariant under arbitrary continuous four-dimensional transformations, but nonseparable in some of the higher dimensions. A third realization might involve the introduction of different manifolds, each with its own metric, for the different interactions.

Proceeding in any of these ways, we might find that from the point of view of the geometry of gravitation all regions of space-time are in fact separable, nonseparability being confined to the microgeometry or to the geometries of other dimensions. My guess, however, is that nonseparability arises from gravitational interactions as well, but that the resulting quantum-gravitational correlations are so weak as not to be evident in most situations.

The second to the last of the just-mentioned alternatives, that involving higher-dimensional geometries, would present to us the aspect of what might be termed "backdoor" connections, higher-dimensional connections between systems that appear distinct within the ordinary four dimensions. One can imagine yet another, far more unusual way to realize "backdoor" connections. This would involve just a four-dimensional manifold, but one with a topology different from that assumed in general relativity. And it would involve identifying the physical systems not with pieces of the manifold, but with various kinds of holes in the manifold. What I have in mind may be represented by a two-dimensional model of a sheet with two "holes" in it that are connected by a "tube." Let the "holes" represent two apparently distinct systems, the apparent "distance" between them over the sheet being relatively large, and let the "tube" represent the connection between them engendered by a previous interaction. What appear to be two distinct "holes" from within the sheet, appear to be merely the "ends" of just a single tubular "channel" from within the "tube." Moreover, the "distance" through the "tube," that is to say the "length" of the "tube," may be made as small as we wish, depending upon the strength of the interaction, by manipulating the metric within the "tube." We can even imagine "tubes" that grow "longer" and "thinner" as the correlations engendered by an interaction grow weaker. And we can also imagine new "tubes" growing as new interactions arise, growing perhaps out of "pipettes"—extremely "thin" "tubes" that we may assume always to connect any two regions of the sheet.

The limit case, perhaps corresponding to such phenomena as pair-creation, may be represented by a "thread"—an infinitely "thin" and sometimes infinitely "short" "pipette"—blossoming into a "tube." There is even room here for deeper layers of structure, for we may imagine both "intratubes" connecting two regions of the wall of a single "tube," and "intertubes" connecting regions of the walls of different "tubes." There is no limit to the depth of structure achievable in this way. And the resulting picture of an infinitely complex "tubular" lattice is really quite a beautiful one. Notice that this suggestion requires our making not only the geometry of the universe dependent upon dynamical considerations, but also the fine details of topological structure as well. I actually find this an attractive feature of the model, because I have long been puzzled as to why in general relativity there is in some respects an *a priori* topological structure, whereas the geometry is merged with the physics. If one can entertain a geometrodynamics, why not also a topodynamics?

Allow me to describe now one final possibility that I can imagine for a fundamental nonseparable theory, the one that is the most radical from an ontological point of view. Earlier, I conceded that some criterion, at least, for the individuation of systems is necessary in order for us even to formulate a physical theory. Perhaps that was too hasty a concession. Perhaps it is possible to adopt a kind of radical ontological holism in which the whole of the forward light cone of any event is regarded as one nonseparable whole because of the pervasiveness of physical interactions, and yet to do this without lapsing into the silence of scientific nirvana. That is to say, maybe we can opt for radical ontological holism and still do some physics.

The possibility of this approach is suggested by a feature of the quantum-mechanical interaction formalism that is too often not emphasized. For while it is true that quantum mechanics assigns to previously interacting systems a single, nonfactorizable joint state, it is also true that once one specifies a "measurement-context," that is to say a set of co-measurable observables for both systems, one can then always construct a factorizable joint state (strictly speaking, a mixture over factorizable joint states) that will give the same predictions for measurements of those observables that would otherwise be given by the nonfactorizable joint state.[30] Of course, this factorizable joint state will always give the wrong predictions for at least one observable not belonging to the set that defines the original context, but this circumstance, which is but a symptom of the underlying nonseparability, does not detract from the fact that, for the observables associated with the specified context, the factorizable state gives all of the right predictions. And one may even pretend, if one wishes, that the separate factors of this joint state are associated with physically separate systems, as long as one remembers that this is really not the case.

The application of contextual factorizability or contextual separability in a fundamental theory should be obvious. The contents of the forward light cone of any event would constitute a single, nonseparable whole, and this single "system" would be assigned a single, nonfactorizable joint state for all properties, including spatio-temporal location. But in any given context, that is, given a specification of the observables we wish to investigate, we can always find a factorizable state description giving us all of the information we need. This factorizable state description would not be the whole truth, but it would represent all of the truth that is accessible in the given context. Or to put the idea differently, the universe is "really" one, but once we put a specific question to it, it falls apart quite naturally into apparent parts.

There you have it. If these metaphysical speculations are a bit woolly-headed, so be it. As I said, it is up to the physicist to decide what to do with them. But there is a limit to my woolly-headedness. About two things I am quite serious. The first is that the problem of nonseparability must be faced squarely. Quantum mechanics is nonseparable, at least when it comes to the individuation of states, but it is not a fundamental theory. At the same time, the nonseparability evinced by quantum mechanics and confirmed by the Bell experiments (assuming that we retain locality) suggests that the route to a satisfactory fundamental theory does not lie through traditional field theories, like general relativity, owing to their manner of individuating systems and states according to the spatio-temporal separability principle. We are thus a long way from finding the kind of fundamental theory that we need.

The second point about which I am serious is that the construction of a satisfactory fundamental theory will require creative imagination of a kind all too rare among contemporary physicists and philosophers of science. What I am talking about is the need for speculative metaphysics, a kind of imagining that by definition carries us beyond the bounds of current physical theory. Here is one place where the philosopher, who should be less inhibited than the physicist, can help to show the way.

---

[30]The relevant theorem is proved in Howard (1979, 382–386).

BOHR ON BELL

HENRY J. FOLSE

Niels Bohr died in 1962, two years before John Bell first opened the way to what appears to be an experimental test of the fundamental issues that Bohr and Einstein had debated a generation earlier. Consequently any reconstruction of how Bohr would "explain" the Bell phenomena must be at least partially speculative, but the exercise can be justified for two reasons. First it can help us to appreciate better the strengths and subtleties of Bohr's framework of complementarity, a position which is at least as widely misunderstood as it is "officially" accepted. Second, although Bohr's vocabulary is far different from the one that physicists and philosophers use today to discuss the Bell inequalities, a proper "translation" of his way of speaking into today's vocabulary can, I believe, reveal Bohr's penetrating insight into the strangeness of physical reality at the microlevel, a strangeness which a philosophy of nature for contemporary physics must finally confront. Bohr still has much to say which has not generally been heard.

Throughout much of the fifty years that Bohr's genius shaped the course of atomic physics, his outlook was in constant interaction with that of Einstein. Thus recent work in uncovering the fundamental presuppositions behind Einstein's rejection of the completeness of quantum theory helps illuminate his opposition to Bohr, and hence indirectly Bohr's own viewpoint. Unfortunately the documentary evidence which has shed much light on the deep thoughts of Einstein does not exist in Bohr's case. Thus what follows must be considered to some extent speculative, not only because I am expressing Bohr's ideas in a new idiom, but because what Bohr has given us is incomplete from a *philosophical* viewpoint, for he never draws out the metaphysical consequences of what he called the "epistemological lesson" taught by the quantum revolution.

It is common to complain about Bohr's obscure and idiosyncratic "jargon" which masks rather than reveals the mysterious nature of the quantum domain. That same complaint is also commonly directed against the cosmologies of Aristotle, Hegel, and Whitehead, for any new and different framework speaks a new and different language, even if the same words are employed. To the outsider such a discourse seems a verbal mumbo-jumbo while to the insider it is likely to appear the clearest way of making the point intended. Bohr's discourse was conditioned by the historical route he personally followed from Planck's discovery of the quantum of action to Heisenberg's formulation of the first consistent quantum mechanics. Today's physicists have not followed that same path; they are not seeking to explain the same phenomena, so it is not surprising that they have adopted a way of speaking very different from Bohr's. Furthermore like all who strive to express a vision new to their age, Bohr's discourse was inevitably retarded by the concrete historical modes of thought of his time. The deep-seated distrust of metaphysics that reigned among scientific thinkers throughout Bohr's prime inevitably acted to suppress expression of the metaphysical consequences of his new outlook, leaving that part of his hoped-for revolution in fragmentary bits and pieces. Nevertheless, even though the speech of physicists in the opening decades of this century barely resembles that of the closing decades, sensitivity to the problematic facing Bohr and Einstein allows at least a partial translation of Bohr's vocabulary into that used in contemporary discussions of the Bell phenomena. The following account of "Bohr on Bell" should be understood in the light of these remarks; a vision of another man's vision, a reconstruction which, for what truth it contains, can give glimpses into the thought that shaped a new order of physics, into the philosophical lessons that the quantum revolution teaches about the description of nature.

1. Bohr's path to EPR

Although it is common to understand Bohr's atomic theory in terms of the model of an orbiting electron which "jumps" from one orbit to another, this does not reflect the way Bohr saw his youthful achievement. Bohr understood his revolutionary break in terms of postulating the nonclassical stability of the "stationary states" of the atomic system from which it changes *discontinuously* when it interacts with the radiation field. This fundamental move was so deeply ingrained into Bohr's outlook that he was more sensitive than most to the fact that this break with the classical scheme—he called it the "quantum postulate"—implied that the description of atomic systems required a deep readjustment of how we are to understand the classical mechanical "pictures" of particles or waves moving through space and time.[1]

---

[1]The "Como Lecture" of 1927 is titled "The Quantum Postulate and the Recent Development of Atomic Theory." The third paragraph begins: "This postulate implies a renunciation as regards the causal space-time co-ordination of atomic process" (Bohr 1934, 52–53).

For the twelve years from his atomic model to the first consistent quantum mechanics Bohr's writings reveal that he fretted constantly over what he called "space-time description." At first, he suspected that the truth contained in the space-time pictures could be preserved at the price of discarding strict conservation of energy, but the pressure of experiment forced him to recognize that one could give either a space-time account or a "causal" account in which the classical conservation principles are respected, but one could not give both simultaneously. This insight, occurring roughly contemporaneously with Heisenberg's derivation of the uncertainly relations, secured his belief that the description of atomic phenomena taught a fundamental epistemological lesson about the use of concepts in the description of nature, a lesson that a new framework was required for the description of atomic phenomena, a framework which he presented as a "rational generalization" of the classical conceptual scheme. This is what he called complementarity (Folse 1985).

Today we think in terms of the complementarity of wave and particle descriptions, but that outlook was derivative, not fundamental in Bohr's thinking. Physicists who heard Bohr's message were not so keen on hearing about fundamental epistemological lessons or a new conceptual framework, but they were anxious to adopt a position on wave-particle dualism. Thus they most commonly took Bohr's admonitions about the limits of visualizability to mean that since the space-time stories we tell cannot be regarded as conceptual snapshots of an observer-independent reality, we have a free license to use one picture on one occasion and the other on another, depending on the phenomenon we are seeking to explain. What the uncertainty principle exposed was that nature could, as it were, never be caught (observed) in the contradiction of behaving like both simultaneously. Thus the inconsistent mixing of concepts was safeguarded from simultaneously telling contradictory stories by a consistent formalism limited by the uncertainty relations.

For this reason, in the early debates the uncertainty relations took the heat from the attack on the quantum description. Disbelievers naturally asked why we are thus limited and defenders replied with thought experiments that showed that because of the discontinuous interaction between the quantum system and the apparatus designed to fix its position or momentum, one could never beat the uncertainty limitations. This emphasis on the fact that there was no way in which *observed phenomena* could beat the uncertainty relations made the new *Copenhagenergeist* totally compatible with Machian phenomenalism and Vienna Circle positivism, both of which appealed to many thinkers of a scientific bent.

The resulting association of these approaches to the description of nature with Bohr's has been a constant source of misunderstanding of complementarity. That it is a misunderstanding should be obvious from the fact that

Bohr's claims about the limitations of space-time descriptions as representations of an observer-independent reality were based on what he believed to be a discovery about how nature really behaved: *atomic systems change state discontinuously*. It had nothing to do with any antimetaphysical argument about the unknowability of a reality behind the phenomena. Bohr's view was not that we know nothing of the real nature of atomic systems, but because their real nature is to change state discontinuously, we cannot represent their behavior in a space-time description. It was not the view that the only real things are the observable phenomena, but it was the view that the only way we can know about the behavior of real atomic systems is through their interactions with the observing systems which are the phenomena that serve as the theory's empirical base. Bohr's view was realistic, but the association with positivism made it seem antirealistic.

2. Bohr on EPR

The turn of the debate from the Solvay encounters of 1927 and 1930 in which the uncertainty relations held the limelight to the EPR argument is in a sense a turn from physical to metaphysical arguments. In the earlier encounters, Einstein sought to prove the *inconsistency* of the formalism with its empirical base by designing a thought experiment in which *physically possible phenomena* revealed more about nature than the theory permitted us to know. In the latter case, the argument urges the *incompleteness* of the formalism on the grounds that any reasonable concept of physical reality (quite apart from its empirical manifestation in physically possible phenomena) allows us to infer that there is more to reality than the theory permits us to know. Of course the use of imagination in the *Gedankenexperiment* is such that the notion of a "physically possible phenomenon" in some sense relies on implicit presuppositions about physical reality. But, even though no experiment was ever performed, Bohr's defense against the earlier attacks was generally perceived as a clear-cut vindication of the consistency of the quantum formalism.

However, this has not been the case with respect to the EPR challenge. Rosenfeld claims that in Bohr's reply to EPR, "Einstein's problem was reshaped and its solution reformulated with such precision and clarity that the weakness in the critics' reasoning became evident, and their whole argumentation, for all its false brilliance, fell to pieces" (Rosenfeld 1964, 129). However, the subsequent history of the discussion hardly supports this exorbitant claim; in fact in Bohr's reply, the two-system experiment proposed by EPR gets scant attention. Bohr devotes most of his attention to making the by then familiar point that the experimental arrangements required for position and momentum measurements are physically exclusive, and thus that both properties are never simultaneously empirically definable. Then he asserts that the EPR experiment involves nothing new:

> My main purpose in repeating these simple, and in substance well-known considerations, is to emphasize that in the phenomena concerned we are not dealing with an incomplete description characterized by the arbitrary picking out of different elements of physical reality at the cost of sacrificing other such elements, but with a rational discrimination between essentially different experimental arrangements and procedures which are suited either for an unambiguous use of the idea of space location or for a legitimate application of the conservation theorem of momentum. . . . Indeed we have in each experimental arrangement suited for the study of proper quantum phenomena not merely to do with an ignorance of the value of certain physical quantities, but with the impossibility of defining these quantities in an unambiguous way.
>
> The last remarks apply equally well to the special problem treated by Einstein, Podolsky and Rosen, which has been referred to above, and which does not actually involve any greater intricacies than the simple examples discussed above. (Bohr 1935, 700)

This outlook is again evident in Bohr's extensive review of his discussions with Einstein; only five pages out of forty are devoted to EPR, whereas twenty are devoted to the Solvay discussions (Schilpp 1949, 210–235). This attitude is especially surprising in light of both the important shift of focus from inconsistency to incompleteness (with the attendant but only dimly recognized shift from physics to metaphysics), and the change in Bohr's style of reasoning from a detailed analysis of extremely concrete imaginary experiments to a very obscure claim about an ambiguity in Einstein's criterion of "physical reality."

Some commentators have seen a significant alteration in Bohr's thought in his reaction to EPR (e.g., Fine 1986, 35). But the point of the quotation above and the historical record show that Bohr himself was certainly not aware of any such substantive change. However, he does admit that in the following years EPR led to some change in expression and emphasis. One way to describe this shift is that presuppositions about physical reality that were covert in the single system *Gedankenexperimenten* of the Solvay conferences were brought to the foreground by introducing two systems in EPR. Certainly this move shifted Bohr's attention from defending against attempts to beat the indeterminacy relations to emphasizing the wholeness ("individuality") of quantum phenomena. For this reason EPR did lead Bohr to a somewhat clearer statement of themes pertinent to understanding the Bell phenomena.

After reviewing the physical impossibility of designing experiments to circumvent the uncertainty relations, a point which he realized Einstein was by now willing to concede, the argument takes a metaphysical turn in Bohr's key objection that "the criterion or reality proposed by Einstein, Podolsky, and Rosen contains an ambiguity as regards the meaning of the expression 'without in any way disturbing a system.' " Bohr admits that "Of course there is in a case like that just considered no question of a mechanical disturbance of the system under investigation during the last critical stage of the measuring procedure" (Bohr 1935, 699). Thus we may be led to expect that there is some other presumably "non-mechanical disturbance" by which the state of system 2 (S2) somehow becomes determined through an observing interaction with system 1 (S1). Indeed, Bohr may seem to refer to such a communication between systems in admitting that "it is therefore clear that a subsequent single measurement either of the position or of the momentum of one of the particles will automatically determine the position or momentum, respectively, of the other particle with any desired accuracy. . . ." How is this state of S2 "automatically determined"? Though "automatically" might suggest instantaneous action-at-a-distance, Bohr's rejection of any "mechanical disturbance" rules out any superluminal *physical* process operating between systems. What he does tells us is that

> even at this stage there is essentially the question of *an influence on the very conditions which define the possible types of predictions regarding the future behavior of the system*. Since these conditions constitute an inherent element of the description of any phenomenon to which the term 'physical reality' can be properly attached, we see that the argumentation of the mentioned authors does not justify their conclusion that quantum mechanical description is essentially incomplete. (Bohr 1935, 699; original emphasis)

This passage probably represents the closest point of contact between Bohr's idiom and current discussion of the Bell phenomena in terms of "locality" and "separability."

In the quotation above, Bohr takes "physical reality" to refer to "phenomena" which he explicitly defines to mean the whole observational interaction in which either position or momentum is measured. As we have seen, his primary point in the Solvay discussions and much of his reply to EPR were to show that these phenomena are mutually exclusive. Thus the "free choice" of the experimenter as to which of these phenomena is to occur shapes the course of "physical reality." However, it is not in the epistemic act of coming to know the result of the measurement that physical reality is thus determined, but rather in choosing to bring about this or that actual physical interaction between the measuring system and the measured object when the particular experiment is performed. Bohr's use of "conditions" here does not refer to the epistemic situation, but to the physical conditions of the particular interaction between the microsystem and the "detector" which actually occurs. Since the "conditions" of the experimental observation are part of the whole phenomenon, and therefore part of the "physical reality" being described, the fact that a position measurement excludes a momentum measurement implies that when position is "defined" the phenomenon in which "momentum" could be "defined" cannot occur, hence the appropriate property cannot be accorded "physical reality." Thus Bohr could only mean that the momentum of S2 is "automatically determined" by a momentum measurement on S1 in the sense that it is determined if we should choose to make that momentum "defined" by performing a momentum measurement on S2, which would make the appropriate phenomenon part of "physical reality." But in that case, of course, the position of S2 would not be defined because the relevant phenomenon would not be part of physical reality.

Einstein attributed Bohr's refusal to ascribe "physical reality" to the unobserved property to be based on a prior adoption of a phenomenalist ontology associated with Machian positivism. Many of Bohr's statements support this interpretation. Nevertheless, no matter how similar his conclusions might have been to those a positivist would have reached, his argument was not based on any positivistic acceptance of a restriction against talking about unobserved objects, but rather on the conviction that in whatever way we characterize physical reality at the atomic level, it must be in a way consonant with his belief that the quantum description provides a complete description of such objects. This belief in the completeness of the quantum description was based on Bohr's persuasive defense of its empirical adequacy in his replies to the earlier attacks.

What Bohr says is that the description of the phenomenon must include the measuring system as part of the phenomenon, because the presence of the exclusive conditions required for a position measurement is essential in order that position be "defined," and similarly for momentum. The exclusive nature of the phenomena in which position and momentum are observed prohibits "defining" the property which is not observed. Now this is very close to the positivist view which denies meaning to terms which cannot be empirically determined.[2] Bohr certainly does hold that we can have empirical warrant for attributing "physical reality" only to the properties of an atomic object that are measured in the phenomenon which actually occurs. Obviously we can never have direct *empirical* grounds *for* attributing any property to an

---

[2]Indeed Fine refers to Bohr's response here as "virtually textbook neopositivism. . . . Here, where it really matters, Bohr invariably lapses into positivist slogans and dogmas" (Fine 1986, 34–35).

unobserved reality, but Bohr never asserts this epistemological tautology as *his* grounds *against* attributing physical reality to the properties of an unobserved object.

The point is that classically we also had *theoretical* grounds for attributing "physical reality" to the observable property even when the system was not observed, for after an interaction the classical mechanical state is well defined in terms of these properties.[3] If the quantum postulate had never broken the classical scheme, Bohr would have had no problems with Einstein's view. However, in the quantum formalism we have no parallel theoretical justification for attributing reality to unmeasured properties, because the quantum state of the system after the measurement interaction is not well defined in terms of these properties. Therefore, if we accept the completeness of the quantum theoretical description—which is the *premise* of Bohr's reply—we must not frame a conception of "physical reality" which attributes more to that reality than does the theory.

Classically, it was possible to define the system's properties apart from the measuring interaction because the separate states of the two systems were always well defined in terms of these properties. However in the quantum framework—owing to the indivisible quantum of action—the "individuality" or wholeness of the interaction implies that the systems do not have separate states that can be precisely defined in the classical terms. Bohr regarded this indivisibility of interaction to be a physical fact, expressed as the quantum postulate, which was the surprising discovery that implied the inapplicability of the classical mode of description.

> The apparent contradiction in [EPR] discloses only an essential inadequacy of the customary viewpoint of natural philosophy for a causal account [i.e., the account permitted by the classical framework] of physical phenomena of the sort with which we are concerned in quantum mechanics. Indeed, the *finite interaction between object and measuring agencies* conditioned by the very existence of the quantum of action entails—because of the impossibility of controlling the action of the object on the measuring instruments if these are to serve their purpose—the necessity of a final renunciation of the classical ideal of causality and a radical revision of our attitude towards the problem of physical reality. (Bohr 1935, 697)

Bohr's reasoning is solidly realistic here: He does not presuppose the meaninglessness of the notion of a reality which produces the phenomenon; indeed, he must presuppose the reality of the microsystem in order to insist that the interaction of that object with the measuring agencies has the character as-

---

[3]This is what cemented the marriage between Paul Teller's "particularism" and the worldview of classical physics; see Teller, this volume.

cribed to it by the quantum postulate. Furthermore he must presuppose that the microsystem has some sort of reality apart from its phenomenal manifestations in order to speak meaningfully of these different phenomena as providing complementary evidence about the same object. This is the whole point of the need to adopt the complementarity viewpoint.

Nevertheless, Bohr is certainly shy of talking about "reality." If he rejects Einstein's way of representing real objects independently of the observer's interaction, but does presuppose that there are such real objects, what can we say about them? In contrast to Einstein's criterion of physical reality, Bohr would say that any concept of physical reality we frame must be one which is in accord with the empirical base on which science rests. The ever-extending exploration of phenomena thus provides empirical grounds for the constant revision of the physicists' concept of physical reality. The discovery of the indivisibility of the interactions at the microlevel implies that observed objects cannot be defined as in states in which they possess simultaneously properties corresponding to the classical state parameters. Therefore a concept of physical reality based on this epistemological lesson must not, in defiance of what our best theory tells us about these objects, presuppose that reality is so constituted. Whatever the atomic entities are, they surely do not bear the same relationship to space and time descriptions we give of them as do the objects of macrocosmic experience. This is the surprising discovery that reminds us of the limitations of the forms of perception for describing objects beyond the range of the experiences which they were developed to describe.

I have said that Bohr's call for a "radical revision of our attitude towards the problem of physical reality" is itself based on the *premise* that the quantum theory is complete. Thus Bohr assumes exactly what Einstein challenges him to prove; his reply simply begs the question. But this *petitio* arises only because of a deeper level of conflict between Bohr and Einstein, which becomes apparent only when we recognize the realist basis of Bohr's position. Bohr saw the "quantum postulate" as expressing a discovery of something new about nature, and the need for revising the concept of physical reality was a consequence of this discovery. Einstein saw the postulate as solely an assumption of the existing theory, not characterizing the nature of reality. Since he had independent grounds for a concept of physical reality which was incompatible with assuming the completeness of the existing theory, he searched for something better. In the next section we will consider how to express the effect of the quantum postulate in the language of "locality" and "separability" in dealing with the Bell phenomena. In the final section we will consider Einstein's independent grounds for his concept of physical reality, and why Bohr's *realist* outlook on science led him not to share Einstein's grounds.

3. Bohr on Bell

Bohr's expression of his disagreement with Einstein in terms of the ambiguity of "without in any way disturbing the system" may make his views seem irrelevant to contemporary discussions of "locality" and "separability" and their relation to the Bell phenomena. Though most physicists may have "accepted" Bohr's view, the subsequent writings on EPR have not adopted his idiom. The philosophical implications of the point of debate were better expressed by Einstein. He correctly understood Bohr to be denying physical reality to the nonobserved properties and for this reason expressed his difference with Bohr in simple terms pregnant with philosophical implications: the Moon has a position when no one is looking. Bohr certainly did regard the relationship of the Moon to the space-time descriptions we give of it to be quite distinct from the relationship of the objects of quantum mechanics to the space-time descriptions we give of them. But Einstein thought it incredible to imagine that the case for atomic systems was seriously different from the Moon. Understanding the grounds for each scientist's outlook may clarify the relation of their debate to the Bell phenomena.

Consider first Bohr's views on the "principle of locality." Much of the discussion of the Bell phenomena suggests an explanation in terms of a superluminal communication between the two subsystems, or a denial of "locality" which asserts a "controllable non-locality" (see Shimony 1984a, 226–227). One can be quite certain that such a transgression of relativistic limits could never be part of Bohr's interpretation. There is no hint in any of his writings or the remembrances of him that would suggest that he saw relativity and quantum theory as in any way incompatible. Indeed, quite the contrary: he always referred to relativity as the great background achievement on his scientific horizon. In lecture after lecture he stressed the parallels between the reasoning and the philosophical lesson of these two great revolutions in twentieth-century thought. Nor was this merely a rhetorical device to win over Einstein, though of course this fact made his failure to do so deeply frustrating. Furthermore, relativity had been used to develop the implications of the old quantum mechanics. And, most crucially, it was through relativistic considerations that Bohr had triumphantly vindicated quantum mechanics in the challenge of the photon in a box *Gedankenexperiment*. Although Bohr gave serious thought to many radical ideas, the prospect of superluminal communication was not one of them.

The Bell phenomena require postulating a superluminal communication only if one presupposes that spatially separated systems exist in separable states which can affect each other causally only by some communication between them. As we have seen, Bohr denies the classical presupposition that space-time description refers to the properties possessed by the object apart from the interaction in which it is observed. Moreover, abandoning the realistic interpretation of the space-time descriptions of isolated systems is neither an *ad hoc* move to escape the EPR paradox nor the result of an antirealistic understanding of science. The belief that the quantum postulate presents a physical barrier against a space-time description of the discontinuous change of state that an atomic system undergoes in its interactions with the radiative field dominated Bohr's outlook already in 1913. He would have been the last person to be surprised by the "news" that the demand for a space-time description of atomic processes leads to paradoxes, for he was the first person to see clearly that this classical expectation must be abandoned.

Bohr does not need to deny locality because he denies the prior assumption of separability, or the belief that "separated systems possess separate real states" (Howard 1985, 173). So there is no need to postulate a superluminal communication between separable systems. The space-time "picture" of the separation between S1 and S2 is not a conceptual snapshot of an independent reality:

> A new epoch in physical science was inaugurated . . . by Planck's discovery of the *elementary quantum of action*, which revealed a feature of *wholeness* inherent in atomic processes, going far beyond the ancient idea of the limited divisibility of matter. Indeed, it became clear that the pictorial descriptions of classical physical theories represents an idealization valid only for phenomena in the analysis of which all actions involved are sufficiently large to permit neglect of the quantum. (Bohr 1963, 2)

After EPR, this emphasis on the claim that only the state of the interacting whole is definable replaces earlier talk of "disturbance" in an observational interaction. Abandoning locality in the sense of introducing a *controllable* nonlocality into the physical world would be a revolution in *physics*: war between relativity and quantum theory. But Bohr's answer to EPR was a call for revolution in *metaphysics*: "a radical revision of our attitude towards the problem of physical reality" (Bohr 1935, 697). But this allows "peaceful coexistence" in physics only by denying the ontological presupposition of separability.[4] Because Bohr's belief in the completeness of the quantum description is based on his acceptance of the quantum postulate, and the quantum postulate in effect denies the principle of separability—quantum theory provides no grounds for framing a concept of physical reality which presupposes that principle. Learning how to describe nature after the quantum revolution requires learning how to describe indivisible interactions with the objects of the quantum theoretical description.

---

[4]See Shimony 1978, 14, and Teller, in this volume.

Now consider Einstein's views on locality and separability. In his review of the EPR experiment in his "Autobiographical Notes," he concludes that one can "escape from" holding that S2 has a definite value of the unmeasured property "only by either assuming that the measurement of S1 (telepathically) changes the real situation of S2 [i.e., denying locality] or by denying independent real situations as such to things which are separated from each other [i.e., denying separability]. Both alternatives appear to me entirely unacceptable" (Schilpp 1949, 85). We have noted that the temptation to deny locality is a consequence of accepting separability. As Don Howard has argued, Einstein's belief in the reality of unobserved properties was a consequence of his acceptance of the principle of *separability*:

> Einstein did believe that all physical systems at all times possess definite, observer-independent properties which are revealed to us by observation. But he did not just *assume* this. Instead, Einstein *grounded* his realism about physical systems and their properties in the deeper assumption of separability, which is important, because the latter assumption is susceptible to revealing kinds of physical and philosophical scrutiny that cannot touch unanalyzed postulates of physical realism. (Howard 1985, 176)

Einstein's belief that we must presuppose that atomic objects do possess spatio-temporal location independently of whether or not they are observed was a consequence of his conviction that separability is what individuates the separate observed and observing systems. As Howard notes, for Einstein "the separability principle is necessary because it provides the only imaginable objective principle for the individuation of physical systems. Needless to say this is an important and provocative claim, because quantum mechanics, interpreted as a theory about individual systems [which is certainly Bohr's view], denies the separability principle" (1985, 191). Einstein believes that the classical state properties have an ontological priority over any epistemological limitations on knowing them because he held for quite independent reasons that an objective science must describe physical reality as it exists apart from the observer, from which, of course, he inferred the incompleteness of the quantum description.

In the next section, I will consider further why Einstein held separability to be essential for individuating the objects of description in physics, but his belief was hardly exceptional. The question to consider here is why did Bohr *not* believe this? He believed that the conditions of knowability act as constraints on the presuppositions we may make about physical reality. This does not imply that he rejected any reference to the objects which cause phenomena on the positivistic grounds that claims about a transphenomenal reality are unverifiable. What he does abandon is the classical view in which the independence of the object is characterized by the space-time description of the state of the "isolated" system. Already, in his atomic model, he recognized that processes inside an atomic system cannot be given an unambiguous space-time description. Although space-time constructs in terms of electron orbits are part of the way in which the stationary states are determined, the atom's interactions with the field in which it gives up or absorbs a quantum cannot be represented in any space-time picture. The limitation here rests on the *physical* postulate of the indivisibility of the quantum of action, not an instrumentalist ban on reference to objects existing independently of observation.

Because space-time description is thus limited, for Bohr space-time separability cannot be the grounds for individuating ontologically independent entities. Hence the "picture" we give of S1 and S2 prior to their interactions with the detectors cannot be regarded as representing the real properties defining separate states for S1 and S2, and hence justifying their separate individuality. This is the way Bohr expresses what Schrödinger first called "entanglement" and which is now referred to as the denial of separability.[5] Each object we observe is given the character it has by the phenomenon in which that object is observed. We cannot speak of choosing to make one or the other of two different observations on the "same" object. The indivisibility of the state of interacting systems is an observation underlies the ambiguity present in the phrase 'disturbing the system' for this implies that different observational interactions constitute different phenomena. Thus the description of these phenomena as different observations of the different properties of a particular observed object in effect refer to different objects.

4.   Separability and the nature of physical reality

In rejecting Einstein's criterion of "physical reality," Bohr leaves himself with the options of either accepting a phenomenalist ontology and banning reference to an independent reality or providing some alternative way of characterizing real atomic objects independently of observational interactions. Bohr's reasoning shows he rejected the first option, but he provides little to indicate his views with respect to the second option. The reason may be that no matter how fascinating such views may be for the philosophy of nature, as the subsequent history of quantum theory seems to show, they are not essential to doing physics.

In the previous section we saw how Bohr's outlook anticipates the current trend to explain the Bell phenomenon by rejecting the principle of

---

[5]Schrödinger 1935a; see Jammer 1974, 211–218, for a discussion of Schrödinger's response to EPR.

separability.[6] The appeal to holism, the "entanglement" of states, as a means of alleviating the paradoxical quality of the Bell phenomena is presaged by Bohr's emphasis on the individuality of interactions to which we must assume the quantum postulate applies. This was not only the basis for his reply to EPR but was part of his outlook from the beginning of his contributions to physics. But because it is so commonly believed that Bohr banned an independent reality from physics for "positivist" phenomenalist reasons, it is appropriate to comment on the status of that reality once separability is denied.

Ultimately Einstein's view was not that there was no way that the quantum description could be considered complete, i.e., that such an interpretation was logically absurd or empirically refutable. What he does assert is that considering it to be complete requires paying a price with respect to our conception of physical reality that he, for one, was not willing to pay. What is this price and why was Einstein unwilling to pay it?

The first relevant point has been made by Fine: the dispute between Bohr and Einstein has nothing to do with the philosopher's debate over whether or not theoretical statements are true because they "correspond" to an ontologically independent order—to which we can never have empirical access—existing "behind" the phenomena. Fine has shown that "realism" for Einstein is a "program" for constructing theories of a certain sort: those which represent conceptually the nature of a world existing apart from the observer. Fine argues that Einstein's "clearest and most succinct statement" on realism is given in his "Autobiographical Notes" for the Schilpp volume: "Physics is the attempt conceptually to grasp reality as it is thought independently of its being observed."[7] Einstein's allegiance to separability clearly derives from this notion of "realism," for it is by presupposing separability that we represent the independence of physical reality from the observer. Howard also "guesses" that Einstein's commitment reflects the need to separate the observer, *qua* physical system, from the observed in order to ground the "objectivity" of the resulting picture of physical reality:

> Like so many realists before him, Einstein speaks of the real world which physics aims to describe as the real "*external*" world, and he does so in such a way as to suggest that the *independence* of the real—its not being dependent in any significant way on ourselves as observers—is grounded in this "externality." For most other realists this talk of "externality" is at best a suggestive metaphor. But for Einstein, it is no metaphor. "Externality" is a relation of spatial separation, and the separability principle, the principle of "the mutually

---

[6]See the papers of Teller, Shimony, Wessels, and Howard in this volume.
[7]Fine 1986, 93–94, quoting Einstein 1949, 81.

independent existence of spatially distant things," asserts that any two systems separated by so much as an infinitesimal spatial interval always possess separate states. Once we realize that observer and observed are themselves just previously interacting physical systems, we see that their independence is grounded by the separability principle along with the independence of all other physical systems. (Howard 1985, 192–193)

Thus Einstein's commitment to the separability principle is not a matter of metaphysical dogmatism but in fact rests on the epistemological presuppositions canonized by the spectator theory of knowledge: the view that objective knowledge of a real external world requires a "conceptual grasp" of an independent reality from which the observer *qua* physical system is, at least in principle, excludable.

If we conclude that Einstein regarded space-time description as the concrete manner in which realist theories represent the independence of physical reality, his version of realism commits him to the belief that all physical entities, be they moons or muons have "a position when nobody is looking." He might have countenanced a non-spatio-temporal alternative to representing an independent reality, if he knew of such, but he could think of no such alternative way to represent the separation of reality existing apart from the observer.[8] Thus the only available contender for a conceptual model of an independent reality is the one which characterizes physical reality as possessing the properties necessary to define separate states.

Although the reality of atomic objects is presupposed by Bohr's interactionist talk, one might suppose that his position is that while the atomic object is real, it is not possible as in Einstein's criterion of realism "conceptually to grasp" that entity "as it is thought independently of its being observed." Bohr's view certainly denies that such a conceptual grasp can employ the space-time mode of description to represent an independent reality. Whatever pictures we conceive of the spatio-temporal evolution of the independently existing "isolated" system is an "abstraction" or "idealization" in Bohr's language and not a conceptual rendering of the characteristics of an independent physical reality. When he unveiled complementarity in 1927, Bohr noted that:

> radiation in free space as well as isolated material particles are abstractions, their properties on the quantum theory being definable and observable only through their interaction with other systems. Nevertheless, these abstractions are . . . indispensable for a description of experience in connection with our ordinary space-time view. . . .

---

[8]Fine 1986, 97–105, and Howard 1985, 190–191.

. . . the main stress had to be laid on the formulation of the laws governing the interaction between the objects which we symbolize by the abstractions of isolated particles and radiation. (Bohr 1934, 56–57, 69)

The pictures we form are "indispensable" but "symbolic" rather than representational. However, does this mean that we can have no "conceptual grasp" of physical reality which in interacting with the detectors produces the phenomena that are interpreted as measurements of position or momentum? Bohr obviously believes there is *something* symbolized by these abstractions about which we can hardly be said to be completely ignorant, for then how could we frame any meaningful symbol?

At times, Einstein seems to attribute a ban on any reference to an independent reality to Bohr. However, actually Einstein does not say that accepting the completeness of the quantum formalism requires abandoning any conceptual grasp of physical reality apart from its phenomenal manifestations; what he does say is much more cautious:

In pre-quantum physics there was no doubt as to how this [i.e., the conceptual grasp on an independent reality] was to be understood. In Newton's theory reality was determined by a material point is space and time; in Maxwell's theory, by the field in space and time. In quantum mechanics it is *not so easily seen*. If one asks: does a φ-function of the quantum theory represent a real factual situation in the same sense in which this is the case of a material system of points or of an electromagnetic field, one hesitates to reply with a simple "yes" or "no"; why? What the φ-function (at a definite time) asserts, is this: What is the probability for finding a definite physical magnitude q (or p) in a definitely given interval, if I measure it at time t? The probability is here to be viewed as an empirically determinable, and therefore certainly as a "real" quantity which I may determine if I create the same φ-function very often and perform a q-measurement each time. But what about the single measured value of q? Did the respective individual system have this q-value even before the measurement? (Schilpp 1949, 79–81)

His point in the rehearsal of the EPR argument which he gives immediately after this passage is *not* that accepting the completeness of the formalism defies realism *per se*, but rather that if one takes this route, one must abandon either locality or separability. Einstein admits it is "thinkable that the system obtains a definite numerical value for q (or p) the measured numerical value only through the measurement itself" (Schilpp 1949, 83). Of course this is a possibility that Einstein rejects, but the important point in this passage is that even Einstein did not deny that the φ-function *could* be considered to represent something "real," even though it was a concept of reality which he would not accept because it could not characterize that reality in a way such that spatially separated entities exist in separate states.

As we have seen, Bohr did not deny locality and assert a "telepathic" superluminal signaling, but what he had already rejected when he accepted the quantum postulate was separability. It is this issue, not the metaphysical battle of realism versus phenomenalism, which marks the disagreement with Einstein. The philosopher's question of whether or not one can make meaningful claims about an independent reality to which an ontological model allegedly corresponds never enters the debate, for Einstein never built his realism on a correspondence between the properties of theoretical constructs and those of an independent reality. What he could not accept was that this reality could not be represented as independent from observational interactions by appeal to the principle of separability.

The "conceptual grasp" quantum theory permits of the state of a system represented as isolated from interaction is defined through the φ-function formalism, but that function presents no "unambiguous" (hence no "objective") way to represent a space-time picture of the careers of physical systems apart from our observations of them. However, it does not follow that we have no knowledge of this reality with which the detectors interact. Bohr surely believed that the objects of the quantum-mechanical descriptions were real physical systems and that the quantum theory provided a great deal of knowledge about the probabilities these systems have to produce the phenomena we cite as the empirical evidence for the theory. As early as 1929 he wrote "at the same time as every doubt regarding the *reality* of atoms has been removed and as we have gained a detailed *knowledge* of the inner structure of atoms, we have been reminded in an instructive manner of the natural limitation of our forms of perception" (Bohr 1934, 102–103, emphasis added). Because Bohr believed both that the entities described by quantum theory are *real* and that the theory gives us *knowledge* of them, his outlook requires a "realist" interpretation of quantum theory, but it is a realist interpretation which denies our wish for a space-time representation of reality as it exists apart from observations of it. Max Born wrote to Bohr that he "disliked thoroughly" the view that "one does not inquire what there is behind the phenomena" and later that "What I meant by 'behind the phenomena' is in mathematical language just 'invariants' in the most general sense of the word. The various aspects of phenomena which we consider in quantum mechanics have also a theory of 'invariants', or in less learned language, common features which do not depend on the aspect, and it is this which I would like to preserve as something beyond our direct experience," Bohr responded, "As you expressed your views in the letter, I agree entirely and had myself the same attitude. . . ."[9] In challenging Einstein's requirements on the nature of physical reality, Bohr does *not* argue that one should

---

[9]See Folse 1985, 247–249, for a fuller account of this dialogue.

"throw out reality"; instead, in effect, he puts forward an argument for the need for a new conception of physical reality, a conception in harmony with what our best theory of the microdomain tells us we can know about that reality.

But this is all he gives us; he does not choose to develop such a conception himself. But, after all, why should he? This is the task of a philosophy of nature, not of physics. Philosophers have long looked upon relativity theory as evidence on behalf of this or that philosophy of nature, but Einstein himself gave us no such a worldview and we do not fault him for it. The same should be understood about Bohr. His defense of the completeness of the quantum description is not based on a positivist rejection of all ontology of science, but on physical reasons expressed in the quantum postulate. As long as that defense is misinterpreted as a defense of an antirealistic view of science, there will be no possibility of developing a philosophy of nature based upon Bohr's interpretation of quantum theory. Understanding that Bohr's interpretation of quantum physics is a kind of realism opens the way to developing a philosophy of nature based on complementarity.[10] It is testimony to the philosophical significance of the Bell phenomena that they call us to a renewed examination of the implications of accepting the completeness of the quantum mechanical description, and thus they invite us to contemplate a philosophy of nature erected upon Bohr's new viewpoint.[11]

---

[10]If we adopt the terminology employed by McMullin (1970, 27–30) (ES = epistemology of science, OS = ontology of science, and PN = philosophy of nature), then we can express the usual view in the following way: A positivist position in ES implies an antirealistic view in OS, which in turn forecloses the possibility of a PN. Now it is certainly true that Bohr never developed a PN, as we would expect if he held an antirealist OS. Since his statements on ES bear a superficial similarity to the conclusions of positivism, we infer he held an antirealist OS and for this reason did not develop a PN. However, I have argued that the arguments given for the positivistic sounding statements in ES are not positivistic arguments, but are in fact based on a realist OS which speaks of the reality of the objects of quantum-mechanical description and their interactions with measuring instruments. Furthermore Bohr's view that different phenomena provide complementary evidence about the same object makes sense only if that object is distinguished from the phenomenal object. The question to be asked, then, is how can philosophers develop a PN based upon Bohr's ES and OS. As to why Bohr himself did not develop such a PN, I think that we can say that quite simply it was not his job, *qua* physicist, to do so.

[11]See Shimony's paper elsewhere in this volume for the view that this century is a "golden age" of metaphysics.

THE EXPLANATION OF DISTANT ACTION:
HISTORICAL NOTES

Ernan McMullin

Can there be science proper in the absence of understanding? How tightly linked *are* the two? Suppose, for example, that statistical correlations are discovered between two classes of events occurring at a distance from one another. The correlations are not mere chance, we can tell, since they can be relied on for accurate statistical prediction. But a causal explanation of the ordinary sort, proposing either a direct causal influence of one on the other or a common causal antecedent of both, can be ruled out. If what happens at *A* makes a difference to what happens at *B,* can this notion of "making a difference" be explicated without involving notions of causal influence, of an action either of one upon the other or of a single system of which both are part? Will it do simply to say that the correlations are brute fact, that no further explanation can (or should) be sought? Can there be understanding in the absence of causal explanation?[1]

Questions like these are familiar to anyone who has been following the debates prompted by the violation of the Bell inequalities in quantum theory coupled with the associated experimental results which suggest that the inequalities do not, in fact, hold. My intention in this essay is not to discuss these debates directly but to ask about historical antecedents. Are debates of this sort entirely novel? Did issues of this kind ever arise before? We shall see that the matter of explaining distant action has often been a controversial one, notably in the wake of the appearance of Newton's *Principia,* the third centenary of whose appearance we have just celebrated. If a formalism is predictively successful in mechanics but lacks explanatory power ought we con-

---

I am indebted to Jim Cushing for many discussions of these issues.

[1]Very similar questions had already been raised in earlier discussions of the EPR paradox; see McMullin (1954) which chronicles that earlier stage of the debate in some detail.

clude that it is deficient as science? Before discussing the Newtonian puzzles, it will be helpful to return briefly to the time when the rules were first laid down.

1. *Science, self-evidence, and explanatory power*

Two themes can be found closely interlinked in the first philosophical fragments from the ancient Greek world. One was that of *explaining,* of taking some feature of the world and making it more intelligible than it had been. This might be done by calling on simple analogies drawn from ordinary experience, the way in which eddies form in a river or mud settles in a lake, for example. The less familiar would thus be explained by reference to something more familiar, what was less intelligible by invoking what was more intelligible. The function of explanation in these cases is to produce understanding; the two terms are indeed correlative though not co-extensive.[2] In the context of *epistēmē,* as the Greeks began to shape the concept that ultimately became our *science,* understanding has to be made both explicit and general in order to be communicable. And this is done through the construction of explanations.

The Ionians asked themselves how the world came to be the way it is. The warrant of the cosmogonies they constructed lay in the coherence of the story they told and in how closely it fitted with ordinary experience. This was the other theme, the concern with proof and evidence, with the quality of the claims made. It found its most famous expression in Plato's allegory of the Cave and the associated metaphor of the Divided Line.[3] Plato draws two crucial distinctions, one between knowledge and opinion and another between the intuitive grasp of principles and the postulation of plausible hypothesis. An hypothesis functions indirectly; it derives its warrant from the degree to which it illuminates something else. The degree of truth attributed to it is measured, therefore, by the quality of the explanation it affords. Plato contrasts this lesser form of knowledge (*dianoia*), lesser precisely because it *is* indirect, with the idea of knowing (*noesis*) arrived at by direct intuition of principles. Once the relations between the Forms are grasped directly—this is, he warns, not easy to achieve—the principles which serve as the foundation of true knowledge are seen to be true in their own right, without need of

---

[2]One might understand (the action of another, for instance) even in the absence of an explanation. 'Understanding' has a primarily psychological reference to a state of mind. 'Explanation' is propositional, explicit, other-directed. An explanation would not, however, count as an explanation unless it could contribute in some way to understanding.

[3]*Republic,* Book VII.

recourse to analogies or to the explanatory merits of hypothesis. Yet this knowledge will only partially serve as guide to the flickering world of the senses, which is not entirely amenable to reason. This is why the testimony of the senses in astronomy can never (in Plato's view) be a sufficient warrant for a theoretical construction of the "real motions" of the heavenly bodies.[4]

Aristotle's world had far less flicker, and so knowledge relates to it in a different and much more direct way. Yet *epistēmē* (Aristotle's term for science) at first sight does not seem to differ all that much from *noēsis*. What marks it off from lesser sorts of knowledge-claim is its necessity and consequently, its stability. It begins from an intuition of principle or of the relation between a particular property and the essence in which it inheres. It progresses by deduction towards necessary conclusions. The premises on which *epistēmē* rests must be self-evident, not in the sense of being obvious, but in the sense of being clear to the trained mind without need of the sort of supporting evidence that hypothesis requires.

What, then, about hypothesis, about indirect knowledge-claims that are warranted in terms of their consequences (by what I am calling their "explanatory power") and not their intrinsic self-evidence? It is surely odd that when Aristotle chooses examples to illustrate his notion of demonstration, he should choose them from astronomy, the science where intuitive self-evidence might seem least likely to be found. How can we demonstrate the *cause* of the odd fact that planets do not twinkle as all other heavenly bodies do? He responds that the nearness of the planets *must* be the cause. But surely it is far from self-evident that the planets *are* near? How do we know this? His response lends itself to different interpretations but the most plausible is that taking the planets to be near gives the best (indeed, he assumes, the only possible) explanation of the known fact that they do not twinkle.[5] But this seems to render the entire construction hypothetical, and to make explanatory power, not self-evidence, the warrant of the causal claim. Generations of commentators have worried over these passages, but it is hard not to see in them some realization on Aristotle's part of the need for a mode of indirect justification that moves backwards from consequence to postulate.

I am beginning my story at this point in order to underline that at this early stage self-evidence tended to blend into explanatory power as the mark of science. Success in "saving the phenomena" was not enough unless it issued in explanation proper, an answer to the question "why?" Aristotle had much to say about the modalities of this "why?" From his general theory of

---

    [4]This leads Plato into many difficulties. See McMullin, "The goals of natural science," *Proceedings American Philosophical Association* 58 (1984): 37–64, see sec. 2.

    [5]For a detailed discussion of *Posterior Analytics* I,13, see McMullin, "Truth and explanatory success," *Proceedings American Catholic Philosophical Association* 59, (1985): 206–231.

nature he concluded that there were just four broad types of explanation; they coincided (he claimed) with the ways people actually *do* answer the question "why?"[6] What counted as explanatory was thus determined by the choice of ontology as well as by a reference to customary practice. "The causes being four, it is the business of the physicist to know about them all, and if he refers his problems back to all of them, he will assign the 'why' in the way proper to his science."[7] Explanation for Aristotle is thus causal *by definition,* since each mode of explanation is linked with a characteristic sort of "cause." To explain a change is to specify one (ultimately all) of the types of "cause" involved. It must be remembered, of course, that the terms 'cause' and 'causal' are much broader here than they would be in modern usage.

2.   *The principle of contact action*

Nature for Aristotle is "a principle of motion and change,"[8] so that (unlike Plato) he sets out to construct a science of motion, i.e., a systematic way of making motion in general, as well as specific kinds of motion, intelligible. The problems with which we are here concerned center on the necessary conditions for (the explanation of) agency. Aristotle admits two primary forms of action: self-motion, which he associates with life, and contact action, whereby one body moves another. Since everything that moves must be moved by something (he devotes a long proof to this), anything which is not self-moved must be moved by contact with something other than itself.

Why is contact needed? He gives two sorts of argument, one *a priori,* based on the meaning of such terms as 'push' and 'pull', the other "based on induction," on experience of the various ways in which motion is, in fact, brought about. He concludes: "It is evident, therefore, that in all locomotion, there is nothing intermediate between mover and moved."[9] Action at a distance is thus excluded; the "principle of locality," as it has recently come to be called, is affirmed. Yet Aristotle's arguments strike the reader as uncommonly weak. The phenomena of magnetism were well known long before his day; many of the Ionian philosophers referred to them. Socrates compares Ion's "divine power" of speaking well on Homer to the power of the "magnetic stone," which not only can attract iron rings but can also impart to the rings a similar power of attraction on their own account.[10] So that Aristotle's

---

[6]*Physics* II, 3.
[7]*Physics* II, 7; 198a 22–24.
[8]*Physics* III, 1; 200b 12. Oxford translation.
[9]*Physics* VII, 2; 244a 16–244b 1.
[10]*Ion,* 533D.

inductive claim that "in every case" one finds mover and moved in contact meets an immediate check. More seriously, the argument claiming the impossibility in principle of a body acting where it is not appears to beg the question.

Why, then, was he so emphatic in his claim? There was obviously for him a fundamental principle of natural order at stake, an unstated presupposition linking action with the position of the agent in space. Agents simply *cannot* push or pull where they themselves are not bodily present. His intuition, schooled by the contact actions of everyday experience, led him to assert the necessary truth of the premise that bodies may not act at a distance without the aid of intermediaries. One can imagine that if the example of the magnet had been pointed out to him, he would have postulated intermediaries of some imperceptible sort to convey the action in a continuous way across the intervening space. But we need not rely on imagination here. Let us take a look at the impact of this principle on Aristotle's astronomy.

The astronomer, Eudoxus, had just discovered an ingenious mathematical way of computing the observed motion of each planet. He postulated a model in which the planet is carried on the equator of a sphere whose axis is in turn carried by another sphere moving at a different (uniform) rate and around a different axis; this sphere is carried on another and this one on another, four in all for each planet (three for the sun and moon). The cluster of spheres for each planet is treated as separate from the other clusters.[11] There is no direct evidence as to whether Eudoxus thought the spheres to be real agents or merely calculational devices.

Aristotle, in any event, was quite definite that the spheres are physically real. If the planet's motion is to be explained by a composition of four circular motions, there have to be four movers intersecting with one another. These movers can only be spheres, the axis of each being carried by the sphere next outside it. Each sphere is not only carried along mechanically by contact with the one above it, it also has its own proper motion of rotation. Since this last can only be a self-motion, he postulates a separate intelligence for each sphere, each moved to a different degree by the desire to emulate the stable state of the Unmoved Mover.

Instead of seven separate clusters of motions, Aristotle has, therefore, a single interconnected system. But this leads to an undesirable consequence. In order to prevent the combined motions of each planetary "cluster" from affecting all the motions below it (which would lose for his system the predictive virtues of the original model), he has to introduce twenty-two extra "counteracting" spheres. The resultant system of fifty-five spheres involves

---

[11]Callippus added seven more spheres to Eudoxus' twenty-six to give a more exact accounting of the motions.

an incredibly complicated set of motions, some mechanically caused, some due to intelligent self-motion. Why are they necessary? Why are there spheres in the first place?

The spheres are not needed for the Eudoxan scheme to work; their warrant does not lie in successful prediction. Rather, they are required by the regulative principles of Aristotle's physics, specifically by the principle that everything that is in motion must be moved by something other than itself, and by the principle of contact action. The nested set of spheres constitute a plausible physical interpretation of the mathematical scheme, "plausible," that is, in the light of these prior principles. The only properties the spheres seem to need are incorruptibility, to ensure that the motions continue perpetually, and something like solidity, to enable them to carry the planets in their observed paths. And these were, in fact, the properties that later Aristotelians attributed to them, though the first created some problems for the second.[12]

The requirement of contact action was frequently discussed in medieval commentaries on Aristotle's *Physics*. Aquinas, for example, mentions apparent violations of the principle, such as the heating of the earth by the sun, and insists that in all such cases there must be a medium of transmission of the action, even though the form the action takes in the medium may be very different from its effect on the object where the action terminates.[13] Even God is bound: "No action of an agent, however powerful it may be, acts at a distance, except through a medium. But it belongs to the supreme power of God that He acts immediately in all things. Hence nothing is distant from Him."[14] Since action at a distance is, in principle, impossible, there cannot be *any* exemption, not even for omnipotence.[15]

---

[12]Because the spheres are incorruptible and because their motions are naturally circular, Aristotle concludes that they must be composed of a fifth element, aether, quite different in kind to the four terrestrial elements. Since they are incorruptible, they must be "exempt from contraries," such as hard and soft (*De Caelo*, 270a 20). Does this mean that the spheres cannot be hard? Aristotle *does* say more than once that they are corporeal: "For this last sphere moves with many others to which it is fixed, each sphere being actually a body." (*De Caelo*, 293a 7–8; see also 288b 5). And they must be capable of carrying the planets along: "the upper bodies are carried on a moving sphere" (289a 29). They must thus be solid, it would seem. Aristotle himself never addresses the question of what kind of "solidity" they can have, if hardness is excluded as a contrary peculiar to the terrestrial realm. For a review of the later medieval discussions of this topic, see Edward Grant, "Celestial orbs in the Middle Ages," *Isis* 78 (1987): 153–173.

[13]*In VIII Libros Physicorum*, VII, lect. 4. Scotus, interestingly, rejects this argument, finding the notion of an intermediary matter incoherent; he goes on to affirm the possibility, in certain special cases at least, of action at a distance. See *Ordinatio*, I, dist. 37, q. 1.

[14]*Summa Theologica*, I, q. 8, a. 1, *ad* 3. Dominican translation.

[15]The issue of action at a distance in Newtonian physics was widely discussed by neo-Thomist writers in the last part of the nineteenth century and the first half of our own. There was virtually unanimous agreement that contact is an absolute requirement for the action of one body

Descartes would have objected to any such limitation; he was reluctant to allow the normal modalities of possibility and impossibility to set bounds on God's power. But he had no doubts about the principle of contact action itself. In fact, he made contact action the sole form of material agency. One looks in vain, however, in the section of his *Principles of Philosophy* entitled "The principles of material objects" for any explicit mention of the impossibility of action at a distance. The reason was simple: the principles of his system entailed this in such an obvious and immediate way that there was no need to devote explicit attention to the issue as the Aristotelians had done.

Among these principles, the one which almost determined the rest of his mechanics was the equating of matter with extension. Since extension is a property, it must be a property of a substance. Hence, Descartes concludes, space must of its nature be a material substance. This leads him into immediate difficulty with the phenomena of density difference: if matter is constituted by extension alone, any two bodies of the same extension must contain the same quantity of matter. Or as he himself puts it: "It is not possible for there to be more matter . . . in a vessel when it is full of lead, gold, or another extremely heavy substance, than when it contains only air, because the quantity of matter . . . is always the same in a given vessel."[16] This is so counterintuitive one wants to know how he could have maintained it.

He presents two sorts of argument. In one we are to remove from the idea of body any property which we can clearly see not to be essential to it. The only property that cannot be lost without losing the idea of body itself is extension.[17] The other argument is only hinted at: among bodily properties, only extension (size, shape) and motion can be grasped by the understanding in terms of clear and distinct (geometrical) ideas. A science of bodies *can*, in fact, be constructed. Hence, it is extension alone that makes a body be a body.[18] In both cases conceivability is the key: what cannot be clearly conceived is excluded.

The possibility of a vacuum is, in principle, rejected; even when a space

---

on another. Henry van Laer reviews this literature and urges the force of the central argument: "Activity outside the quantitative limits of the acting body would be an accident without a substantial support and therefore absurd." See "The possibility of action at a distance," chap. 4 of *Philosophico-Scientific Problems* (Pittsburgh: Duquesne University Press, 1953), p. 75. Many neo-scholastic writers also argued more concretely that the hypothesis of action at a distance is unable to account for the influence of distance upon action (as in the inverse-square law) and the finite time of propagation of light.

[16]*Principles of Philosophy*, trans. V. R. Miller and R. Miller (Dordrecht: Reidel, 1983), II, 18; p. 48.

[17]*Principles*, II, 4; II, 11. He mentions properties like hardness, but does *not* mention density.

[18]*Principles*, I, 69.

appears to be quite empty, there must in fact be a material substance present. This body which "does not affect the senses" he calls aether. It is composed of "extremely tiny scrapings of matter" which are flexible enough to fill all of the interstices of body without remainder.[19] He goes on to speculate about the roles such an aether might play in the transmission of light, in magnetic action, and so on. But there is never any attempt to test this theory by means of the empirical consequences drawn from it. In Descartes's mind, the assertion that such a substance must exist rests primarily on the first principles of his mechanics. The speculative discussions of light and magnetism might, in a very loose sense, have been regarded as additional corroboration, but Descartes's assurance about the aether in no sense rested upon such claims to specific explanatory virtues.

The other factor in Cartesian mechanics is motion. God must be the primary cause of all motion in the universe; since He is constant in his manner of acting, Descartes reasons, He must maintain a constant quantity of motion in the universe, this last being defined as the product of quantity of matter (for him, extension) and speed (taken as a scalar quantity). He chooses this as the conserved quantity presumably because he views it as the sole resource for causal action, for the bringing about of change in the material order.[20] The motion, once communicated, keeps being redistributed; what one body gains, another loses. He goes on to specify the rules of contact action governing the redistribution and ends: "These things require no proof, because they are obvious in themselves."[21] They were, indeed, far from obvious, in part because they employed notions, like force, for which no quantitative measure had been given. More seriously, several of them were clearly at odds with experience. But the most serious objection, of course, as Leibniz in particular would later urge, was that Cartesian bodies lack the means of making contact with one another in the first place. There can be no contact action between volumes as such; if change of motion ensues, it can *only* be because the primary cause, God, wills to bring it about.

The understanding of motion that Descartes purports to give rests, then, not at all on any appeal to experiment or even to the most generalized sort of experience. Indeed, in the French edition of the *Principles,* the section on the rules of contact action ends: "The demonstrations of this are so certain that, even if experience were to show us the opposite, we should nevertheless be obliged to place more trust in our reason than in our senses."[22] The presump-

---

[19]*Principles,* III, 49.

[20]There was, of course, a famous difficulty here about the operation of soul in the material order, a problem that Descartes does not address in the *Principles.*

[21]*Principles, II, 52.*

[22]There is reason to suppose that Descartes himself was responsible for this emendation. See the Miller translation, p. 69.

tion here is that human reason has the inherent capacity to determine what counts as properly explanatory in mechanics, and that this is independent of any inductive testimony that might seem to count against it. This holds only for mechanics; Descartes is clear that in other parts of natural science, where an appeal to underlying causes must be made, experiment will be needed in order to discover which of the many possible causes is the one actually realized.[23] Mechanics is unique in this respect, in part because it requires— indeed, in his view, permits—no such causes, and also because it lends itself more readily to the assumption that its concepts can be given the clarity and distinctness of geometry.

To summarize our story thus far, both Aristotle and Descartes laid down in advance what kinds of agency may be called upon in the causal explanation of motion. But where Aristotle presupposed that the basic concepts of mechanics are learned from our everyday experience of motion, Descartes attributed to them an *a priori* status; it is difficult to determine from the texts, however, just what kind of priority the ideas are supposed to have.[24] Aristotle and Descartes each postulated theoretical entities (the spheres, aether) to supplement the causal agencies responsible for motion. The warrant in each case was not the predictive accuracy of the account provided, but the way in which the principle of contact action was preserved and in that sense a proper "understanding" of motion given.

3. Kepler's dynamics of planetary motion

Meanwhile, what were the astronomers doing? The *Almagest* and the *De Revolutionibus* were approximately equivalent in their predictions of planetary movements. But Copernicus was emphatic in claiming a "naturalness" for his own system that that of Ptolemy had so notably lacked. He devoted little attention, however, to the dynamical question of *why* planets move as they do. He retained spheres as carriers, but since he had not been able to dispense with epicycles, it was not at all clear how the spheres were to function. They needed neither external force nor immanent intelligence to keep them in motion, he thought; the natural motion of a spherical body simply *is* circular.[25]

---

[23]*Discourse on Method,* Part *VI.* See McMullin, "Conceptions of science in the Scientific Revolution," in *Reappraisals of the Scientific Revolution,* ed. D. Lindberg and R. Westman (Cambridge: Cambridge University Press, 1989).

[24]See, for example, "Reason in Cartesian science," chap. 3 in Desmond M. Clarke, *Descartes' Philosophy of Science* (Manchester: Manchester University Press, 1982).

[25]Alexandre Koyré argues that for Copernicus their spherical shape is a *sufficient* cause for the uniform motions of rotation of the heavenly spheres and of the planets, including the earth (*The Astronomical Revolution* [Ithaca, N.Y. Cornell University Press, 1973], pp. 58, 113).

Copernicus saw a link between the tendency of bodies to fall and the spherical shapes that the planets (and sun) take on. Parts of the same body, if separated, have a "natural desire" to rejoin one another.[26] This is a tendency of one towards another; natural place plays no part. Is it an attraction? Are there forces at work? This is the sort of ambiguity that will appear again in Newton's theory; it is unlikely that Copernicus would have been troubled by it, since the kind of dynamic context that prompts the question was not yet at hand. The notion of a "natural inclination" of "like" things towards like, was not new;[27] it seemed to him a sufficient explanation of falling motion, as well as of the spherical shapes from which rotatory motions derived.

If Copernicus paid little attention to the question of what it is that makes the planets move, it was evidently the question that above all others Kepler wished to solve. Brahe's computations of the orbit of the comet of 1577 "destroyed the reality of the spheres," as Kepler more than once reminded his readers,[28] and hence made the assumption underlying the traditional explanation untenable. But his aim was much larger than merely finding an alternative to the spheres. As he set out to refigure the planetary motions, he hoped to discover a single mathematical formalism that would be common to all of the motions and which could then show the way to a new physics linking earth and sky. The kinematics of celestial motions was to furnish the basis for the new dynamics. This was a truly revolutionary aim; though he did not succeed in it, its formulation was achievement enough.[29]

In his first work, the *Mysterium Cosmographicum* of 1596, Kepler put his finger on the main reason for preferring Copernicus's system over that of Ptolemy. It predicted more accurately, but what was:

> much more important, things which arouse our astonishment in the work of the other are given a reasonable explanation by Copernicus, and thereby he destroys the source of our astonishment which lies in the ignorance of causes.[30]

Copernicus has more commonly been supposed to take the sun as the primary cause of planetary motions. Koyré argues, however, that although the sun is the source of light, heat, and hence of life for Copernicus, it "plays no part whatsoever" in the *mechanics* of his system (p. 65). The issue hinges, in part, on the sense in which geometrical figures are "causes" in a broadly neo-Platonic worldview like the Copernican one.

[26]*De Revolutionibus*, I, 9.

[27]Max Jammer traces the origins of this notion to the Stoic tradition (*Concepts of Force* [Cambridge, Mass: Harvard University Press, 1957], p. 72). Oresme gives express formulation to it: "*omnia similia habent quandem inclinationem naturalem ad invicem*" (*Quodlibeta*, q. 22); Jammer, p. 68.

[28]See, for example, *Astronomia Nova*, chap. 33.

[29]See E. J. Dijksterhuis, *The Mechanization of the World Picture* (Oxford: Clarendon, 1961), pp. 309ff.

[30]*Mysterium Cosmographicum*, chap. 1; *Gesammelte Werke*, ed. W. von Dyck and Max Caspar (Munchen, 1938–1959), vol. 1, p. 14. Quoted by Koyré, *Astronomical Revolution*, p. 129. I am indebted to Koyré for the translations of the passages from Kepler quoted below.

Kepler lists the various phenomena that appear *ad hoc* in the Ptolemaic system and shows how they follow without need for further assumption, once the earth be set in motion. The motions of the planets he thought to be probably due to an *anima motrix* (moving soul) in the sun, whose force (*virtus*) weakens with distance.[31] The more distant planets will move more slowly, therefore, than the nearby ones. And the sun will have to be at the center, not slightly off center as it had been for Copernicus. But that was as far as he could carry the argument.

And then came his long labors on Brahe's observations of Mars and the final triumph of the *Astronomia Nova* of 1609, where circles were replaced, at last, by ellipses and the varying orbital speeds of the planets were determined. The title of the new work, *A New Astronomy Based on Causes or: A Celestial Physics Drawn from Commentaries on the Motions of the Planet Mars,* conveys Kepler's own assessment of what he had accomplished. What he draws from his new knowledge of the motion of Mars is that planets are accelerated or retarded according as they are nearer to, or farther away from, the sun. He argues that the cause of this variation cannot lie in the planet itself; the only plausible conclusion is that the source of motion lies in the sun.[32] But how does it operate?

Analogies with light and with magnetic force immediately suggested themselves; these are the two other obvious contexts where bodies appear to affect one another at a distance. Just as light propagates as an intangible *species* or form which (unlike heat or odor) does not affect the intervening spaces, so the motive force which moves the planets is an immaterial *species* of the motive force that lies in the sun itself. Since both types of action propagate in time and space, they obey geometrical laws. Even though they are immaterial in their mode of transmission, passing instantaneously through space, they can act upon material bodies.[33] The effect of the sun's motive force on a distant body depends on three factors, just as in the case of magnetism: the original intensity of the force (which in turn depends on the size and density of the active body),[34] the distance between them (it falls off proportionally to distance),[35] and the bulk (*moles*) of the body acted upon.[36]

---

[31]*Mysterium,* chap. 20; *Werke,* vol. 1, p. 70.

[32]*Astronomia Nova,* Introduction; *Werke,* vol. 3, p. 23.

[33]*Astronomia Nova,* chap. 33; *Werke,* vol. 3, p. 240.

[34]He later wrote: "The Sun is the densest body in the whole universe, evidence whereof is its immense and manifold force, which could not but have a proportionate cause." *Epitome Astronomiae Copernicanae, Werke,* vol. 7, p. 283; Koyré, p. 353.

[35]When estimating how the motive force might vary with distance, Kepler usually assumes that it acts along a plane sheet instead of on a surface perpendicular to incidence; hence it comes out as inversely proportional to distance, instead of to the square of the distance.

[36]*Moles* (bulk) depends on both volume and density, and hence is similar to the "quantity of matter" of late medieval natural philosophy. In the *Epitome,* he later gave a precise formulation

Kepler was, at this point, developing a dynamics entirely different from the Aristotelian one. Bodies do not have an intrinsic gravity or weight, nor do they tend to move naturally to a particular place. They have a natural *inertia,* a resistance to motion which is proportionate to their bulk. An external force acting in the direction of the motion is required to *keep* them in motion, i.e., to overcome their continuing inertia. (There will be echoes of this conception later on in Newton's *vis inertiae.*) The difference in periodic times of the planets is thus due not only to their different distances from the sun, but also to their different bulks.

But now there was a problem. How could forces radiating outwards from the sun produce the orbital motions of the planets? There had to be a force acting in the direction of the motions. But where was it to come from? His solution was to set the sun in rotation, swinging the immaterial *species* it emits into a sort of "whirlpool," as he calls it, whose intensity depends on distance as well as on the original intensity of emission, a whirlpool which urges the planets directly onwards in their paths. It was not at all clear, however, how such an agency *could* keep the planets at relatively constant distances, nor how exactly its effects could be calculated. The vortex was also Descartes' choice a few years later for propelling the planets along their paths. But where Kepler's was an immaterial swirl of forms somehow conveying the force of the sun, Descartes's involved a material aether, operating strictly by transfer of pressure. The constraints on what constituted an acceptable explanation of motion were obviously very different for the two.

The notion of attraction is conspicuously absent from the story so far. Kepler *did,* in fact, develop what he called a "true theory of gravity" in the *Astronomia Nova.* And it involved real attractions, not just tendencies: "It is the earth which attracts the stone, rather than the stone which tends to the earth."[37] But they operate only between *like* bodies. If two stones are placed near each other anywhere in the universe, they would draw together, just as magnets would. Since earth and moon are alike, they attract one another; they would, in fact, draw together if the earth (and presumably the moon) were not "held on its course by an animal or some other such force."[38] The strength of the attraction is proportional to the bulk of the attracting body; its effect on the body attracted is, likewise, proportional to *its* bulk since, like any other force, it has to overcome the inertia of that body. The earth's attraction keeps the waters of the sea from rising to the moon, and the moon's attraction shows itself in the phenomenon of the tides. There is every reason to think, Kepler

---

of the concept, noting that the abundance (*copia*) of matter in, or mass of, a planet is to be calculated by multiplying density and volume.

[37]*Harmonia Nova,* Introduction; *Werke,* vol. 3, p. 23; Koyré, p. 194.

[38]This force would have to act in the direction away from the sun; Kepler is obviously at somewhat of a loss as to how to account for it.

concludes, that the range of the earth's attraction extends to the moon "and far beyond."

But no matter how far beyond, it will not affect planets or sun since they are unlike in nature to earth and to one another. Even where he saw an affinity sufficient to extend the action of gravity from earth to moon, he could not see this action as in any way responsible for the moon's motion. Kepler has not yet made the crucial step. Much less could he postulate a universal force of which terrestrial gravity is only one manifestation; the evident lack of affinity between earth and sun stood in the way.

There was one further problem: Why do the planets move at constantly varying distances from the sun? The whirlpool metaphor could not explain this. Kepler labored in the *Astronomia Nova* to find a solution. Eventually, he hit on an ingenious idea. The work of Gilbert had already shown the earth to be a vast magnet. What if all the other planets were magnets too? Then the manner in which their distances from the sun systematically change could be due to alternating repulsions and attractions, as their poles alter position relatively to the sun. As always, his guiding principle was that some "natural" forces would be found to underlie the regularities of observed celestial motions.

Later, in the *Epitome Astronomiae Copernicanae* of 1618–1621, he tried out a different and much more radical suggestion. Perhaps "the Sun possesses the active and energetic faculty to attract, or repel, or hold a planet."[39] The source of the forces of attraction and repulsion is thus transferred from the planets to the sun. In the earlier work, these latter had been regarded as too dissimilar in nature to give rise to any form of attraction. Now he uses the analogy of pieces of iron attracted by a magnet. There is a sufficient similarity of substance, he thinks, to allow the sun to attract planets. Because it possesses something like magnetic poles, repulsion will occur when its "hostile part" is presented to the planet. But these magnetic forces are intended to account only for the regular variation in planetary distances from the sun. To account for the primary orbital motions, he retains the idea of a swirl of immaterial *species* which pushes the planets onwards. Were the sun to cease rotating, he says, the planets would either fall inwards towards it or recede outwards to the stars.

In the *Epitome,* he sums up the arguments that had led him to the new celestial dynamics. The solid spheres are no longer necessary because the inertia of matter will cause the planets to remain wherever they are in space, unless moved by motive forces. Though these forces are "far removed from our understanding and without positive example," that is, without precise parallel in our everyday experience, certain things at least are clear:

We should not have recourse either to a mind which would make the planets

---

[39]*Epitome, Werke,* vol. 7, p. 300; Koyré, p. 297.

rotate by the dictates of reason, as it were by order; or to a soul which would preside over the motion of revolution. . . . The motion of the primary planets around the Sun is to be attributed solely and uniquely to the body of the Sun placed in the middle of the Universe.[40]

The rotation of the sun still requires the action of a soul; Kepler cannot think of any other agency that could bring about this crucial motion on which all other celestial motions depend. While allowing this single role still to an *anima motrix,* he reminds the reader that soul of itself cannot impart a motion of translation any more than intelligence can. Nor would intelligences be able to take into account the many things that would be needed in order to compute elliptical orbits:

> On the contrary, the elliptical shape of the planetary orbits, and the laws of motion by which such a figure is traced, reveal more of the nature of balance and material necessity than of the determination by a mind.[41]

Kepler's celestial dynamics could claim in its support that it did, in a general way at least, account for the newly discovered kinematical laws of planetary motion. It also depended upon plausible physical analogies with light, magnetism, and gravity. His reliance on these analogies helps to explain why it is that he never adverted to the issues regarding contact action which had so often preoccupied earlier natural philosophers. The simplest form of contact action, the carrier sphere, had been eliminated. The other traditional solution, the soul or intelligence that would guide the planet in a natural motion akin to that of the living being, could not account for translatory motion, only for the motion of rotation. That left only forces, attractions, immaterial *species* instantaneously transmitted, to convey the causal action of the sun outwards to the planets. It was obvious to Kepler that the sun *had* to be the source of the motions of those bodies which revolved in such an intricate array around it. He did not expect the manner of this action to be plain to us; the analogies with familiar but also not very well understood phenomena like magnetism were effective in showing that an explanation for the mysterious phenomenon of planetary motion *could* legitimately be hoped for, and perhaps to warn that an extension of the familiar categories of mechanical explanation would almost surely prove to be necessary.

4. *Before the* Principia

Though Descartes's writings had served to introduce Newton to mechanics, he very soon reacted against what he termed the ''fictions'' of his

---

[40]*Epitome, Werke,* vol. 7, p. 297; Koyré, p. 290.
[41]*Epitome, Werke,* vol. 7, p. 295; Koyré, p. 288.

predecessor. In a short unfinished tract, *De gravitatione,* written shortly after his student days, he attacks some of the fundamental principles of Cartesian mechanics. He takes it as obvious that matter cannot be defined in terms of extension alone but requires also the separate quality of impenetrability; since matter and void can now be distinguished, the Cartesian doctrine of the plenum can be dismissed. Force he defines as "the causal principle of motion and rest";[42] though it is not clear where forces are to be found, they are definitely taken to be real causes, and hence a way opens up to define "real" (i.e., absolute) motion. And gravity is detached from the earth; it is a "force in a body impelling it to descend" to a particular point in a space now absolute in character. In another paper from about the same time, he makes the acceleration of a falling body a measure of the force of gravity acting upon it and goes on to calculate the ratio (4000:1) between this acceleration at the earth's surface and the "endeavour of the moon to recede from the centre of the earth." He then uses Kepler's third law to show that for each planet its "endeavour to recede" from the sun is inversely proportional to the square of the distance between them. It seems likely that it would have occurred to him that the force countering the moon's "endeavour to recede" was, in fact, the *same* force of gravity which causes bodies on earth to fall.[43]

Granted that forces are the causal principles responsible for planetary motion, how do they operate? How do they relate to the "endeavour to recede"? Newton had three very different sorts of resource to draw upon in answer to questions such as these. The mechanical philosophy would suggest an aether of some kind as a medium of transmission of contact action; it was not clear, however, that the new notion of force could be contained within the restrictive bounds of mechanical explanation as it had been defined by the Cartesians. His reading of the Cambridge neo-Platonist, Henry More, on the other hand, would lead him to think of spirit, not body, as the source of new motion in the world. In the *De gravitatione,* he argues that as we move our bodies by an act of will alone, so God must be thought to move the bodies He creates. Newton makes his theological motivation clear: "We find almost no other reason for atheism than this notion of bodies having, as it were, a complete, absolute and independent reality in themselves."[44] His opposition to this aspect of the mechanical philosophy would remain a lifelong one. And

---

[42]A. R. and M. B. Hall, *Unpublished Scientific Papers of Isaac Newton* (Cambridge: Cambridge University Press, 1962), p. 148.

[43]Since he took the moon to be 60 earth radii distant from the earth, the inverse proportionality would give a ratio of 3600:1 between the two "endeavors," close to the ratio directly derived from the accelerations. In later years, he more than once claimed to have hit upon this crucial generalization in the 1660s. But the "endeavor" here, in the case of the moon, is still one of *receding.*

[44]*Unpublished Scientific Papers of Newton,* p. 144.

the openness of the world to the operation of spirit at all levels, and the corresponding passivity of matter in its own right, will become an article of faith, despite the problems it later creates for him.

The third source of analogies which might have illuminated for Newton the operation of force in the world was the alchemical writings into which he plunged at this time. The alchemists saw the world as suffused with active principles of all sorts, incessantly transforming one kind into another. In some notes from around 1669, Newton contrasts mechanical with "vegetable" action and argues that the earth can best be understood as a "great vegetable" within which a "prodigious active principle," a "subtle spirit," perpetually works:

> There is therefore besides the sensible changes wrought in the textures of the grosser matter a more subtle secret and noble way of working in all vegetation which makes its products distinct from all others, and the immediate seat of these operations is not the whole bulk of matter, but rather an exceedingly subtle and unimaginably small portion of matter diffused through the mass which, if it were separated, there would remain but a dead and inactive earth.[45]

One can see Newton trying out in these early writings different paradigms of natural action. The contact action to which the mechanical philosophy would have limited him seemed entirely insufficient to explain the phenomena of the living world or magnetism or light or chemical transformation. In the "Hypothesis of Light and Colors" of 1675, he suggests that light is "something or other capable of exciting vibrations in the aether,"[46] which sounds mechanical enough. But then he goes on to say that the aether cannot be considered to be a homogeneous sort of matter. Besides the main "phlegmatic body" of the aether, it must contain various other "aetherial spirits" just as air contains a variety of different exhalations:

> For the electric and magnetic effluvia and the gravitating principle seem to argue such variety. Perhaps the whole frame of nature may be nothing but various contextures of some certain aetherial spirits or vapors, condensed as it were by precipitation. . . . Thus perhaps may all things be originated from aether.[47]

He goes on to argue that the Cartesian attempt to explain this wide variety of actions by invoking a single property, the "subtlety" or size and

---

[45] "On the vegetation of metals," Burndy mss. 16, f. 6v 6; spelling modernized. Cited by R. S. Westfall, *Never At Rest* (Cambridge: Cambridge University Press, 1980), p. 307.

[46] Communicated to the Royal Society in a letter to Oldenburg, *The Correspondence of Isaac Newton*, ed. H. W. Turnbull (Cambridge: Cambridge University Press, 1959), vol. 1, p. 363.

[47] *Ibid.*, p. 364.

shape of the constituent particles, simply cannot work. The ways in which simple substances combine or refuse to combine (acid and salts, oil and water) show that there must be some "secret principle of unsociableness" at work. In short, we ought never measure the modes of natural action by *our* abilities to understand them:

> God, who gave animals motion beyond our understanding is, without doubt, able to implant other principles of motion in bodies which we may understand as little. Some would readily grant this may be a spiritual one; yet a mechanical one might be shown, did not I think it better to pass it by.[48]

In 1679, Hooke wrote Newton for his opinion on an hypothesis Hooke had formulated some years before, one that he predicted would reduce all celestial motions to a single rule, thus leading to "the true perfection of astronomy." All celestial bodies, he suggests, have an "attraction or gravitating power" toward their centers which not only makes nearby bodies fall towards those centers but also affects, to a degree dependent on their distance, "all other celestial bodies that are within the sphere of their activity."[49] The sixty years since Kepler had proposed an attractive power between sun and planets had not been a hospitable time among astronomers for such powers. And Hooke had made the step that Kepler could not by supposing that the attraction is responsible for keeping the planet in its orbit. For Newton, this was challenge, indeed, and the challenge became sharper when, despite Newton's professed lack of interest, Hooke in successive letters suggested that the attraction might vary inversely as the square of the distance[50] and asked what form the orbit would take in such a case. Further, he wondered what the "physical reason" for an attraction varying in this way might be.[51] Thus began the celebrated process that led to the writing of the *Principia*. Newton had none of the scruples about attraction that an orthodox mechanical philosopher would have had.[52] But to find a "physical reason" for the attraction was going to prove more taxing than he could at that point have guessed.

---

[48]*Ibid.*, p. 370.

[49]*Attempt to Prove the Motion of the Earth,* written in 1674, and published in his *Lectiones Cutlerianae* in 1679; see R. T. Gunther, *Early Science in Oxford* (Oxford: Oxford University Press, 1931), vol. 8, pp. 27–28.

[50]*Correspondence,* Jan. 6, 1680; vol. 2, p. 309.

[51]*Correspondence,* Jan. 17, 1680; vol. 2, p. 313.

[52]Westfall remarks: "Newton's philosophy of nature underwent a profound conversion in 1679–80 under the combined influence of alchemy and the cosmic problem of orbital mechanics, two unlikely partners which made common cause on the issue of action at a distance" (*Never at Rest,* p. 390). They did not, however, persuade him that unqualified action at a distance is possible.

5. *The* Principia

It was all very well to show that an inverse-square attraction entails an elliptical orbit for the planets, and conversely, that an elliptical orbit requires an inverse-square attraction to one focus of the ellipse. What was more difficult, and what took Newton months of effort and many successive trial drafts, was to construct a conceptual network relating the key ideas of motion, mass, and force that he was using. What Hooke's questions had helped him see was that the operation of an attractive force on a planet in a circular orbit would not alter the planet's speed; its effect, instead, is to alter the direction of motion in the direction of the force. R. S. Westfall sums up:

> Only the principle of inertia, which abandoned the concept of the force of a body's motion, would allow him to treat changes of direction in the same terms as changes of speed. Only the principle of inertia would allow the unique leap of the imagination which made the new dynamics possible, the recognition of the dynamic identity of uniform circular motion and uniformly accelerated motion in a straight line. Heretofore in the history of mechanics, these two motions had been treated as irreducible opposites. What Newton perceived were the possibilities that opened up if he treated the two as dynamically the same.[53]

But even in the final version of the eight Definitions and three Axioms or Laws that prefaced the first edition of the *Principia,* he still had not a secure grip on the principle of inertia. He still invokes a force (*vis insita*) to account for the persistence of uniform motion.[54] And he had not sufficiently clearly distinguished between force and impulse. Proposition I, for example, makes use of the parallelogram of forces to compound two incommensurables, one of them a discrete impulse, the other a continuously acting force. But these were the awkwardnesses that one could expect in so intricate a realignment, faults that would gradually be eliminated.

Of far greater concern was the central concept on which the entire planetary dynamics depended, the concept of attraction. Newton realized full well that it would be unacceptable to devotees of the mechanical philosophy. But, more to the point, he was not entirely comfortable with it himself. There was more than a suggestion of action at a distance about it, and this notion Newton could not accept. In a much-quoted letter to Bentley in 1693, he wrote:

> It is inconceivable that inanimate brute matter should, without the mediation of

---

[53]*Ibid.,* p. 416.

[54]See McMullin, *Newton on Matter and Activity* (Notre Dame, Ind.: University of Notre Dame Press, 1978), pp. 33–42.

something else which is not material, operate upon and affect other matter
without mutual contact. . . . That gravity should be innate, inherent, and essential to matter, so that one body may act upon another at a distance and through a
vacuum without the mediation of anything else by and through which their
action or force may be conveyed from one to another is to me so great an
absurdity that I believe no man who has in philosophical matters any competent
faculty of thinking can ever fall into it. Gravity must be caused by an agent
acting constantly according to certain laws, but whether this agent be material or
immaterial is a question I have left to the consideration of my readers.[55]

He could scarcely have been more emphatic. Attraction was *not* to be
interpreted as action at a distance. There had to be some intermediate sort of
agency, but of what sort, Newton could not determine. To block off awkward
questions, he had a strategy to hand, one that (as we have seen) had ancient
roots in planetary astronomy. In Definition VIII, at the opening of the
Principia, he makes a disclaimer:

I . . . use the words, 'attraction', 'impulse', or propensity of any sort towards a
center, promiscuously and indifferently, one for another, considering those
forces not physically but mathematically; wherefore the reader is not to imagine
that by those words I anywhere take upon me to define the kind or the manner of
any action, the causes or the physical reason thereof, or that I attribute forces, in
a true and physical sense, to certain centers (which are only mathematical
points) when at any time I happen to speak of centers as attracting. . .[56]

What are we to make of this and other similar disavowals? What did he
mean by ''considering forces not physically but mathematically''? The
''mathematicians'' in earlier astronomy were those who focused on a descriptive account of planetary motions, expressed in geometrical terms, leaving
aside questions of cause. What Newton had realized was that his reworking of
the concept of force had, for the first time, made it possible to express motion
in terms that are at once ''mathematical'' and quasi-causal. From the observational standpoint, to say that a force is operating is to say that an accelerated
motion of a certain sort is occurring. Force could thus be treated ''mathematically'' as a tendency to move something in a particular way. He clarifies
this later in Book I:

I here use the word, 'attraction', in general for any endeavor whatever made by
bodies to approach to each other, whether that endeavor arise from the action of
the bodies themselves, as tending to each other or agitating each other by spirits

[55]Newton to Bentley, February 25, 1692/3; *Correspondence,* vol. 3, 253–254.
[56]Definition VIII, *Principia,* Motte-Cajori translation (Berkeley, Calif.: University of California Press, 1962), pp. 5–6.

emitted; or whether it arises from the action of the aether or the air, or of any medium whatever, whether corporeal or incorporeal, in any manner impelling bodies placed therein toward each other.[57]

Here 'attraction' is taken to designate the endeavor or tendency only, prescinding entirely from the manner in which that tendency is brought about. It is not necessary to inquire into causes in order to give a unified account of the effects. Are we to see here a precedent for later-day instrumentalism, as commentators have sometimes suggested? By no means. First of all, Newton was not entirely setting aside the quest for causes. Taking gravity as a tendency is not to put an end to causal inquiry:

> In mathematics, we are to investigate the quantities of forces . . . then, when we enter upon physics, we compare those proportions with the phenomena of Nature, that we may know what conditions of those forces answer to the several kinds of attractive bodies. And this preparation being made, we argue more safely concerning the physical species, causes, and proportions of the forces.[58]

This is not quite the same way of dividing "mathematics" and "physics" as the one prompting the earlier distinction, because it would label as "physics" those parts of the *Principia* where the "phenomena of Nature," Kepler's laws, for example, are explained by means of the new "mathematics." Newton's use of these terms is not entirely consistent if one follows them through his works.[59] But it is clear in this passage that he is anticipating a future stage of inquiry when one can "argue safely" about the causes of gravitational motion, presumably by finding empirical ways to test rival causal hypotheses of the sort he has already hinted at.

But there is another and deeper reason to reject the characterization of Newton as proto-instrumentalist. Had he been asked how his own "mathematics" differed from that of Ptolemy, he would assuredly not have restricted himself to the criterion of "saving the appearances," any more than Copernicus had. What excited him was the fact that he had discovered a single *physical* hypothesis that would unify the motions of earth and sky:

> Hitherto we have explained the phenomena of the heavens and of our sea by the power of gravity, but have not yet assigned the cause of this power. This is certain, that it must proceed from a cause that penetrates to the very center of the sun and planets, without suffering the least diminution of its force; that operates not according to the quantity of the surfaces of the particles upon which it acts

---

[57]*Ibid.*, p. 192.

[58]*Ibid.*, p. 192.

[59]See Anita Pampusch, "'Experimental', 'Metaphysical', and 'Hypothetical' philosophy in Newton's methodology," *Centaurus* 18 (1974): 289–300.

(as mechanical causes used to do), but according to the quantity of the solid matter which they contain, and propagates its virtue on all sides to immense distances.[60]

No hesitation here: it is "certain" that gravitational motion must proceed from a cause that "penetrates," or as he had already put it in Definition VIII, a "cause, without which those motive forces would not be propagated through the spaces round about," whatever that cause may be. And that same cause operates also on the smaller scale. In some early draft-material for the *Opticks,* written in the 1690s, Newton's optimism shows through:

> Hypothesis 2. As all the great motions in the world depend upon a certain kind of force (which in this earth we call gravity) whereby great bodies attract one another at great distances; so all the little motions in the world depend upon certain kinds of forces whereby minute bodies attract or dispel one another at little distances.
>
> . . . And if Nature be most simple and fully consonant to herself, she observes the same method in regulating the motions of smaller bodies which she doth in regulating those of the greater.[61]

It is this conviction of the "consonance of Nature to herself" that carries him along. In the *Principia,* he had had to hold back:

> This principle of nature being very remote from the conceptions of philosophers I forbore to describe it in that book [lest it] should be accounted an extravagant freak and so prejudice my readers against all of those things which were the main design of the book. And yet I hinted . . . both in the preface and in the book itself where I speak of the [refraction] of light and of the elastic power of the air.[62]

But now that the *Principia* has received the "approbation of the mathematicians" that he had hoped it would, he can at last use "plain words." The truth of the physical hypothesis about the universality of forces as agents at all levels he cannot prove, "but I think it very probable because a great part of the phenomena of nature do easily flow from it," and he goes on to list a dozen chemical phenomena with which his alchemical researches had made him familiar, as well as another dozen drawn from the phenomena of light, heat, and gases.[63] He is speaking here as a natural philosopher; his assurance that a single sort of cause underlies all these phenomena rests on the belief that the phenomena "do easily flow" from this supposition. The notion of "flow-

---

[60]General Scholium, added to the second edition of the *Principia* in 1713, Motte-Cajori translation, p. 546.

[61]Cambridge University Library, Add. Mss., 3970, f.338r.

[62]Add. Mss, 3970, f.338. Words missing due to damage to the manuscript.

[63]*Ibid.*

ing from'' was, as we can now clearly see, dangerously ambiguous because of the dual status of force as cause and effect. But it was, in the end, the causal consonance of the whole and not just the success of the ''mathematics,'' that separated the *Principia* from the *Almagest* and guaranteed for Newton its significance as science.

6.   *After the* **Principia**: *The gravity dilemma*

But others were not so easily convinced of this. The problem was not merely that the distinction between ''mathematics'' and a ''physics'' still to come seemed to many like a ruse, but that Newton had, to all appearances, made a future physics impossible, a physics acceptable to mechanical philosophers, at least. To see why this was the case, why the problem of explaining distant gravitational motions seemed more, not less, difficult in the aftermath of the *Principia,* let us lay out the logic of the ''gravity dilemma.'' Though the escape routes were not as tightly sealed as they appear to be in the case of the Bell theorem and the associated experimental results, the challenge to assumptions about explanation must have seemed no less then than it once again does now.

First, it was agreed by all that action at a distance was to be excluded as nonexplanatory. There was no theory of special relativity to support this ''locality'' principle, but it did not seem to be needed. The near-unanimity of natural philosophers all the way back to Aristotle ensured that the unintelligibility of the notion of action at a distance would remain unquestioned.

Second, the *Principia* seemed to have excluded the possibility of a mechanical aether that could convey action from planet to sun. Newton had showed in Book II of the *Principia* that such an aether would inevitably disturb the planet's motion and would cause it to spiral inwards. To retain a mechanical aether, one would either have to attribute to it some very strange properties or else challenge the entire structure of the new dynamics and produce a credible alternative.

Third, it would not be enough to talk in terms of innate tendencies to motion on the part of the planet. The measure of the attraction on a planet was given not only by the mass of the planet itself but by the mass of the distant sun as well. Though one could for practical purposes abstract from this latter factor and assume the sun to be at rest, it could not be set aside in translating the scheme into explanatory terms. Change the mass of that distant body and the motion of the planet will be different. So some form of ''influence'' is involved, an influence impossible to understand in contact terms of vortices or the like. A pressure transmitted through a Cartesian aether would not be able to carry ''information'' of the required sort.

So how *is* planetary motion to be explained? It was an impasse for many of the readers of the *Principia* at least. If the new mechanics were accepted, one either had to allow action at a distance or else introduce new and illegitimate-seeming modes of bridging the spaces between planet and sun. In a letter to Newton in 1693, Leibniz tried to avoid this conclusion:

> You have made the astonishing discovery that Kepler's ellipses result simply from the conception of attraction or gravitation and passage in a planet. And yet I would incline to believe that all these are caused or regulated by the motion of a fluid medium, on the analogy of gravity and magnetism as we know it here. Yet this solution would not at all detract from the value and truth of your discovery.[64]

Newton was, however, uncompromising. A fluid medium would *disturb* the motions of the planets, not sustain them, he responded. Besides no such medium is needed; gravity is enough:

> I have myself concluded that all other causes are to be rejected and that the heavens are to be stripped as far as may be of all matter, lest the motions of planets and comets be hindered or rendered irregular.[65]

If, he goes on, someone can explain gravity by a "subtle matter" of a kind that does not resist planetary motion (and thus is not "material" in the standard mechanical sense), he will be happy to hear about it. But he wanted to make it quite clear that talk of a fluid medium would not help *unless* it were recognized that it could not be a material medium of the traditional sort.

Leibniz, on the other hand, could be pardoned for being unhappy with this response. To dispense with all causes save gravity is to trade on the ambiguity between gravity as cause and gravity as effect that Newton constantly fell back on. Here, in response to Leibniz's invocation of a medium as cause, he proposes gravity as a sufficient alternative. But elsewhere, as we have seen, he recognizes that this leaves him open to the objection of admitting action at a distance and so he takes gravity instead as an "endeavor," capable of being certified "mathematically" but itself now requiring a cause. But this cause must somehow bridge the intervening spaces. How?

7. The alternatives: A "mechanical" aether

It may be instructive to chronicle very briefly in closing, some of the responses of the natural philosophers to the gravity dilemma. These responses fall very roughly into three classes. First, there were those who maintained

---

[64]Leibniz to Newton, March 1692/3, *Correspondence*, vol. 3, p. 258.
[65]Newton to Leibniz, October 16, 1693, *Correspondence*, vol. 3, p. 287.

that the traditional notion of an aether could on no account be dispensed with and proposed some means of salvaging it. Second, there were those who insisted that a mechanical aether of any sort was no longer tenable and that some other means of bridging the spaces, "whether material or spiritual," had to be found. Finally, there was at least one who argued that no physical explanation of the correlations between planetary motions and the sun's position and mass was really needed and made use of the occasion to propose an instrumentalist account of natural science generally. Analogies between these alternatives and the responses to the quantum dilemma chronicled in this book are not hard to discern.

In the first camp were the Cartesians and others like Leibniz and Huygens[66] who, though critical of one or other principle of Cartesian physics, were assured that an aether *must* be retained at all costs as the only "intelligible" mode of explanation of distant action. Huygens's response to the *Principia,* his *Discourse on the Cause of Gravity* (1690), may be taken as a paradigm of this solution. The immediate focus of the work is on terrestrial gravity, which had always posed more of a problem to the Cartesians than planetary motions did. It was plausible to suppose that vortices of invisible matter might whirl the planets in their paths. But how about accelerated downward motion towards the center of the earth? Towards the earth's axis of rotation, perhaps, but to its *center?* Where would the aether *go?*

In the opening lines of the *Discourse,* Huygens makes his position plain. If we are to find an *intelligible* cause of gravity, we must not introduce different kinds of matter (like Descartes's three kinds of element); size, shape, and motion are sufficient to characterize the entire material world. No "inclinations" are needed. And so he proposes a complicated mechanism of aether-particles moving very rapidly in concentric shells around the earth, whose effect on heavy bodies is to force them inwards, i.e., to make them "fall." The warrant for this is the intelligibility of the *terms* used in the explanation; since size and shape cannot explain the apparent tendency of bodies to fall, it *must* be due to motion, and the motion here can only be of an invisible medium, since no motions are actually observed other than the motion of fall itself. When it comes to planetary motions, Hugyens nonchalantly remarks that "nothing prevents" a similar explanation from being given,[67] he does not actually produce one, however, though he dismisses Descartes's vortices as inadequate for the task allotted them.

---

[66]See A. Koyré, "Huygens and Leibniz on universal attraction," and "Attraction an occult quality?" in *Newtonian Studies* (Chicago: University of Chicago Press, 1965), pp. 115–138, 139–148.

[67]*Discours de la Cause de la Pesanteur,* published as an appendix to the *Traité de la Lumière* (Leiden: 1690), p. 160.

He allows that he is impressed by Newton's discovery of the inverse-square law and has no hesitation in saying that he believes Newton's "hypothesis of gravity" is true since it brings together so many previously disparate classes of phenomena. But, of course, though he does not say so, it is understood that it is incomplete since it still has to be *explained* in terms of an aether. Huygens calculates the speed at which the aether must circulate around the earth in order to give to falling bodies the acceleration they actually have; it comes out as seventeen times the rotational velocity of the earth at the equator. But the difference between this and Newton's procedure is plain: Newton postulates a very general hypothesis of universal attraction from which all sorts of testable consequences follow. Huygens can only infer, after the fact, the velocity of the aether, according to his hypothesis, from the value of gravity at each point. But nothing further follows from this. The warrant for his theory in no way derives, therefore, from the predictions it enables him to make. His aether is thus reminiscent of the spheres of Aristotle. As a "physics," it stands apart from the "mathematics." This is precisely the split that Newton's work had made unacceptable; though Newton usually recognizes that "mathematics" alone cannot provide a *sufficient* warrant for a "physics," he rejects hypotheses that cannot be tested observationally in their own right.[68]

But what does Huygens make of the obvious objection that the *Principia* has shown that an aether would render the planetary system unstable? He attempts to evade the difficulty by distinguishing between two kinds of rarity in aetherial matter, one due to the distance between the particles, the other deriving from the rarity of the particles themselves, which are like honeycombs touching one another. An aether of the former kind falls prey to Newton's argument, but not the latter, he claims. It would have "very little resistance" to motion through it because of the voids within the particles. But, of course, "very little" is still too much. What Newton had shown was not (as Huygens here says) that interplanetary space "contains only a very rare matter,"[69] but that if it contains any resisting matter at all, the planet would gradually spiral inwards. And it is far from clear why the honeycomb particles would offer so little resistance, nor, on the other hand, how they could still function to drive falling bodies downwards.

This rather desperate attempt to save the original aether hypothesis at all costs appears, however, like great good sense in comparison with some later efforts like those of Bernoulli and Lesage. In his *New Celestial Physics,* written nearly half a century after the *Principia,* John Bernoulli explains

---

[68]Newton was far from consistent in his prescriptions for the methods of natural philosophy, but this is a large topic on its own account. See, for example, McMullin, "Conceptions of science in the Scientific Revolution."

[69]*Discours*, p. 161.

gravity by a "central torrent" of aetherial matter which pours ceaselessly towards the center of each planet. Because the matter is "perfectly liquid," being infinitely divided, it sets up no resistance to motion through it since its parts lack inertia, being infinitely small.[70] Its subtilty indeed "surpasses imagination" (a dangerous comment in a narrative that claims as its warrant "clear and distinct ideas"!).[71] Yet this matter without inertia is also the agent responsible for planetary motions. Whewell comments sardonically on:

> the overwhelming thought of the whole universe filled with torrents of an invisible but material and tangible substance, rushing in every direction in infinitely prolonged straight lines and with immense velocity. Whence can such matter come, and whither can it go? Where can be its perpetual and infinitely distant fountain, and where the ocean into which it pours itself when its infinite course is ended? . . . The central torrent of Bernoulli [is] an explanation far more inconceivable than the thing explained.[72]

8. The alternatives: "Nonmechanical" natural agency

A second way to approach the gravity dilemma is to look for alternatives to an aether, causal agencies other than media functioning through contact action. Recall Newton's own language of "active principles," "active spirits," "material or immaterial," "corporeal or incorporeal," which would surely encourage the greatest latitude in the quest. His disciple, Samuel Clarke, might be as good an example as any of someone who took these hints seriously. In his famous debate with Leibniz, who held it to be a "supernatural thing that bodies should attract one another at a distance without any intermediate cause,"[73] Clarke responded that such action at a distance would

---

[70]*Essai d'une Nouvelle Physique Celeste (1735), Opera Omnia* (Lausanne, 1742), vol. 3, p. 277. The *Essai* was awarded the prize of the Royal Academy of Sciences of Paris in 1734, suggesting how much of a hold Cartesian physics still had in France.

[71]*Ibid.*, pp. 284, 270.

[72]William Whewell, "On the transformation of hypotheses in the history of science," in *On the Philosophy of Discovery* (London: 1860), p. 499. Whewell was troubled by the discontinuities so evident in the history of science, where even long-accepted theories are sometimes rejected in favor of other quite different ones. What he labored to show was that the discontinuities were not as great as they appeared since the hypothesis under attack is gradually modified until it becomes indistinguishable from its rival. Bernoulli's "central torrent" has no function, he claims, other than to produce the results already certified in Newtonian mechanics. Whewell's attempt to smooth out the history of science will not quite do, however, since Bernoulli's theory is not just *ad hoc,* it is full of inconsistencies, and the Newtonian results cannot be quantitatively derived from it. So the theories do not *really* merge.

[73]*The Leibniz-Clarke Correspondence,* ed. H. G. Alexander (Manchester: Manchester University Press, 1956), p. 43.

be not just a miracle but a contradiction, "for 'tis supposing something to act where it is not." But, he goes on:

> the means by which two bodies attract each other may be invisible and intangible, and of a different nature from mechanism; and yet, acting regularly and constantly, may well be called natural, being much less wonderful than animal motion, which yet is never called a miracle. If the word 'natural forces' means here 'mechanical', then all animals, and even men, are as mere machines as a clock. But if the word does not mean mechanical forces, then gravitation may be effected by regular and natural powers, though they be not mechanical.[74]

Leibniz, predictably, sniffed at such a response. As far as he was concerned, to say it was "invisible, intangible, not mechanical" was equivalent to calling it: "inexplicable, unintelligible, precarious, groundless, and unexampled."[75] Clarke's response, of course, was that explanation ought not be bound into the rigid framework of the mechanical philosophy. If 'mechanical' had to mean intelligible in terms of size, shape, and motion alone, then not only was gravitation nonmechanical but so also was all living motion. Boyle had, years before, criticized the strict reductionism of those who would allow explanation only in terms of such categories as size, shape, and motion. Though broadly sympathetic to the mechanical philosophy, he thought it absurd to exclude explanations in terms of such familiar qualities as spring of the air, or gravity ("that known affection of almost all bodies here below"), unless the further causes of *these* qualities could themselves be immediately specified in narrowly mechanical terms.[76] But the strongest reaction to the damaging restrictions imposed by the mechanical philosophy had, of course, always come from those whose imaginations were formed by the richer cosmologies of neo-Platonism and alchemy.

What Clarke is arguing for is that it is sufficient that something act in a regular way for it to be called "natural" and to be understood in natural terms. There is no reason why explanation has to be reduced to the sparse categories of mechanism, as though all the permissible explanatory modalities could be laid down in advance and drawn, at that, from so impoverished an exemplar as clockwork. Here, of course, was the "great divide" in seventeenth-century philosophy. The primary/secondary distinction, which had been canonical among most natural philosophers from Galileo's time onwards, was being challenged by the gravity dilemma. The earlier list of primary qualities was simply insufficient in Clarke's eyes to handle the prob-

---

[74]*Ibid.*, p. 53.

[75]*Ibid.*, p. 94.

[76]"Some considerations touching experimental essays in general" (1661), *The Works of the Honourable Robert Boyle*, ed. Thomas Birch, 6 vol. (London: 1772), vol. 1, p. 309.

lems of gravitational motion in a coherent way. The list would have to be enlarged, and the strait-jacket imposed by clear and distinct ideas discarded.

Clarke could not, of course, translate any of this into action. Talk of "active principles" or "elastic spirits" infusing the spaces around magnetic bodies, and perhaps the entire interplanetary spaces as well, would be dismissed as a return to the despised occult qualities of the scholastics until some way was found to anchor it in "mathematics." It could serve as heuristic but it would remain "hypothesis" in the sense proscribed by Newton, as long as it rested only on broad explanatory analogies like the Cartesian aether. There had to be a way to specify it in "mathematical," that is, in measurable terms. Clarke could not have guessed that the task would take so long. But when Faraday and Maxwell began to fill up those troublesome spaces with their electromagnetic fields, they were anchoring Newton's active principles in a way of which he would have approved.[77]

9.   The alternatives: No physical explanation possible

In 1720, the topic of the prize-essay contest of the Royal Academy of Sciences in Paris was "The cause of motion." One of the entrants was George Berkeley who at the time was in Lyons His essay, *De motu*, offers the most succinct statement of yet another possible response to the gravity dilemma. It draws on the positions already laid down by him in the *Treatise Concerning the Principles of Human Knowledge* of 1710 in order to respond to the general question as to how motion is to be explained causally and, more particularly, how gravitational motions are to be treated. He begins by deploring the custom of natural philosophers of using metaphors like *attraction* to explain. A term like 'gravity' suggests a cause, but if gravity is a cause, it obviously lies beyond the reach of the senses. It is therefore an occult quality and, as such, must be excluded. Only that which is subject to sense and fully intelligible to reason is admissible in natural philosophy.

Terms like 'gravity' and 'attraction' are perfectly acceptable for the purposes of "reckoning," or prediction; this was their function, he claims, for Newton. They describe, at best, effects, not causes. Indeed, it is not:

> the business of physics or mechanics to establish efficient causes, but only the rules of impulsions or attractions, and, in a word, the laws of motions, and from the established laws to assign the solution, not the efficient cause, of particular phenomena.[78]

[77]See Mary Hesse, *Forces and Fields* (Edinburgh: Nelson, 1961), chap. 8.

[78]*De Motu. The Works of George Berkeley*, ed. A. A. Luce and T. E. Jessop (Edinburgh: Nelson, 1952), vol. 4, p. 40.

By 'solution' here, he means effective prediction. A motion can be said to be "explained mechanically," he goes on, when it is derived deductively from general "laws," or descriptions. "That is the sole mark at which the physicist must aim."[79] There is, however, a weaker sense of 'cause' which is applicable in physics:

> The physicist studies the series or successions of sensible things, noting by what laws they are connected, in what order, what precedes as cause, and what follows as effect. And the body in motion is the cause of motion in the other, and impresses motion on it, draws it also or impels it. In this sense, second corporeal causes ought to be understood, no account being taken of the actual seat of the forces or of the active powers or of the real cause in which they are.[80]

If one is satisfied merely with regularities of succession between observables, he says, one can call what comes before "cause" and what comes after "effect." But this is a secondary sense, and there ought be no mistaking "second" causes of this sort with *real* causes, that is, with the agents actually responsible for the occurrence of the effect. There is no need for the real cause to show itself in the limited way that second causes do, no reason to think the cause to be necessarily a "sensible thing," manifesting its causal tie by means of an invariable temporal sequence between one class of such things and another. (He would have been astonished at the success that Hume's erection of second causes into the [only] real causes would have in a later world.) In this sense, and this sense only, the sun can be said to "attract" a planet, or to be the "cause" of its motion. When Newton propounds the inverse-square law to describe the operation of gravity, there is nothing more for the physicist to say, there is no need to seek in aethers or elastic spirits (both excluded in principle by Berkeley's empiricism) a further underlying cause of the motion.

Does this mean that gravity has no real cause? By no means. But we must go on to metaphysics to discover it. "Reason and experience advise us," he says (this had been a central theme of the *Principles*), "that there is nothing active except mind or soul."[81] And so he will ultimately find in God the real cause of the correlations discovered in natural philosophy. These correlations are brute fact requiring, indeed permitting, no further explanation from the point of view of physics proper:

> Metaphysical principles and real efficient causes of the motion and existence of bodies or of corporeal attributes in no way belong to mechanics or experiment,

---

[79]*Ibid.*, p. 41.

[80]*Ibid.*, p. 51. He makes a similar point in the *Treatise Concerning the Principles of Human Knowledge*, par. 105.

[81]*De Motu*, p. 41.

nor throw light on them, except insofar as by being known beforehand, they may serve to define the limits of physics, and in that way to remove imported difficulties and problems.[82]

Berkeley's solution of the gravity dilemma is, then, to reject intermediates of whatever sort, whether mechanical or nonmechanical, as well as action at a distance of one body on another. This leaves him with unexplained correlations between the motions of distant bodies. To deal with them, he has to ascend to another order of explanation entirely. Mind (ultimately God) is immediately responsible for both the being *and* the motion of all bodies, the *only* real cause of planetary motions. Thus Newton's mechanics and the gravity dilemma both served Berkeley's purposes well. The mechanics provided a plausible context for his instrumentalism, precisely because of the ambiguous status of force/gravity as cause/effect. And the resolution of the gravity dilemma along strictly empiricist lines excluded all forms of cross action between the bodies, leaving a choice between only two alternatives: allowing the correlations to remain *entirely* unexplained, or turning to metaphysics and recognizing a common cause who is God.

10. The explanation of distant action

We have traced three different ways in which the natural philosophers of Newton's day responded to the gravity dilemma. It may be worth recalling that all three of these responses are explored in Newton's *own* work. This is obvious in the case of the second one, the one typified by Clarke. But in notes from the 1690s, he frequently draws on the analogy of the way in which the human mind moves the human body to suggest that space may be the "sensorium of an immaterial, living, thinking being" whose mind may be directly responsible for the regular way in which bodies move.[83] But the difficulties (to be later urged by Leibniz against Clarke) in the way of this solution were sufficiently obvious to a natural philosopher. It meant giving up on the quest for physical causes in the natural order, and though the gravity dilemma might ultimately enforce this solution, Newton, unlike Berkeley, was not one to welcome it despite his firm theological conviction of God's ultimacy in the order of explanation.

More significant, perhaps, was his return to a quasi-mechanical aether in the third edition of the *Opticks,* despite his earlier assurance that any such

---

[82]*Ibid.,* p. 42.

[83]Addendum crossed out in a draft for Query 31 of the *Opticks,* Add. Mss., 3970 f. 252v. For other similar passages, see J. E. McGuire and P. M. Rattansi, "Newton and the pipes of Pan," *Proc. Royal Society,* NR, 21 (1966): 108–143.

aether had to be "in exile from the nature of things."[84] It was for light, in particular, that he needed an aether. He had, after all, shown the periodicity of many optical phenomena, and how could this be explained without a medium that could sustain vibration? Queries 17–24, inserted in 1717, postulate a medium filling the "celestial spaces" whose density differences (its density *increases* with distance from solid bodies) could be responsible for gravitational motion, and whose extraordinary elasticity/density ratio of 49.10^{10} might prevent it from setting up a resistance to planetary motion. It was an implausible suggestion, as Newton must have known. This aether is not, in the Cartesian sense, strictly mechanical since it operates by elastic forces of repulsion between its particles, not by contact action. But it may have appeared less open to challenge because its forces are short-range, like those of magnetism, and the threat of action at a distance may thus have seemed less immediate. But it could not possibly move the planets unless it had some *vis inertiae* on its own account, and if it had, no matter how little, this would be sufficient to destabilize the planetary orbits. So it could not solve the gravity dilemma.

Newton obviously remained open to every explanatory lead, no matter how initially unpromising. His estimations of likelihood were rooted in a far richer metaphysics and thus a far wider range of possible agencies, than the mechanical philosophy permitted. He did not allow himself to be bound by an intelligibility that rested exclusively on the intuitively simplest forms of action. His sustained and imaginative campaign to solve the gravity dilemma did not, in the end, succeed. The problem of understanding distant gravitational action, of moving beyond the neutral terms of the "mathematician," proved far more difficult than he could ever have anticipated. It required a profound transformation not only of the concepts of action, space, and time, but also of the notion of explanation itself.

And, as the essays in this book eloquently testify, that transformation still continues.

---

[84]Draft for a second edition of the *Principia*, dating from the 1690s; Gregory Mss., 247, f. 14a.

Reference List

Achinstein, Peter. (1983) *The Nature of Explanation*. Oxford: Oxford University Press.

Aspect, A., J. Dalibard, and G. Roger. (1982) "Experimental tests of Bell's inequalities using time-varying analyzers." *Physical Review Letters* 49:1804–1807.

Aspect, A., P. Grangier, and G. Roger. (1981) "Experimental tests of realistic local theories via Bell's Theorem." *Physical Review Letters* 47: 460–463.

——— (1982) "Experimental realization of Einstein-Podolsky-Rosen-Bohm *Gedankenexperiment*: A new violation of Bell's inequalities." *Physical Review Letters* 49: 91–94.

Ballentine, L. E. (1987) "Resource letter IQM-2: Foundations of quantum mechanics since the Bell inequalities." *American Journal of Physics* 55: 785–792.

Ballentine, L. E., and J. Jarrett. (1987) "Bell's theorem: Does quantum mechanics contradict relativity?" *American Journal of Physics* 55: 696–701.

Belinfante, F. J. (1973) *A Survey of Hidden-Variable Theories*. Oxford: Pergamon Press.

Bell, J. S. (1964) "On the Einstein-Podolsky-Rosen paradox." *Physics* 1: 195–200; reprinted in Bell (1987).

——— (1966) "On the problem of hidden variables in quantum mechanics." *Reviews of Modern Physics* 38: 447–452; reprinted in Bell (1987).

——— (1971) "Introduction to the hidden variable question." In *Foundations of Quantum Mechanics*, ed. B. d'Espagnat. New York: Academic Press, 171–181; reprinted in Bell (1987).

——— (1975) "On wave-packet reduction in the Coleman-Hepp model." *Helvetica Physica Acta* 48: 93–98; reprinted in Bell (1987).

——— (1981) "Bertlmann's socks and the nature of reality." *Journal de Physique*, Colloque C2, 42(3): 41–61; reprinted in Bell (1987).

——— (1984) "*Be*ables for quantum field theory." CERN Preprint, CERN-TH.4035/84; reprinted in Bell (1987).

——— (1987) *Speakable and Unspeakable in Quantum Mechanics*. Cambridge: Cambridge University Press.

Bell, J. S., A. Shimony, M. A. Horne, and J. F. Clauser. (1985) "An exchange on local beables." *Dialectica* 39: 85–110; reprinted from *Epistemological Letters*, 9(1976), 13(1976), 15(1977), 18(1978).

Beltrametti, E. G., and G. Cassinelli. (1981) *The Logic of Quantum Mechanics.* Reading, Mass.: Addison-Wesley.

Bohm, David. (1951) *Quantum Theory.* Englewood Cliffs, N.J.: Prentice-Hall.

——— (1952) "A suggested interpretation of the quantum theory in terms of 'hidden' variables, I." *Physical Review* 85: 166–179.

——— (1957) *Causality and Chance in Modern Physics.* New York: Harper.

Bohm, D., and Y. Aharonov. (1957) "Discussion of experimental proof for the paradox of Einstein, Podolsky and Rosen." *Physical Review* 108: 1070–1076.

Bohm, D., and B. J. Hiley. (1981) "Nonlocality in quantum theory understood in terms of Einstein's nonlinear field approach." *Foundations of Physics* 11: 529–546.

——— (1984) "Quantum potential model for the quantum theory." In Kamefuchi et al. (1984), 231–252.

Bohm, D., B. J. Hiley, and P. N. Kaloyerou. (1987) "An ontological basis for the quantum theory." *Physics Reports* 144: 321–375.

Bohr, N. (1913) "On the constitution of atoms and molecules." *Philosophical Magazine* 26: 1–25.

——— (1934) *Atomic Theory and the Description of Nature.* Cambridge: Cambridge University Press.

——— (1935) "Can quantum-mechanical description of physical reality be considered complete?" *Physical Review* 38: 696–702.

——— (1949) "Discussion with Einstein on epistemological problems in atomic physics." In Schilpp (1949), 199–241.

——— (1963) *Essays 1958–1962 on Atomic Physics and Human Knowledge.* New York: Random House.

Born, M. (1926) "Zur Quantenmechanik der Stossvorgänge." *Zeitschrift für Physik* 37: 863–867; *38*: 803–827.

——— (1949) *Natural Philosophy of Cause and Chance.* Oxford: Oxford University Press.

———, editor. (1969) *Albert Einstein—Hedwig und Max Born. Briefwechsel, 1916–1955.* Munich: Nymphenburger.

———, editor. (1971) *The Born-Einstein Letters,* trans. I. Born. New York: Walker: translation of Born (1969).

Brown, Harvey, and Rom Harré, editors. (1988) *Philosophical Foundations of Quantum Field Theory.* Oxford: Oxford University Press.

Bub, J. (1977) "Von Neumann's projection postulate as a possibility conditionalization rule in quantum mechanics." *Journal of Philosophical Logic* 6: 381–390.

——— (1979) "The measurement problem of quantum mechanics." In *Problems in the Philosophy of Physics,* Bologna: Societa Italiana de Fisica, 100–104.

Cartwright, Nancy. *Causality.* Forthcoming.

Clauser, J. F., and M. A. Horne. (1974) "Experimental consequences of objective local theories." *Physical Review* D10: 526–535.

Clauser, J. F., and A. Shimony. (1978) "Bell's Theorem: Experimental tests and implications." *Reports on Progress in Physics* 41: 1881–1927.

Costa De Beauregard, O. (1983) "Running backwards the Mermin device: Causality in EPR correlations." *American Journal of Physics* 51: 513–516.

Cushing, J. T. (1986) "Causality as an overarching principle in physics." In *PSA 1986*, ed. A. Fine and P. Machamer. East Lansing, Mich.: Philosophy of Science Association, vol. 1, 3–11.

———— (1988) "Foundational problems in and methodological lessons from quantum field theory." In Brown and Harré (1988), 25–39.

Davies, P. C. W., and J. R. Brown, editors. (1986) *The Ghost in the Atom: A Discussion of the Mysteries of Quantum Physics*. Cambridge: Cambridge University Press.

d'Espagnat, Bernard. (1976) *The Conceptual Foundations of Quantum Mechanics*, 2d ed. Reading, Mass.: Benjamin.

———— (1979) "The quantum theory and reality." *Scientific American* 241(5): 158–181 (European edition, 128–140).

———— (1984) "Nonseparability and the tentative descriptions of reality." *Physics Reports* 110: 201–264.

Dewitt, B. S., and N. Graham. (1973) *The Many-Worlds Interpretation of Quantum Mechanics*. Princeton: Princeton University Press.

Earman, John. (1986a) *A Primer on Determinism*. Dordrecht: Reidel.

———— (1986b) "Locality, non-locality and action-at-a-distance." In *Theoretical Physics in the 100 Years since Kelvin's Baltimore Lectures*, ed. P. Achinstein and R. Kargon. Cambridge, Mass.: MIT Press, 449–490.

Eberhard, P. (1977) "Bell's theorem without hidden variables." *Nuovo Cimento* 38B: 75 80.

———— (1978) "Bell's theorem and the different concepts of locality." *Nuovo Cimento* 46B: 392–419.

Einstein, A. (1917) "Zur Quantentheoric der Strahlung." *Physikalische Zeitschrift* 18: 121–128; translated as "On the quantum theory of radiation," in van der Waerden (1967), 63–67.

———— (1919) "Einstein on his theory: Time, space and gravitation." *Times*, London, 28 November 1919, 13; reprinted as "What is the theory of relativity?" in *Ideas and Opinions*, ed. C. Seelig and S. Bargmann. New York: Crown, 1954, 227–232.

———— (1936) "Physik und Realität." *Journal of the Franklin Institute* 221: 313–347.

———— (1948) "Quantenmechanik und Wirklichkeit." *Dialectica* 2: 320–324.

———— (1949) "Autobiographical Notes." In Schilpp (1949), 3–95.

Einstein, A., B. Podolsky, and N. Rosen. (1935) "Can quantum-mechanical description of physical reality be considered complete?" *Physical Review* 47: 777 780.

Everett, H. (1957) "Relative state formulation of quantum mechanics." *Reviews Modern Physics* 29: 454–462.

Feynman, R. P. (1982) "Simulating physics with computers." *International Journal of Theoretical Physics* 21: 467–488.

Fine, A. (1973) "Probability and the interpretation of quantum mechanics." *British Journal for the Philosophy of Science* 24: 1–37.

———— (1981) "Correlations and physical locality." In *PSA 1980*, ed. P. Asquith and R. Giere. East Lansing, Mich.: Philosophy of Science Association, vol. 2, 535–562.

———— (1982a) "Joint distributions, quantum correlations, and commuting observables." *Journal of Mathematical Physics* 23: 1306–1310.

———— (1982b) "Hidden variables, joint probability, and the Bell inequalities." *Physical Review Letters* 48: 291–295.

———— (1982c) "Antinomies of entanglement: The puzzling case of the tangled statistics." *Journal of Philosophy* 79: 733–747.

———— (1982d) "Some local models for correlation experiments." *Synthèse* 50: 279–294.

———— (1984a) "What is Einstein's statistical interpretation, or: Is it Einstein for whom Bell's theorem tolls?" *Topoi* 3: 23–36; reprinted in Fine (1986).

———— (1984b) "Einstein's realism." In *Science and Reality*, ed. J. Cushing et al. Notre Dame, Ind.: University of Notre Dame Press, 106–133; reprinted in Fine (1986).

———— (1986) *The Shaky Game: Einstein, Realism and the Quantum Theory.* Chicago: University of Chicago Press.

Folse, Henry J. (1985) *The Philosophy of Niels Bohr: The Framework of Complementarity.* Amsterdam: North Holland.

———— (1987) "Causality and reality in quantum physics." Paper delivered at the annual meeting of the Metaphysical Society of America, New York, March 1987.

French, S., "Individuality, supervenience, and Bell's theorem." *Philosophical Studies.* Forthcoming.

Friedman, Michael. (1983) *Foundations of Space-Time Theories: Relativistic Physics and Philosophy of Science.* Princeton, N.J.: Princeton University Press.

Ghirardi, G. C., A. Rimini, and T. Weber. (1980) "A general argument against superluminal transmission through the quantum-mechanical measurement process." *Lettere al Nuovo Cimento* 27: 293–298.

———— (1986) "Unified dynamics for microscopic and macroscopic systems." *Physical Review* D34: 470–491.

Gisin, N. (1981) "A simple non-linear dissipative quantum evolution equation." *Journal of Physics* A14: 2259–2267.

———— (1984) "Quantum measurements and stochastic processes." *Physical Review Letters* 52: 1657–1660.

Gisin, N., and C. Piron. (1981) "Collapse of the wave-packet without mixture." *Letters in Mathematical Physics* 5:379–385.

Gleason, A. M. (1957) "Measures on the closed sub-spaces of Hilbert spaces." *Journal of Mathematics and Mechanics* 6: 885–893.

Glymour, C. (1985) "Explanation and realism." In *Images of Science*, ed. P. Churchland and C. A. Hooker. Chicago: University of Chicago Press, 99–117.

Greenberger, Daniel M., editor. (1986) *New Techniques and Ideas in Quantum Measurement Theory.* Annals of the New York Academy of Sciences, vol. 480.

Halpin, J. (1986) "Stalnaker's conditional and Bell's problem." *Synthese* 69: 325–340.

Heisenberg, Werner. (1958) *Physics and Philosophy.* New York: Harper & Row.

Hellman, G. (1982) "Stochastic Einstein-locality and the Bell theorems." *Synthese* 53: 461–504.

―――― (1987) "EPR, Bell, and collapse: A route around 'stochastic' variables." *Philosophy of Science* 54: 639–657.

Herbert, N., and J. Karush. (1978) "Generalization of Bell's theorem." *Foundations of Physics* 8: 313–317.

Heywood, P., and M. Redhead. (1983) "Nonlocality and the Kochen-Specker paradox." *Foundations of Physics* 13: 481–499.

Howard, Don. (1979) *Complementarity and Ontology: Niels Bohr and the Problem of Scientific Realism in Quantum Physics*. Dissertation, Boston University.

―――― (1984) "Realism and conventionalism in Einstein's philosophy of science: The Einstein-Schlick correspondence." *Philosophia Naturalis* 21: 616–629.

―――― (1985) "Einstein on locality and separability." *Studies in History and Philosophy of Science* 16: 171–201.

―――― (1986) "What makes a classical concept classical? Toward a reconstruction of Niels Bohr's philosophy of physics." Boston University preprint.

―――― (1987) "Locality, separability, and the physical implications of the Bell experiments." Boston University preprint.

―――― (1988) "Einstein's conventionalism." In preparation.

Hoyer, U., editor. (1981) *Niels Bohr: Collected Works*. Amsterdam: North Holland, vol. 2.

Hughes, R. I. G. (1981) "Quantum logic." *Scientific American* 243 (10): 202–213.

―――― (1989) *The Structure and Interpretation of Quantum Mechanics*. Cambridge, Mass.: Harvard University Press.

Jammer, Max. (1974) *The Philosophy of Quantum Mechanics*. New York: Wiley.

Jarrett, Jon P. (1983) *Bell's Theorem, Quantum Mechanics, and Local Realism*. Dissertation, University of Chicago.

―――― (1984) "On the physical significance of the locality conditions in the Bell arguments." *Noûs* 18: 569–589.

―――― (1986) "Does Bell's theorem apply to theories that admit time-dependent states?" In Greenberger (1986), 428–437.

Jauch, Josef M. (1968) *Foundations of Quantum Mechanics*. Reading, Mass.: Addison-Wesley.

Kamefuchi, S., et al., editors. (1984) *Foundations of Quantum Mechanics in the Light of New Technology*. Tokyo: Physical Society of Japan.

Kitchener, Richard, editor. (1988) *The World View of Modern Physics: Does It Need a New Metaphysics?* Ithaca, N.Y.: SUNY Press.

Kochen, S., and E. P. Specker. (1967) "The problem of hidden variables in quantum mechanics." *Journal of Mathematics and Mechanics* 17: 59 87.

Kraus, K. (1985) "Quantum theory, causality, and EPR experiments." In Lahti and Mittelstaedt (1985), 461–480.

Krips, Henry. (1987) *The Metaphysics of Quantum Theory*. Oxford: Oxford University Press.

Lahti, Peter, and Peter Mittelstaedt, editors. (1985) *Symposium on the Foundations of Modern Physics*. Singapore: World Scientific.

Lewis, D.(1980) "A subjectivist's guide to objective chance." In *Studies in Inductive Logic and Probability*, ed. R. Jeffrey. Berkeley, Calif.: University of California Press; reprinted in Lewis (1986), 83–113.

———— (1986) *Philosophical Papers*. Oxford: Oxford University Press, vol. 2.

Lo, T. K., and A. Shimony. (1981) "Proposed molecular test of local hidden-variable theories." *Physical Review* 23A: 3003–3012.

Lüders, G. (1951) "Über die Zustandsänderung durch den Messprozess," *Annalen der Physik* 8: 323–328.

Margenau, H. (1936) "Quantum mechanical description." *Physical Review* 49: 240–242.

McMullin, Ernan. (1954) *The Quantum Principle of Uncertainty*. Dissertation, University of Louvain.

———— (1970) "The history and philosophy of science: A taxonomy." In *Minnesota Studies in the Philosophy of Science*, ed. R. Stuewer. Minneapolis: University of Minnesota Press, vol. 5, 12–67.

———— (1984) "Stability and change in science." *New Ideas in Psychology* 2: 9–19.

———— (1989) "Newton and scientific realism." To appear in McMullin, *Rationality, Realism, and the Growth of Knowledge*. Dordrecht: Reidel.

Mermin, N. D. (1981a) "Quantum mysteries for anyone." *Journal of Philosophy* 78: 397–408; reprinted in this volume.

———— (1981b) "Bringing home the atomic world: Quantum mysteries for anyone." *American Journal of Physics* 49: 940–943.

———— (1985) "Is the moon there when nobody looks? Reality and the quantum theory." *Physics Today* 38(4): 38–47.

———— (1986) "The EPR experiment: Thoughts about the 'loophole'." In Greenberger (1986) 422–427.

Moses, Lincoln. (1986) *Think and Explain with Statistics*. Reading, Mass.: Addison-Wesley.

Murdoch, Dugald. (1987) *Niels Bohr's Philosophy of Physics*. Cambridge: Cambridge University Press.

Newton, Isaac. (1726) *Mathematical Principles of Natural Philosophy*, translated A. Motte (1729); ed. F. Cajori (1934). Berkeley, Calif.: University of California Press.

Page, D. (1982) "The Einstein-Podolsky-Rosen physical reality is completely described by quantum mechanics." *Physics Letters* 91A: 57–60.

Pais, A. (1979) "Einstein and the quantum theory." *Reviews of Modern Physics* 51: 863–914.

Pattee, H. H. (1967) "Quantum mechanics, heredity, and the origin of life." *Journal of Theoretical Biology* 17: 410–420.

Pearle, P. (1976) "Reduction of the state-vector by a non-linear Schrödinger equation." *Physical Review* D13: 857–868.

———— (1979) "Toward explaining why events occur." *International Journal of Theoretical Physics* 18: 489–518.

———— (1982) "Might God toss coins?" *Foundations of Physics* 12: 249–263.

———— (1986a) "Models for reduction." In Penrose and Isham (1986), 84–108.

———— (1986b) "Suppose the state-vector is real: The description and consequences of dynamical reduction." In Greenberger (1986), pp. 539–539.

Penrose, Roger, and Christopher J. Isham, editors. (1986) *Quantum Concepts in Space and Time*. Oxford: Oxford University Press.

Peres, A. (1978) "Unperformed experiments have no results." *American Journal of Physics* 46: 745–747.

——— (1985) "Einstein, Gödel, Bohr." *Foundations of Physics* 15: 201–205.

——— (1988) "Schrödinger's immortal cat." *Foundations of Physics* 18: 57–76.

Peres, A., and W. H. Zurek. (1982) "Is quantum theory universally valid?" *American Journal of Physics* 50: 807–810.

Przibram, K., editor. (1967) *Letters on Wave Mechanics*. New York: Philosophical Library.

Putnam, H. (1969) "Is logic empirical?" *Boston Studies in the Philosophy of Science* 5: 199–215.

——— (1982) "Why reason can't be naturalized." *Synthese* 52: 3–23.

Quine, W. V. O. (1951) "Two dogmas of empiricism." *Philosophical Review* 60: 20–43.

Redhead, M. L. G. (1983) "Nonlocality and peaceful coexistence." In *Space, Time and Causality*, ed. R. Swinburne. Dordrecht: Reidel, 151–189.

——— (1986) "Relativity and quantum mechanics: Conflict or peaceful coexistence?" in Greenberger (1986), 14–20.

——— (1987a) "Undressing Baby Bell." Cambridge University reprint.

——— (1987b) *Incompleteness, Nonlocality, and Realism: A Prolegomenon to the Philosophy of Quantum Mechanics*. Oxford: Clarendon Press.

Reichenbach, Hans. (1924) *Axiomatik der relativistischen Raum-Zeit-Lehre*, Braunschweig: Vieweg.

——— (1956) *The Direction of Time*. Berkeley, Calif.: University of California Press.

Rosenfeld, Leon. (1964) "Niels Bohr in the thirties." In *Niels Bohr: His Life and Work as Seen by his Friends and Colleagues*, ed. S. Rozental. Amsterdam: North Holland, 114–136; English edition, 1967.

Salmon, W. (1978) "Why ask 'Why'?" *Proceedings American Philosophical Association* 51: 683–705.

——— (1984) *Scientific Explanation and the Causal Structure of the World*. Princeton, N.J.: Princeton University Press.

Schilpp, Paul, editor. (1949) *Albert Einstein: Philosopher-Scientist*. Evanston, Ill.: Library of Living Philosophers.

Schlick, M. (1936) "Sind die Naturgesetze Konventionen?" In *Actes du Congrès Internationale de Philosophie Scientifique 1935*, Vol. 4, *Induction et Probabilité*. 8–17 Paris: Hermann.

Schrödinger, E. (1935a) "Discussion of probability relations between separated systems." *Proceedings Cambridge Philosophical Society* 31: 553–563.

——— (1935b) "Die gegenwärtige Situation in der Quantenmechanik," *Naturwissenschaften* 23: 823–881.

Selleri, Franco, editor. (1988) *Quantum Mechanics versus Local Realism: The Einstein-Podolsky-Rosen Paradox*. New York: Plenum.

Shimony, A. (1971) "Perception from an evolutionary point of view." *Journal of Philosophy* 68: 571–583.

——— (1978) "Metaphysical problems in the foundations of quantum mechanics." *International Philosophical Quarterly* 18: 3–17.

————— (1984a) "Controllable and uncontrollable non-locality." In Kamefuchi et al. (1984), 225–230.

————— (1984b) "Contextual hidden variables theories and Bell's inequalities." *British Journal for the Philosophy of Science* 35: 25–45.

————— (1986) "Events and processes in the quantum world." In Penrose and Isham (1986), 182–203.

————— (1988) "The reality of the quantum world." *Scientific American* 258(1): 46–53 (European edition, 36–43).

Skyrms, Brian. (1980) *Causal Necessity.* New Haven, Conn.: Yale University Press.

————— (1982) "Counterfactual definiteness and local causation." *Philosophy of Science* 49: 43–50.

Stairs, A. (1983) "On the logic of pairs of quantum systems." *Synthese* 56: 437–460.

————— (1984) "Sailing into Charybdis: van Fraassen on Bell's theorem." *Synthese* 61: 351–359.

————— (1988) "Locality and logic." Forthcoming.

Stapp, H. P. (1968) "Correlation experiments and the non-validity of ordinary ideas about the physical world." Lawrence Berkeley Lab. Report LBL-5333.

————— (1971) "S-Matrix interpretation of quantum theory." *Physical Review* D3: 1303–1320.

————— (1972) "The Copenhagen interpretation." *American Journal of Physics* 40: 1098–1116.

————— (1977) "Are superluminal connections necessary?" *Nuovo Cimento* 40B: 191–205.

————— (1979) "Whiteheadian approach to quantum theory and the generalized Bell's theorem." *Foundations of Physics* 9: 1–25.

————— (1985a) "Bell's theorem and the foundations of quantum physics." *American Journal of Physics* 53: 306–317.

————— (1985b) "Consciousness and values in the quantum universe." *Foundations of Physics* 15: 35–47.

————— (1987) "Transcending Newton's legacy." Lawrence Berkeley Laboratory Report LBL-24322.

————— (1988a) "Quantum theory and the physicist's conception of nature: Philosophical implications of Bell's theorem." In Kitchener (1988), 38–48.

————— (1988b) "Quantum non-locality." *Foundations of Physics* 18: 427–488.

————— (1988c) "Are faster-than-light influences necessary?" In Selleri (1988).

Stein, H. (1970) "Is there a problem of interpreting quantum mechanics?" *Noûs* 4: 93–103.

————— (1972) "On the conceptual structure of quantum mechanics." In *Paradigms and Paradoxes: The Philosophical Challenge of the Quantum Domain,* ed. R. G. Colodny. Pittsburgh: University of Pittsburgh Press, 367–438.

Suppes, P. (1966) "The probabilistic argument for a non-classical logic in quantum mechanics." *Philosophy of Science* 33: 14–21.

————— (1984) *Probabilistic Metaphysics.* Oxford: Blackwell.

Swinburne, Richard, editor. (1983) *Space, Time and Causality.* Dordrecht: Reidel.

Teller, P. (1983) "The projection postulate as a fortuitous approximation." *Philosophy of Science* 50: 413–431.

———— (1986) "Relational holism and quantum mechanics." *British Journal for the Philosophy of Science* 37: 71–81.

Toulmin, Stephen. (1961) *Foresight and Understanding*. New York: Harper and Row.

van der Waerden, B. L., editor. (1967) *Sources of Quantum Mechanics*. Amsterdam: North-Holland Publishing Co.

van Fraassen, Bas. (1974) "The Einstein-Podolsky-Rosen paradox." *Synthese* 29: 291–309.

———— (1980) *The Scientific Image*. Oxford: Clarendon Press.

———— (1981) "Baby Bell," preprint.

———— (1982a) "The Charybdis of realism: Epistemological implications of Bell's inequality." *Synthese* 52: 25–38; reprinted in this volume.

———— (1982b) "Rational belief and the common cause principle." In *What? Where? When? Why?* ed. R. McLaughlin. Dordrecht: Reidel, 193–209.

———— (1984) "Glymour on evidence and explanation." In *Testing Scientific Theories,* ed. J. Earman. Minneapolis: University of Minnesota Press, 165–176.

———— (1985) "EPR: When is a correlation not a mystery?" In Lahti and Mittelstaedt (1985), 113–128.

von Neumann, John. (1932) *Mathematische Grundlagen der Quantenmechanik*. Berlin: Springer; translated R. T. Beyer. (1955) *Mathematical Foundations of Quantum Mechanics*. Princeton: Princeton University Press.

Wessels, L. (1985) "Locality, factorability and the Bell inequalities." *Nous* 19: 481–519.

Wheeler, J. A., and W. H. Zurek, editors. (1983) *Quantum Theory and Measurement*. Princeton: Princeton University Press.

Wigner, E. P. (1970) "On hidden variables and quantum mechanical probabilities." *American Journal of Physics* 38: 1005–1009.

INDEX OF NAMES

Articles by multiple authors are listed under the first author only.

312